I0486258

Entropy & Negentropy

" The End & The Beginning "

Edited by Paul F. Kisak

Contents

Chapter 1

Introduction to entropy

This article is a non-technical introduction to the subject. For the main encyclopedia article, see Entropy.

The idea of "irreversibility" is central to the understanding of **entropy**. Everyone has an intuitive understanding of irreversibility (a dissipative process) - if one watches a movie of everyday life running forward and in reverse, it is easy to distinguish between the two. The movie running in reverse shows impossible things happening - water jumping out of a glass into a pitcher above it, smoke going down a chimney, water in a glass freezing to form ice cubes, crashed cars reassembling themselves, and so on. The intuitive meaning of expressions such as "you can't unscramble an egg", "don't cry over spilled milk" or "you can't take the cream out of the coffee" is that these are irreversible processes. There is a direction in time by which spilled milk does not go back into the glass.

In thermodynamics, one says that the "forward" processes – pouring water from a pitcher, smoke going up a chimney, etc. – are "irreversible": they cannot happen in reverse, even though, on a microscopic level, no laws of physics would be violated if they did. All real physical processes involving systems in everyday life, with many atoms or molecules, are irreversible. For an irreversible process in an isolated system, the thermodynamic state variable known as entropy is always increasing. The reason that the movie in reverse is so easily recognized is because it shows processes for which entropy is decreasing, which is physically impossible. In everyday life, there may be processes in which the increase of entropy is practically unobservable, almost zero. In these cases, a movie of the process run in reverse will not seem unlikely. For example, in a 1-second video of the collision of two billiard balls, it will be hard to distinguish the forward and the backward case, because the increase of entropy during that time is relatively small. In thermodynamics, one says that this process is practically "reversible", with an entropy increase that is practically zero. The statement of the fact that the entropy of the Universe never decreases is found in the second law of thermodynamics.

In a physical system, **entropy** provides a measure of the amount of thermal energy that *cannot* be used to do work. In some other definitions of entropy, it is a measure of how evenly energy (or some analogous property) is distributed in a system. *Work* and *heat* are determined by a process that a system undergoes, and only occur at the boundary of a system. *Entropy* is a function of the state of a system, and has a value determined by the state variables of the system.

The concept of entropy is central to the second law of thermodynamics. The second law determines which physical processes can occur. For example, it predicts that the flow of heat from a region of high temperature to a region of low temperature is a spontaneous process – it can proceed along by itself without needing any extra external energy. When this process occurs, the hot region becomes cooler and the cold region becomes warmer. Heat is distributed more evenly throughout the system and the system's ability to do work has decreased because the temperature difference between the hot region and the cold region has decreased. Referring back to our definition of entropy, we can see that the entropy of this system has increased. Thus, the second law of thermodynamics can be stated to say that the entropy of an isolated system always increases, and such processes which increase entropy can occur spontaneously. The entropy of a system increases as its components have the range of their momentum and/or position increased.

The term *entropy* was coined in 1865 by the German physicist Rudolf Clausius, from the Greek words *en-*, "in", and *trope* "a turning", in analogy with *energy*.[1]

1.1   Explanation

The concept of thermodynamic entropy arises from the second law of thermodynamics. By this law of entropy increase it quantifies the reduction in the capacity of a system for change, for example heat always flows from a region of higher temperature to one with lower temperature until temperature becomes uniform or determines whether a thermodynamic process may occur.

Entropy is calculated in two ways, the first is the **entropy change** (ΔS) to a system containing a sub-system which undergoes heat transfer to its surroundings (inside the system of interest). It is based on the macroscopic relationship between heat flow into the sub-system and the temperature at which it occurs summed over the boundary of that sub-system. The second calculates the **absolute entropy** (S) of a system based on the microscopic behaviour of its individual particles. This is based on the natural logarithm of the number of microstates possible in a particular macrostate (W or Ω) called the thermodynamic probability. Roughly the probability of the system being in that state. In this sense it effectively defines entropy independently from its effects due to changes which may involve heat, mechanical, electrical, chemical energies etc. but also includes logical states such as information.

Following the formalism of Clausius, the first calculation can be mathematically stated as:[2]

$$\delta S = \frac{\delta q}{T}.$$

Where δS is the increase or decrease in entropy, δq is the heat added to the system or subtracted from it, and T is temperature. The equal sign indicates that the change is reversible . If the temperature is allowed to vary, the equation must be integrated over the temperature path. This calculation of entropy change does not allow the determination of absolute value, only differences. In this context, the Second Law of Thermodynamics may be stated that for heat transferred over any valid process for any system, whether isolated or not,

$$\delta S \geq \frac{\delta q}{T}.$$

The second calculation defines entropy in absolute terms and comes from statistical mechanics. The entropy of a particular macrostate is defined to be Boltzmann's constant times the natural logarithm of the number of microstates corresponding to that macrostate, or mathematically

$$S = k_B \ln \Omega.$$

Where S is the entropy, kB is Boltzmann's constant, and Ω is the number of microstates.

The macrostate of a system is what we know about the system, for example the temperature, pressure, and volume of a gas in a box. For each set of values of temperature, pressure, and volume there are many arrangements of molecules which result in those values. The number of arrangements of molecules which could result in the same values for temperature, pressure and volume is the number of microstates.

The concept of energy is related to the first law of thermodynamics, which deals with the conservation of energy and under which the loss in heat will result in a decrease in the internal energy of the thermodynamic system. Thermodynamic entropy provides a comparative measure of the amount of this decrease in internal energy of the system and the corresponding increase in internal energy of the surroundings at a given temperature. A simple and more concrete visualization of the second law is that energy of all types changes from being localized to becoming dispersed or spread out, if it is not hindered from doing so. Entropy change is the quantitative measure of that kind of a spontaneous process: how much energy has flowed or how widely it has become spread out at a specific temperature.

The concept of entropy has been developed to describe any of several phenomena, depending on the field and the context in which it is being used. Information entropy takes the mathematical concepts of statistical thermodynamics into areas of probability theory unconnected with heat and energy.

Ice melting provides an example of entropy increasing

1.2   Example of increasing entropy

Main article: Disgregation

Ice melting provides an example in which entropy increases in a small system, a thermodynamic system consisting of the surroundings (the warm room) and the entity of glass container, ice, water which has been allowed to reach thermodynamic equilibrium at the melting temperature of ice. In this system, some heat (δQ) from the warmer surroundings at 298 K (77 °F, 25 °C) transfers to the cooler system of ice and water at its constant temperature (T) of 273 K (32 °F, 0 °C), the melting temperature of ice. The entropy of the system, which is $\delta Q/T$, increases by $\delta Q/273 K$. The heat δQ for this process is the energy required to change water from the solid state to the liquid state, and is called the enthalpy of fusion, i.e. ΔH for ice fusion.

It is important to realize that the entropy of the surrounding room decreases less than the entropy of the ice and water increases: the room temperature of 298 K is larger than 273 K and therefore the ratio, (entropy change), of $\delta Q/298 K$ for the surroundings is smaller than the ratio (entropy change), of $\delta Q/273 K$ for the ice and water system. This is always true in spontaneous events in a thermodynamic system and it shows the predictive importance of entropy: the final net entropy after such an event is always greater than was the initial entropy.

As the temperature of the cool water rises to that of the room and the room further cools imperceptibly, the sum of the $\delta Q/T$ over the continuous range, "at many increments", in the initially cool to finally warm water can be found by calculus. The entire miniature 'universe', i.e. this thermodynamic system, has increased in entropy. Energy has spontaneously become more dispersed and spread out in that 'universe' than when the glass of ice and water was introduced and became a 'system' within it.

1.3   Origins and uses

Originally, entropy was named to describe the "waste heat," or more accurately, energy loss, from heat engines and other mechanical devices which could never run with 100% efficiency in converting energy into work. Later, the term came to acquire several additional descriptions, as more was understood about the behavior of molecules on the microscopic level. In the late 19th century, the word "disorder" was used by Ludwig Boltzmann in developing statistical views of entropy using probability theory to describe the increased molecular movement on the microscopic level. That was before quantum behavior came to be better understood by Werner Heisenberg and those who followed. Descriptions of thermodynamic (heat) entropy on the microscopic level are found in statistical thermodynamics and statistical mechanics.

For most of the 20th century, textbooks tended to describe entropy as "disorder", following Boltzmann's early conceptualisation of the "motional" (i.e. kinetic) energy of molecules. More recently, there has been a trend in chemistry and physics textbooks to describe entropy as energy dispersal.[3] Entropy can also involve the dispersal of particles, which are themselves energetic. Thus there are instances where both particles and energy disperse at different rates when substances are mixed together.

The mathematics developed in statistical thermodynamics were found to be applicable in other disciplines. In particular, information sciences developed the concept of information entropy where a constant replaces the temperature which is inherent in thermodynamic entropy.

1.4   Heat and entropy

At a microscopic level, kinetic energy of molecules is responsible for the temperature of a substance or a system. "Heat" is the kinetic energy of molecules being transferred: when motional energy is transferred from hotter surroundings to a cooler system, faster-moving molecules in the surroundings collide with the walls of the system which transfers some of their energy to the molecules of the system and makes them move faster.

- Molecules in a gas like nitrogen at room temperature at any instant are moving at an average speed of nearly 500 miles per hour (210 m/s), repeatedly colliding and therefore exchanging energy so that their individual speeds are always changing. Assuming an ideal-gas model, average kinetic energy increases linearly with temperature, so the average speed increases as the square root of temperature.

- Thus motional molecular energy ('heat energy') from hotter surroundings, like faster-moving molecules in a flame or violently vibrating iron atoms in a hot plate, will melt or boil a substance (the system) at the temperature of its melting or boiling point. That amount of motional energy from the surroundings that is required for melting or boiling is called the phase-change energy, specifically the enthalpy of fusion or of vaporization, respectively. This phase-change energy breaks bonds between the molecules in the system (not chemical bonds inside the molecules that hold the atoms together) rather than contributing to the motional energy and making the molecules move any faster – so it does not raise the temperature, but instead enables the molecules to break free to move as a liquid or as a vapor.

- In terms of energy, when a solid becomes a liquid or a vapor, motional energy coming from the surroundings is changed to 'potential energy' in the substance (phase change energy, which is released back to the surroundings when the surroundings become cooler than the substance's boiling or melting temperature, respectively). Phase-change energy increases the entropy of a substance or system because it is energy that must be spread out in the system from the surroundings so that the substance can exist as a liquid or vapor at a temperature above its melting or boiling point. When this process occurs in a 'universe' that consists of the surroundings plus the system, the total energy of the 'universe' becomes more dispersed or spread out as part of the greater energy that was only in the hotter surroundings transfers so that some is in the cooler system. This energy dispersal increases the entropy of the 'universe'.

The important overall principle is that *"Energy of all types changes from being localized to becoming dispersed or spread out, if not hindered from doing so. Entropy (or better, entropy change) is the quantitative measure of that kind of a spontaneous process: how much energy has been transferred/T or how widely it has become spread out at a specific temperature.*

1.4.1 Classical calculation of entropy

When entropy was first defined and used in 1865 the very existence of atoms was still controversial and there was no concept that temperature was due to the motional energy of molecules or that "heat" was actually the transferring of that motional molecular energy from one place to another. Entropy change, ΔS, was described in macroscopic terms that could be directly measured, such as volume, temperature, or pressure. However, today the classical equation of entropy, $\Delta S = \frac{q_{rev}}{T}$ can be explained, part by part, in modern terms describing how molecules are responsible for what is happening:

- ΔS is the change in entropy of a system (some physical substance of interest) after some motional energy ("heat") has been transferred to it by fast-moving molecules. So, $\Delta S = S_{final} - S_{initial}$.

- Then, $\Delta S = S_{final} - S_{initial} = \frac{q_{rev}}{T}$, the quotient of the motional energy ("heat") q that is transferred "reversibly" (rev) to the system from the surroundings (or from another system in contact with the first system) divided by T, the absolute temperature at which the transfer occurs.

  - "Reversible" or "reversibly" (rev) simply means that T, the temperature of the system, has to stay (almost) exactly the same while any energy is being transferred to or from it. That's easy in the case of phase changes, where the system absolutely must stay in the solid or liquid form until enough energy is given to it to break bonds between the molecules before it can change to a liquid or a gas. For example in the melting of ice at 273.15 K, no matter what temperature the surroundings are – from 273.20 K to 500 K or even higher, the temperature of the ice will stay at 273.15 K until the last molecules in the ice are changed to liquid water, i.e., until all the hydrogen bonds between the water molecules in ice are broken and new, less-exactly fixed hydrogen bonds between liquid water molecules are formed. This amount of energy necessary for ice melting per mole has been found to be 6008 joules at 273 K. Therefore, the entropy change per mole is $\frac{q_{rev}}{T} = \frac{6008J}{273K}$, or 22 J/K.

  - When the temperature isn't at the melting or boiling point of a substance no intermolecular bond-breaking is possible, and so any motional molecular energy ("heat") from the surroundings transferred to a system raises its temperature, making its molecules move faster and faster. As the temperature is constantly rising, there is no longer a particular value of "T" at which energy is transferred. However, a "reversible" energy transfer can be measured at a very small temperature increase, and a cumulative total can be found by adding each of many small temperature intervals or increments. For example, to find the entropy change $\frac{q_{rev}}{T}$ from 300 K to 310 K, measure the amount of energy transferred at dozens or hundreds of temperature increments, say from 300.00 K to 300.01 K and then 300.01 to 300.02 and so on, dividing the q by each T, and finally adding them all.

- Calculus can be used to make this calculation easier if the effect of energy input to the system is linearly dependent on the temperature change, as in simple heating of a system at moderate to relatively high temperatures. Thus, the energy being transferred "per incremental change in temperature" (the heat capacity, C_p), multiplied by the integral of $\frac{dT}{T}$ from $T_{initial}$ to T_{final} , is directly given by $\Delta S = C_p \ln \frac{T_{final}}{T_{initial}}$.

1.5   Introductory descriptions of entropy

Traditionally, 20th century textbooks have introduced entropy as order and disorder so that it provides "a measurement of the disorder or randomness of a system". It has been argued that ambiguities in the terms used (such as "disorder" and "chaos") contribute to widespread confusion and can hinder comprehension of entropy for most students. A more recent formulation associated with Frank L. Lambert describing entropy as energy dispersal.[4]

1.6   See also

- Entropy (energy dispersal)

- Second law of thermodynamics

- Statistical mechanics

- Thermodynamics

1.7   References

[1] "etymonline.com:entropy". Retrieved 2009-06-15.

[2] I. Klotz, R. Rosenberg, *Chemical Thermodynamics - Basic Concepts and Methods*, 7th ed., Wiley (2008), p. 125

[3] welcome to entropysite.

[4] welcome to entropysite.

1.8   Further reading

- Goldstein, Martin and Inge F. (1993). *The Refrigerator and the Universe: Understanding the Laws of Energy*. Harvard Univ. Press. ISBN 9780674753259. chapters=4-12 touch on entropy

Chapter 2

History of entropy

The concept of **entropy** developed in response to the observation that a certain amount of functional energy released from combustion reactions is always lost to dissipation or friction and is thus not transformed into useful work. Early heat-powered engines such as Thomas Savery's (1698), the Newcomen engine (1712) and the Cugnot steam tricycle (1769) were inefficient, converting less than two percent of the input energy into useful work output; a great deal of useful energy was dissipated or lost. Over the next two centuries, physicists investigated this puzzle of lost energy; the result was the concept of entropy.

In the early 1850s, Rudolf Clausius set forth the concept of the thermodynamic system and posited the argument that in any irreversible process a small amount of heat energy δQ is incrementally dissipated across the system boundary. Clausius continued to develop his ideas of lost energy, and coined the term *entropy*.

Since the mid-20th century the concept of entropy has found application in the field of information theory, describing an analogous loss of data in information transmission systems.

2.1 Classical thermodynamic views

In 1803, mathematician Lazare Carnot published a work entitled *Fundamental Principles of Equilibrium and Movement*. This work includes a discussion on the efficiency of fundamental machines, i.e. pulleys and inclined planes. Lazare Carnot saw through all the details of the mechanisms to develop a general discussion on the conservation of mechanical energy. Over the next three decades, Lazare Carnot's theorem was taken as a statement that in any machine the accelerations and shocks of the moving parts all represent losses of *moment of activity*, i.e. the useful work done. From this Lazare drew the inference that perpetual motion was impossible. This *loss of moment of activity* was the first-ever rudimentary statement of the second law of thermodynamics and the concept of 'transformation-energy' or *entropy*, i.e. energy lost to dissipation and friction.[1]

Lazare Carnot died in exile in 1823. During the following year Lazare's son Sadi Carnot, having graduated from the École Polytechnique training school for engineers, but now living on half-pay with his brother Hippolyte in a small apartment in Paris, wrote *Reflections on the Motive Power of Fire*. In this book, Sadi visualized an ideal engine in which any heat (i.e., caloric) converted into work, could be reinstated by reversing the motion of the cycle, a concept subsequently known as thermodynamic reversibility. Building on his father's work, Sadi postulated the concept that "some caloric is always lost" in the conversion into work, even in his idealized reversible heat engine, which excluded frictional losses and other losses due to the imperfections of any real machine. He also discovered that this idealized efficiency was dependent only on the temperatures of the heat reservoirs between which the engine was working, and not on the types of working fluids. Any real heat engine could not realize the Carnot cycle's reversibility, and was condemned to be even less efficient. This loss of usable caloric was a precursory form of the increase in entropy as we now know it. Though formulated in terms of caloric, rather than entropy, this was an early insight into the second law of thermodynamics.

2.2   1854 definition

*Rudolf Clausius - originator of the concept of **entropy***

In his 1854 memoir, Clausius first develops the concepts of *interior work*, i.e. that "which the atoms of the body exert upon each other", and *exterior work*, i.e. that "which arise from foreign influences [to] which the body may be exposed", which may act on a working body of fluid or gas, typically functioning to work a piston. He then discusses the three categories into which heat Q may be divided:

1. Heat employed in increasing the heat actually existing in the body.

2. Heat employed in producing the interior work.

3. Heat employed in producing the exterior work.

Building on this logic, and following a mathematical presentation of the *first fundamental theorem*, Clausius then presented the first-ever mathematical formulation of entropy, although at this point in the development of his theories he called it "equivalence-value", perhaps referring to the concept of the mechanical equivalent of heat which was developing at the time rather than entropy, a term which was to come into use later.[2] He stated:[3]

> the *second fundamental theorem* in the mechanical theory of heat may thus be enunciated:
> If two transformations which, without necessitating any other permanent change, can mutually replace one another, be called equivalent, then the generations of the quantity of heat Q from work at the temperature T, has the *equivalence-value*:
>
> $$\frac{Q}{T}$$
>
> and the passage of the quantity of heat Q from the temperature T_1 to the temperature T_2, has the equivalence-value:
>
> $$Q\left(\frac{1}{T_2} - \frac{1}{T_1}\right)$$
>
> wherein T is a function of the temperature, independent of the nature of the process by which the transformation is effected.

In modern terminology, we think of this equivalence-value as "entropy", symbolized by S. Thus, using the above description, we can calculate the entropy change ΔS for the passage of the quantity of heat Q from the temperature T_1, through the "working body" of fluid (see heat engine), which was typically a body of steam, to the temperature T_2 as shown below:

If we make the assignment:

$$S = \frac{Q}{T}$$

Then, the entropy change or "equivalence-value" for this transformation is:

$$\Delta S = S_{final} - S_{initial}$$

which equals:

$$\Delta S = \left(\frac{Q}{T_2} - \frac{Q}{T_1}\right)$$

and by factoring out Q, we have the following form, as was derived by Clausius:

$$\Delta S = Q\left(\frac{1}{T_2} - \frac{1}{T_1}\right)$$

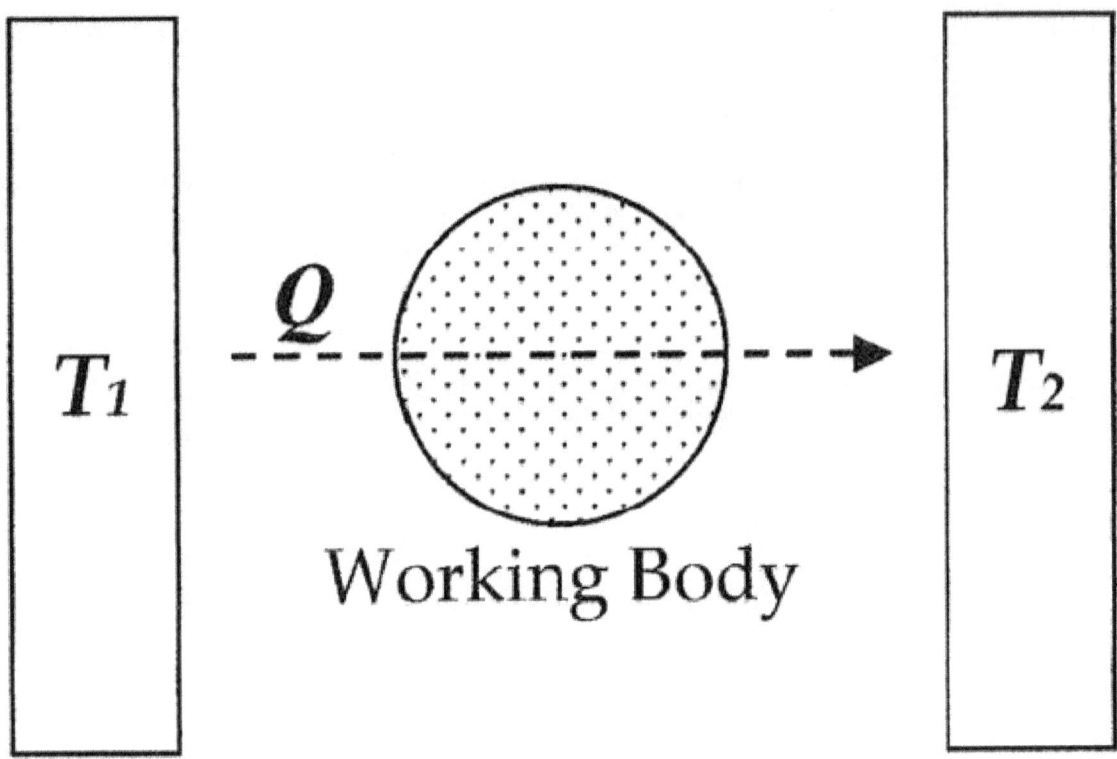

Diagram of Sadi Carnot's heat engine, 1824

2.3   1856 definition

In 1856, Clausius stated what he called the "second fundamental theorem in the mechanical theory of heat" in the following form:

$$\int \frac{\delta Q}{T} = -N$$

where N is the "equivalence-value" of all uncompensated transformations involved in a cyclical process. This equivalence-value was a precursory formulation of entropy.[4]

2.4   1862 definition

Main article: disgregation

In 1862, Clausius stated what he calls the "theorem respecting the equivalence-values of the transformations" or what is now known as the second law of thermodynamics, as such:

> *The algebraic sum of all the transformations occurring in a cyclical process can only be positive, or, as an extreme case, equal to nothing.*

Quantitatively, Clausius states the mathematical expression for this theorem is as follows. Let δQ be an element of the heat given up by the body to any reservoir of heat during its own changes, heat which it may absorb from a reservoir being here reckoned as negative, and T the absolute temperature of the body at the moment of giving up this heat, then the equation:

$$\int \frac{\delta Q}{T} = 0$$

must be true for every reversible cyclical process, and the relation:

$$\int \frac{\delta Q}{T} \geq 0$$

must hold good for every cyclical process which is in any way possible. This was an early formulation of the second law and one of the original forms of the concept of entropy.

2.5 1865 definition

In 1865, Clausius gave irreversible heat loss, or what he had previously been calling "equivalence-value", a name:[5][6]

Although Clausius did not specify why he chose the symbol "S" to represent entropy, it is arguable that Clausius chose "S" in honor of Sadi Carnot, to whose 1824 article Clausius devoted over 15 years of work and research. On the first page of his original 1850 article "On the Motive Power of Heat, and on the Laws which can be Deduced from it for the Theory of Heat", Clausius calls Carnot the most important of the researchers in the theory of heat.[7]

2.6 Later developments

In 1876, physicist J. Willard Gibbs, building on the work of Clausius, Hermann von Helmholtz and others, proposed that the measurement of "available energy" ΔG in a thermodynamic system could be mathematically accounted for by subtracting the "energy loss" $T\Delta S$ from total energy change of the system ΔH. These concepts were further developed by James Clerk Maxwell [1871] and Max Planck [1903].

2.7 Statistical thermodynamic views

Main article: statistical thermodynamics

In 1877, Ludwig Boltzmann developed a statistical mechanical evaluation of the entropy S, of a body in its own given macrostate of internal thermodynamic equilibrium. It may be written as:

$$S = k_B \ln \Omega$$

where

kB denotes Boltzmann's constant and

Ω denotes the number of microstates consistent with the given equilibrium macrostate.

Boltzmann himself did not actually write this formula expressed with the named constant kB, which is due to Planck's reading of Boltzmann.[8]

Boltzmann saw entropy as a measure of statistical "mixedupness" or disorder. This concept was soon refined by J. Willard Gibbs, and is now regarded as one of the cornerstones of the theory of statistical mechanics.

Erwin Schrödinger made use of Boltzmann's work in his book *What is Life?*[9] to explain why living systems have far fewer replication errors than would be predicted from Statistical Thermodynamics. Schrödinger used the Boltzmann equation in a different form to show increase of entropy

$$S = k_B \ln D$$

where D is the number of possible energy states in the system that can be randomly filed with energy. He postulated a local decrease of entropy for living systems when (1/D) represents the number of states that are prevented from randomly distributing, such as occurs in replication of the genetic code.

$$-S = k_B \ln(1/D)$$

Without this correction Schrödinger claimed that statistical thermodynamics would predict one thousand mutations per million replications, and ten mutations per hundred replications following the rule for square root of n, far more mutations than actually occur.

Schrödinger's separation of random and non-random energy states is one of the few explanations for why entropy could be low in the past, but continually increasing now. It has been proposed as an explanation of localized decrease of entropy[10] in radiant energy focusing in parabolic reflectors and during dark current in diodes, which would otherwise be in violation of Statistical Thermodynamics.

2.8   Information theory

An analog to *thermodynamic entropy* is **information entropy**. In 1948, while working at Bell Telephone Laboratories electrical engineer Claude Shannon set out to mathematically quantify the statistical nature of "lost information" in phone-line signals. To do this, Shannon developed the very general concept of information entropy, a fundamental cornerstone of information theory. Although the story varies, initially it seems that Shannon was not particularly aware of the close similarity between his new quantity and earlier work in thermodynamics. In 1949, however, when Shannon had been working on his equations for some time, he happened to visit the mathematician John von Neumann. During their discussions, regarding what Shannon should call the "measure of uncertainty" or attenuation in phone-line signals with reference to his new information theory, according to one source:[11]

According to another source, when von Neumann asked him how he was getting on with his information theory, Shannon replied:[12]

In 1948 Shannon published his famous paper *A Mathematical Theory of Communication*, in which he devoted a section to what he calls Choice, Uncertainty, and Entropy.[13] In this section, Shannon introduces an *H function* of the following form:

$$H = -K \sum_{i=1}^{k} p(i) \log p(i).$$

where K is a positive constant. Shannon then states that "any quantity of this form, where K merely amounts to a choice of a unit of measurement, plays a central role in information theory as measures of information, choice, and uncertainty."

Then, as an example of how this expression applies in a number of different fields, he references R.C. Tolman's 1938 *Principles of Statistical Mechanics*, stating that "the form of *H* will be recognized as that of entropy as defined in certain formulations of statistical mechanics where *pi* is the probability of a system being in cell *i* of its phase space... *H* is then, for example, the *H* in Boltzmann's famous H theorem." As such, over the last fifty years, ever since this statement was made, people have been overlapping the two concepts or even stating that they are exactly the same.

Shannon's information entropy is a much more general concept than statistical thermodynamic entropy. Information entropy is present whenever there are unknown quantities that can be described only by a probability distribution. In a series of papers by E. T. Jaynes starting in 1957,[14][15] the statistical thermodynamic entropy can be seen as just a particular application of Shannon's information entropy to the probabilities of particular microstates of a system occurring in order to produce a particular macrostate.

2.9 Popular use

The term entropy is often used in popular language to denote a variety of unrelated phenomena. One example is the concept of **corporate entropy** as put forward somewhat humorously by authors Tom DeMarco and Timothy Lister in their 1987 classic publication *Peopleware*, a book on growing and managing productive teams and successful software projects. Here, they view energy waste as red tape and business team inefficiency as a form of entropy, i.e. energy lost to waste. This concept has caught on and is now common jargon in business schools.

In another example, entropy plays the main villain in Isaac Asimov's short story The Last Question (first copyrighted in 1956). The story plays with the idea that when tampering with the Second law of thermodynamics, entropy must always increase.

2.10 Terminology overlap

When necessary, to disambiguate between the statistical thermodynamic concept of entropy, and entropy-like formulae put forward by different researchers, the statistical thermodynamic entropy is most properly referred to as the **Gibbs entropy**. The terms *Boltzmann–Gibbs entropy* or *BG entropy*, and *Boltzmann–Gibbs–Shannon entropy* or *BGS entropy* are also seen in the literature.

2.11 See also

- Entropy
- Enthalpy
- Thermodynamic free energy

2.12 References

[1] Mendoza, E. (1988). *Reflections on the Motive Power of Fire – and other Papers on the Second Law of Thermodynamics by E. Clapeyron and R. Clausius*. New York: Dover Publications. ISBN 0-486-44641-7.

[2] *Mechanical Theory of Heat*, by Rudolf Clausius, 1850-1865

[3] Published in *Poggendoff's Annalen*, December 1854, vol. xciii. p. 481; translated in the *Journal de Mathematiques*, vol. xx. Paris, 1855, and in the *Philosophical Magazine*, August 1856, s. 4. vol. xii, p. 81

[4] Clausius, Rudolf. (1856). "*On the Application of the Mechanical theory of Heat to the Steam-Engine*." as found in: Clausius, R. (1865). The Mechanical Theory of Heat – with its Applications to the Steam Engine and to Physical Properties of Bodies. London: John van Voorst, 1 Paternoster Row. MDCCCLXVII.

[5] Laidler, Keith J. (1995). *The Physical World of Chemistry*. Oxford University Press. pp. 104–105. ISBN 0-19-855919-4.

[6] OED, Second Edition, 1989, "*Clausius (Pogg. Ann. CXXV. 390), assuming (unhistorically) the etymological sense of energy to be 'work-contents' (werk-inhalt), devised the term entropy as a corresponding designation for the 'transformation-contents' (verwandlungsinhalt) of a system*"

[7] Clausius, Rudolf (1850). *On the Motive Power of Heat, and on the Laws which can be deduced from it for the Theory of Heat*. Poggendorff's *Annalen der Physick*, LXXIX (Dover Reprint). ISBN 0-486-59065-8.

[8] Partington, J.R. (1949), *An Advanced Treatise on Physical Chemistry*, volume 1, *Fundamental Principles, The Properties of Gases*, London: Longmans, Green and Co., p. 300.

[9] Schrödinger, Erwin (2004). *What is Life?* (11th reprinting ed.). Cambridge: Canto. p. 72 - 73. ISBN 0-521-42708-8.

[10] "Random and Non Random States".

[11] M. Tribus, E.C. McIrvine, "Energy and information", *Scientific American*, 224 (September 1971).

[12] Avery, John (2003). *Information Theory and Evolution*. World Scientific. ISBN 981-238-400-6.

[13] C.E. Shannon, "A Mathematical Theory of Communication", *Bell System Technical Journal*, vol. 27, pp. 379-423, 623-656, July, October, 1948, Eprint, PDF

[14] E. T. Jaynes (1957) Information theory and statistical mechanics, *Physical Review* **106**:620

[15] E. T. Jaynes (1957) Information theory and statistical mechanics II, *Physical Review* **108**:171

2.13   External links

- Max Jammer (1973). *Dictionary of the History of Ideas*: Entropy

Chapter 3

Entropy

This article is about entropy in thermodynamics. For other uses, see Entropy (disambiguation).
For a more accessible and less technical introduction to this topic, see Introduction to entropy.
Do not confuse this with Enthalpy

In thermodynamics, **entropy** (usual symbol S) is a measure of the number of specific ways in which a thermodynamic system may be arranged, commonly understood as a measure of disorder. According to the second law of thermodynamics the entropy of an isolated system never decreases; such a system will spontaneously proceed towards thermodynamic equilibrium, the configuration with maximum entropy. Systems that are not isolated may decrease in entropy, provided they increase the entropy of their environment by at least that same amount. Since entropy is a state function, the change in the entropy of a system is the same for any process that goes from a given initial state to a given final state, whether the process is reversible or irreversible. However, irreversible processes increase the combined entropy of the system and its environment.

The change in entropy (ΔS) of a system was originally defined for a thermodynamically reversible process as

$$\Delta S = \int \frac{dQ_{rev}}{T}$$

where T is the absolute temperature of the system, dividing an incremental reversible transfer of heat into that system (dQ). (If heat is transferred out the sign would be reversed giving a decrease in entropy of the system.) The above definition is sometimes called the macroscopic definition of entropy because it can be used without regard to any microscopic description of the contents of a system. The concept of entropy has been found to be generally useful and has several other formulations. Entropy was discovered when it was noticed to be a quantity that behaves as a function of state, as a consequence of the second law of thermodynamics.

Entropy is an extensive property. It has the dimension of energy divided by temperature, which has a unit of joules per kelvin ($J\ K^{-1}$) in the International System of Units (or $kg\ m^2\ s^{-2}\ K^{-1}$ in terms of base units). But the entropy of a pure substance is usually given as an intensive property — either entropy per unit mass (SI unit: $J\ K^{-1}\ kg^{-1}$) or entropy per unit amount of substance (SI unit: $J\ K^{-1}\ mol^{-1}$).

The *absolute* entropy (S rather than ΔS) was defined later, using either statistical mechanics or the third law of thermodynamics.

In the modern microscopic interpretation of entropy in statistical mechanics, entropy is the amount of additional information needed to specify the exact physical state of a system, given its thermodynamic specification. Understanding the role of thermodynamic entropy in various processes requires an understanding of how and why that information changes as the system evolves from its initial to its final condition. It is often said that entropy is an expression of the disorder, or randomness of a system, or of our lack of information about it. The second law is now often seen as an expression of the fundamental postulate of statistical mechanics through the modern definition of entropy.

3.1 History

Rudolf Clausius (1822–1888), originator of the concept of entropy

Main article: History of entropy

The analysis which led to the concept of entropy began with the work of French mathematician Lazare Carnot who in his 1803 paper *Fundamental Principles of Equilibrium and Movement* proposed that in any machine the accelerations and shocks of the moving parts represent losses of *moment of activity*. In other words, in any natural process there exists an inherent tendency towards the dissipation of useful energy. Building on this work, in 1824 Lazare's son Sadi Carnot published *Reflections on the Motive Power of Fire* which posited that in all heat-engines, whenever "caloric" (what is now known as heat) falls through a temperature difference, work or motive power can be produced from the actions of its fall from a hot to cold body. He made the analogy with that of how water falls in a water wheel. This was an early insight into the second law of thermodynamics.[1] Carnot based his views of heat partially on the early 18th century "Newtonian hypothesis" that both heat and light were types of indestructible forms of matter, which are attracted and repelled by other matter, and partially on the contemporary views of Count Rumford who showed (1789) that heat could be created by friction as when cannon bores are machined.[2] Carnot reasoned that if the body of the working substance, such as a body of steam, is returned to its original state at the end of a complete engine cycle, that "no change occurs in the condition of the working body".

The first law of thermodynamics, deduced from the heat-friction experiments of James Joule in 1843, expresses the concept of energy, and its conservation in all processes; the first law, however, is unable to quantify the effects of friction and dissipation.

In the 1850s and 1860s, German physicist Rudolf Clausius objected to the supposition that no change occurs in the working body, and gave this "change" a mathematical interpretation by questioning the nature of the inherent loss of usable heat when work is done, e.g. heat produced by friction.[3] Clausius described entropy as the *transformation-content*, i.e. dissipative energy use, of a thermodynamic system or working body of chemical species during a change of state.[3] This was in contrast to earlier views, based on the theories of Isaac Newton, that heat was an indestructible particle that had mass.

Later, scientists such as Ludwig Boltzmann, Josiah Willard Gibbs, and James Clerk Maxwell gave entropy a statistical basis. In 1877 Boltzmann visualized a probabilistic way to measure the entropy of an ensemble of ideal gas particles, in which he defined entropy to be proportional to the logarithm of the number of microstates such a gas could occupy. Henceforth, the essential problem in statistical thermodynamics, i.e. according to Erwin Schrödinger, has been to determine the distribution of a given amount of energy E over N identical systems. Carathéodory linked entropy with a mathematical definition of irreversibility, in terms of trajectories and integrability.

3.2 Definitions and descriptions

Any method involving the notion of entropy, the very existence of which depends on the second law of thermodynamics, will doubtless seem to many far-fetched, and may repel beginners as obscure and difficult of comprehension.

Willard Gibbs, *Graphical Methods in the Thermodynamics of Fluids*[4]

There are two related definitions of entropy: the thermodynamic definition and the statistical mechanics definition. Historically, the classical thermodynamics definition developed first. In the classical thermodynamics viewpoint, the system is composed of very large numbers of constituents (atoms, molecules) and the state of the system is described by the average thermodynamic properties of those constituents; the details of the system's constituents are not directly considered, but their behavior is described by macroscopically averaged properties, e.g. temperature, pressure, entropy, heat capacity. The early classical definition of the properties of the system assumed equilibrium. The classical thermodynamic definition of entropy has more recently been extended into the area of non-equilibrium thermodynamics. Later, the thermodynamic properties, including entropy, were given an alternative definition in terms of the statistics of the motions of the microscopic constituents of a system — modeled at first classically, e.g. Newtonian particles constituting a gas, and later quantum-mechanically (photons, phonons, spins, etc.). The statistical mechanics description of the behavior of a system is necessary as the definition of the properties of a system using classical thermodynamics become an increasingly unreliable method of predicting the final state of a system that is subject to some process.

3.2.1 Function of state

There are many thermodynamic properties that are functions of state. This means that at a particular thermodynamic state (which should not be confused with the microscopic state of a system), these properties have a certain value. Often, if two properties of the system are determined, then the state is determined and the other properties' values can also be determined. For instance, a gas at a particular temperature and pressure has its state fixed by those values, and has a particular volume that is determined by those values. As another instance, a system composed of a pure substance of a single phase at a particular uniform temperature and pressure is determined (and is thus a particular state) and is at not only a particular volume but also at a particular entropy.[5] The fact that entropy is a function of state is one reason it is useful. In the Carnot cycle, the working fluid returns to the same state it had at the start of the cycle, hence the line integral of any state function, such as entropy, over the cycle is zero.

3.2.2   Reversible process

Entropy is defined for a reversible process and for a system that, at all times, can be treated as being at a uniform state and thus at a uniform temperature. Reversibility is an ideal that some real processes approximate and that is often presented in study exercises. For a reversible process, entropy behaves as a conserved quantity and no change occurs in total entropy. More specifically, total entropy is conserved in a reversible process and not conserved in an irreversible process.[6] One has to be careful about system boundaries. For example, in the Carnot cycle, while the heat flow from the hot reservoir to the cold reservoir represents an increase in entropy, the work output, if reversibly and perfectly stored in some energy storage mechanism, represents a decrease in entropy that could be used to operate the heat engine in reverse and return to the previous state, thus the *total* entropy change is still zero at all times if the entire process is reversible. Any process that does not meet the requirements of a reversible process must be treated as an irreversible process, which is usually a complex task. An irreversible process increases entropy.[7]

Heat *transfer* situations require two or more non-isolated systems in thermal contact. In irreversible heat transfer, heat energy is irreversibly transferred from the higher temperature system to the lower temperature system, and the combined entropy of the systems increases. Each system, by definition, must have its own absolute temperature applicable within all areas in each respective system in order to calculate the entropy transfer. Thus, when a system at higher temperature T_H transfers heat dQ to a system of lower temperature T_C, the former loses entropy dQ/T_H and the latter gains entropy dQ/T_C. Since $T_H > T_C$, it follows that $dQ/T_H < dQ/T_C$, whence there is a net gain in the combined entropy.

3.2.3   Carnot cycle

The concept of entropy arose from Rudolf Clausius's study of the Carnot cycle.[8] In a Carnot cycle, heat Q_H is absorbed at temperature T_H from a 'hot' reservoir (an isothermal process), and given up as heat Q_C to a 'cold' reservoir (isothermal process) at T_C. According to Carnot's principle, work can only be produced by the system when there is a temperature difference, and the work should be some function of the difference in temperature and the heat absorbed (Q_H). Carnot did not distinguish between Q_H and Q_C, since he was using the incorrect hypothesis that caloric theory was valid, and hence heat was conserved (the incorrect assumption that Q_H and Q_C were equal) when, in fact, $Q_H > Q_C$.[9] Through the efforts of Clausius and Kelvin, it is now known that the maximum work that a system can produce is the product of the Carnot efficiency and the heat of the hot reservoir:

Equation 1:

$$W = \left(\frac{T_H - T_C}{T_H} \right) Q_H = \left(1 - \frac{T_C}{T_H} \right) Q_H$$

In order to derive the Carnot efficiency, $1 - \frac{T_C}{T_H}$ (a number less than one), Kelvin had to evaluate the ratio of the work output to the heat absorbed during the isothermal expansion with the help of the Carnot-Clapeyron equation which contained an unknown function, known as the Carnot function. The possibility that the Carnot function could be the temperature as measured from a zero temperature, was suggested by Joule in a letter to Kelvin. This allowed Kelvin to establish his absolute temperature scale.[10] It is also known that the work produced by the system is the difference between the heat of the hot reservoir and the heat of the cold reservoir:

Equation 2:

$$W = Q_H - Q_C$$

Since the latter is valid over the entire cycle, this gave Clausius the hint that at each stage of the cycle, work and heat would not be equal, but rather their difference would be a state function that would vanish upon completion of the cycle. The state function was called the internal energy and it became the first law of thermodynamics.[11]

Now equating the two expressions gives

$$\frac{Q_H}{T_H} - \frac{Q_C}{T_C} = 0$$

or

$$\frac{Q_H}{T_H} = \frac{Q_C}{T_C}$$

This implies that there is a function of state which is conserved over a complete cycle of the Carnot cycle. Clausius called this state function *entropy*. One can see that entropy was discovered through mathematics rather than through laboratory results. It is a mathematical construct and has no easy physical analogy. This makes the concept somewhat obscure or abstract, akin to how the concept of energy arose.

Clausius then asked what would happen if there should be less work produced by the system than that predicted by Carnot's principle. The right-hand side of the first equation would be the upper bound of the work output by the system, which would now be converted into an inequality $W < \left(1 - \frac{T_C}{T_H}\right) Q_H$ When the second equation is used to express the work as a difference in heats, we get $Q_H - Q_C < \left(1 - \frac{T_C}{T_H}\right) Q_H$ or $Q_C > \frac{T_C}{T_H} Q_H$ So more heat is given off to the cold reservoir than in the Carnot cycle. If we denote the entropies by $S_i = Q_i/T_i$ for the two states, then the above inequality can be written as a decrease in the entropy

$$S_H - S_C < 0$$

or

$$S_H < S_C$$

In other words, the entropy that leaves the system is greater than the entropy that entered the system, implying that some irreversible process prevented the cycle from outputting the maximum amount of work as predicted by the Carnot equation.

The Carnot cycle and efficiency are invaluable as they define the upper bound of the possible work output and the efficiency of any classical thermodynamic system. Other cycles such as the Otto cycle, Diesel cycle, Brayton cycle etcetera can be analyzed from the standpoint of the Carnot cycle. Any machine or process that is claimed to produce an efficiency greater than the Carnot efficiency is not viable as it would violate the second law of thermodynamics. For very small numbers of particles in the system, statistical thermodynamics must be used. The efficiency of devices such as photovoltaic cells require an analysis from the standpoint of quantum mechanics.

3.2.4 Classical thermodynamics

Main article: Entropy (classical thermodynamics)

The thermodynamic definition of entropy was developed in the early 1850s by Rudolf Clausius and essentially describes how to measure the entropy of an isolated system in thermodynamic equilibrium with its parts. Clausius created the term entropy as an extensive thermodynamic variable that was shown to be useful in characterizing the Carnot cycle. Heat transfer along the isotherm steps of the Carnot cycle was found to be proportional to the temperature of a system (known as its absolute temperature). This relationship was expressed in increments of entropy equal to the ratio of incremental heat transfer divided by temperature, which was found to vary in the thermodynamic cycle but eventually return to the same value at the end of every cycle. Thus it was found to be a function of state, specifically a thermodynamic state of the system. Clausius wrote that he "intentionally formed the word Entropy as similar as possible to the word Energy", basing the term on the Greek ἡ τροπή *trope*, "transformation".[12][note 1]

While Clausius based his definition on a reversible process, there are also irreversible processes that change entropy. Following the second law of thermodynamics, entropy of an isolated system always increases. The difference between an isolated system and closed system is that heat may *not* flow to and from an isolated system, but heat flow to and from a closed system is possible. Nevertheless, for both closed and isolated systems, and indeed, also in open systems, irreversible thermodynamics processes may occur.

According to the Clausius equality, for a reversible cyclic process: $\oint \frac{\delta Q_{rev}}{T} = 0$. This means the line integral $\int_L \frac{\delta Q_{rev}}{T}$ is path-independent.

So we can define a state function S called entropy, which satisfies $dS = \frac{\delta Q_{rev}}{T}$.

To find the entropy difference between any two states of a system, the integral must be evaluated for some reversible path between the initial and final states.[13] Since entropy is a state function, the entropy change of the system for an irreversible path will be the same as for a reversible path between the same two states.[14] However, the entropy change of the surroundings will be different.

We can only obtain the change of entropy by integrating the above formula. To obtain the absolute value of the entropy, we need the third law of thermodynamics, which states that $S = 0$ at absolute zero for perfect crystals.

From a macroscopic perspective, in classical thermodynamics the entropy is interpreted as a state function of a thermodynamic system: that is, a property depending only on the current state of the system, independent of how that state came to be achieved. In any process where the system gives up energy ΔE, and its entropy falls by ΔS, a quantity at least $TR \, \Delta S$ of that energy must be given up to the system's surroundings as unusable heat (TR is the temperature of the system's external surroundings). Otherwise the process will not go forward. In classical thermodynamics, the entropy of a system is defined only if it is in thermodynamic equilibrium.

3.2.5   Statistical mechanics

The statistical definition was developed by Ludwig Boltzmann in the 1870s by analyzing the statistical behavior of the microscopic components of the system. Boltzmann showed that this definition of entropy was equivalent to the thermodynamic entropy to within a constant number which has since been known as Boltzmann's constant. In summary, the thermodynamic definition of entropy provides the experimental definition of entropy, while the statistical definition of entropy extends the concept, providing an explanation and a deeper understanding of its nature.

The interpretation of entropy in statistical mechanics is the measure of uncertainty, or *mixedupness* in the phrase of Gibbs, which remains about a system after its observable macroscopic properties, such as temperature, pressure and volume, have been taken into account. For a given set of macroscopic variables, the entropy measures the degree to which the probability of the system is spread out over different possible microstates. In contrast to the macrostate, which characterizes plainly observable average quantities, a microstate specifies all molecular details about the system including the position and velocity of every molecule. The more such states available to the system with appreciable probability, the greater the entropy. In statistical mechanics, entropy is a measure of the number of ways in which a system may be arranged, often taken to be a measure of "disorder" (the higher the entropy, the higher the disorder).[15][16][17] This definition describes the entropy as being proportional to the natural logarithm of the number of possible microscopic configurations of the individual atoms and molecules of the system (microstates) which could give rise to the observed macroscopic state (macrostate) of the system. The constant of proportionality is the Boltzmann constant.

Specifically, entropy is a logarithmic measure of the number of states with significant probability of being occupied:

$$S = -k_B \sum_i p_i \ln p_i,$$

where k_B is the Boltzmann constant, equal to 1.38065×10^{-23} J/K. The summation is over all the possible microstates of the system, and p_i is the probability that the system is in the i-th microstate.[18] This definition assumes that the basis set of states has been picked so that there is no information on their relative phases. In a different basis set, the more general expression is

$$S = -k_B \text{Tr} \left(\hat{\rho} \ln(\hat{\rho}) \right).$$

where $\hat{\rho}$ is the density matrix, Tr is trace (linear algebra) and ln is the matrix logarithm. This density matrix formulation is not needed in cases of thermal equilibrium so long as the basis states are chosen to be energy eigenstates. For most practical purposes, this can be taken as the fundamental definition of entropy since all other formulas for S can be mathematically derived from it, but not vice versa.

In what has been called *the fundamental assumption of statistical thermodynamics* or *the fundamental postulate in statistical mechanics*, the occupation of any microstate is assumed to be equally probable (i.e. $P_i = 1/\Omega$, where Ω is the number of microstates); this assumption is usually justified for an isolated system in equilibrium.[19] Then the previous equation reduces to

$$S = k_B \ln \Omega.$$

In thermodynamics, such a system is one in which the volume, number of molecules, and internal energy are fixed (the microcanonical ensemble).

The most general interpretation of entropy is as a measure of our uncertainty about a system. The equilibrium state of a system maximizes the entropy because we have lost all information about the initial conditions except for the conserved variables; maximizing the entropy maximizes our ignorance about the details of the system.[20] This uncertainty is not of the everyday subjective kind, but rather the uncertainty inherent to the experimental method and interpretative model.

The interpretative model has a central role in determining entropy. The qualifier "for a given set of macroscopic variables" above has deep implications: if two observers use different sets of macroscopic variables, they will observe different entropies. For example, if observer A uses the variables U, V and W, and observer B uses U, V, W, X, then, by changing X, observer B can cause an effect that looks like a violation of the second law of thermodynamics to observer A. In other words: the set of macroscopic variables one chooses must include everything that may change in the experiment, otherwise one might see decreasing entropy![21]

Entropy can be defined for any Markov processes with reversible dynamics and the detailed balance property.

In Boltzmann's 1896 *Lectures on Gas Theory*, he showed that this expression gives a measure of entropy for systems of atoms and molecules in the gas phase, thus providing a measure for the entropy of classical thermodynamics.

3.2.6 Entropy of a system

Entropy is the above-mentioned unexpected and, to some, obscure integral that arises directly from the Carnot cycle. It is reversible heat divided by temperature. It is, remarkably, a function of state and it is fundamental and very useful.

In a thermodynamic system, pressure, density, and temperature tend to become uniform over time because this equilibrium state has higher probability (more possible combinations of microstates) than any other; see statistical mechanics. As an example, for a glass of ice water in air at room temperature, the difference in temperature between a warm room (the surroundings) and cold glass of ice and water (the system and not part of the room), begins to be equalized as portions of the thermal energy from the warm surroundings spread to the cooler system of ice and water. Over time the temperature of the glass and its contents and the temperature of the room become equal. The entropy of the room has decreased as some of its energy has been dispersed to the ice and water. However, as calculated in the example, the entropy of

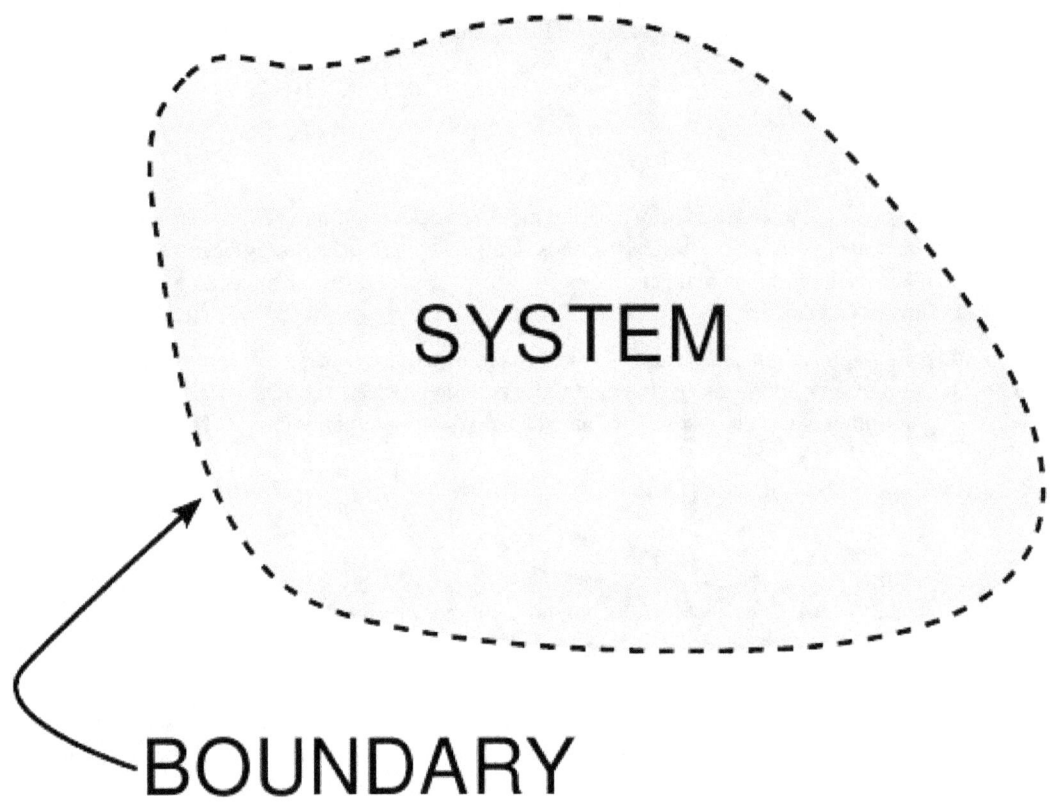

A thermodynamic system

the system of ice and water has increased more than the entropy of the surrounding room has decreased. In an isolated system such as the room and ice water taken together, the dispersal of energy from warmer to cooler always results in a net increase in entropy. Thus, when the "universe" of the room and ice water system has reached a temperature equilibrium, the entropy change from the initial state is at a maximum. The entropy of the thermodynamic system is a measure of how far the equalization has progressed.

Thermodynamic entropy is a non-conserved state function that is of great importance in the sciences of physics and chemistry.[15][22] Historically, the concept of entropy evolved in order to explain why some processes (permitted by conservation laws) occur spontaneously while their time reversals (also permitted by conservation laws) do not; systems tend to progress in the direction of increasing entropy.[23][24] For isolated systems, entropy never decreases.[22] This fact has several important consequences in science: first, it prohibits "perpetual motion" machines; and second, it implies the arrow of entropy has the same direction as the arrow of time. Increases in entropy correspond to irreversible changes in a system, because some energy is expended as waste heat, limiting the amount of work a system can do.[15][16][25][26]

Unlike many other functions of state, entropy cannot be directly observed but must be calculated. Entropy can be calculated for a substance as the standard molar entropy from absolute zero (also known as absolute entropy) or as a difference in entropy from some other reference state which is defined as zero entropy. Entropy has the dimension of energy divided by temperature, which has a unit of joules per kelvin (J/K) in the International System of Units. While these are the same units as heat capacity, the two concepts are distinct.[27] Entropy is not a conserved quantity: for example, in an isolated

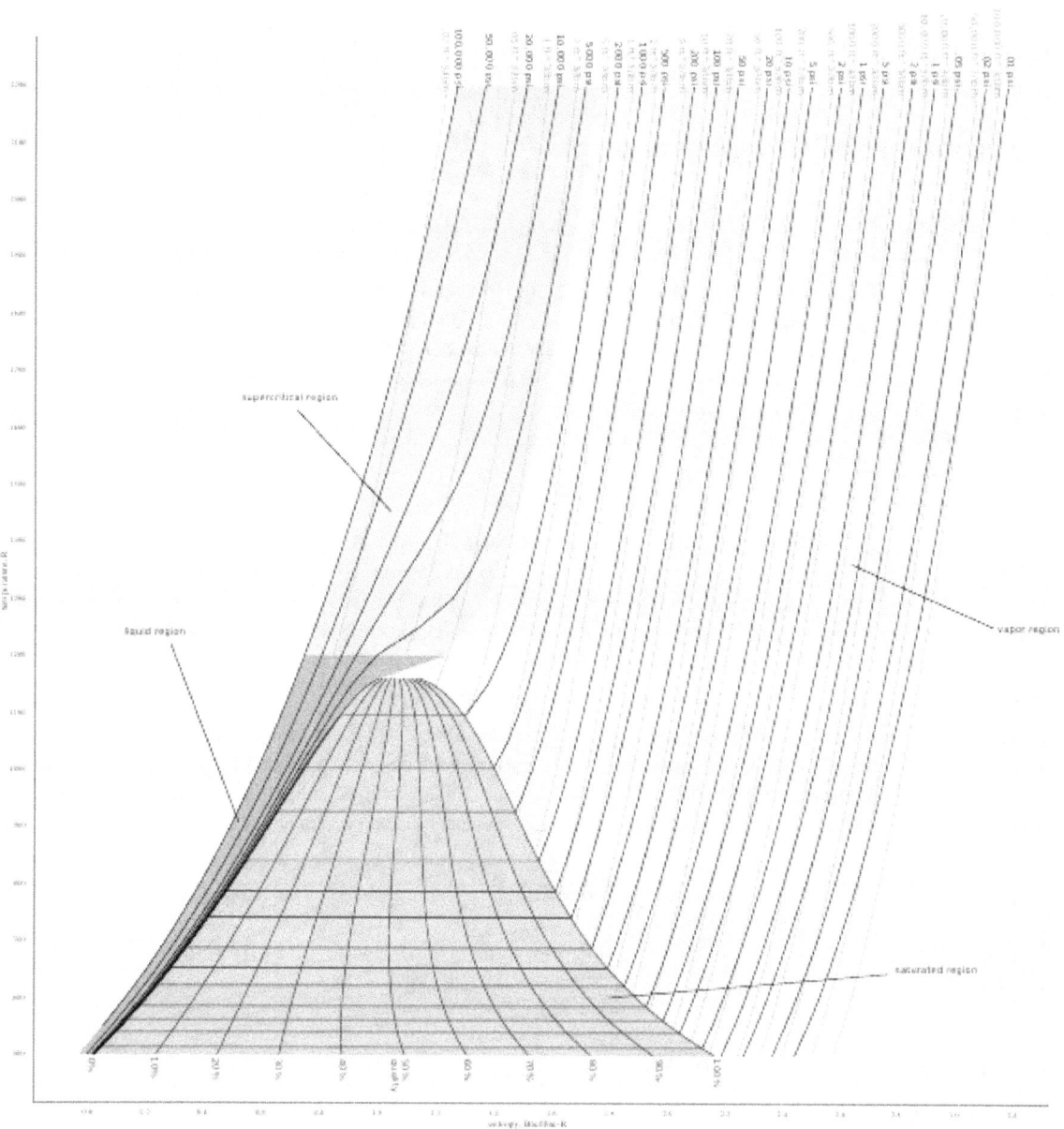

A temperature–entropy diagram for steam. The vertical axis represents uniform temperature, and the horizontal axis represents specific entropy. Each dark line on the graph represents constant pressure, and these form a mesh with light gray lines of constant volume. (Dark-blue is liquid water, light-blue is boiling water, and faint-blue is steam. Grey-blue represents supercritical liquid water.)

system with non-uniform temperature, heat might irreversibly flow and the temperature become more uniform such that entropy increases. The second law of thermodynamics, states that a closed system has entropy which may increase or otherwise remain constant. Chemical reactions cause changes in entropy and entropy plays an important role in determining in which direction a chemical reaction spontaneously proceeds.

One dictionary definition of entropy is that it is "a measure of thermal energy per unit temperature that is not available for useful work". For instance, a substance at uniform temperature is at maximum entropy and cannot drive a heat engine. A substance at non-uniform temperature is at a lower entropy (than if the heat distribution is allowed to even out) and some of the thermal energy can drive a heat engine.

A special case of entropy increase, the entropy of mixing, occurs when two or more different substances are mixed. If the substances are at the same temperature and pressure, there will be no net exchange of heat or work – the entropy change will be entirely due to the mixing of the different substances. At a statistical mechanical level, this results due to the change in available volume per particle with mixing.[28]

3.3   Second law of thermodynamics

Main article: Second law of thermodynamics

The second law of thermodynamics requires that, in general, the total entropy of any system will not decrease other than by increasing the entropy of some other system. Hence, in a system isolated from its environment, the entropy of that system will tend not to decrease. It follows that heat will not flow from a colder body to a hotter body without the application of work (the imposition of order) to the colder body. Secondly, it is impossible for any device operating on a cycle to produce net work from a single temperature reservoir; the production of net work requires flow of heat from a hotter reservoir to a colder reservoir, or a single expanding reservoir undergoing adiabatic cooling, which performs adiabatic work. As a result, there is no possibility of a perpetual motion system. It follows that a reduction in the increase of entropy in a specified process, such as a chemical reaction, means that it is energetically more efficient.

It follows from the second law of thermodynamics that the entropy of a system that is not isolated may decrease. An air conditioner, for example, may cool the air in a room, thus reducing the entropy of the air of that system. The heat expelled from the room (the system), which the air conditioner transports and discharges to the outside air, will always make a bigger contribution to the entropy of the environment than will the decrease of the entropy of the air of that system. Thus, the total of entropy of the room plus the entropy of the environment increases, in agreement with the second law of thermodynamics.

In mechanics, the second law in conjunction with the fundamental thermodynamic relation places limits on a system's ability to do useful work.[29] The entropy change of a system at temperature T absorbing an infinitesimal amount of heat δq in a reversible way, is given by $\delta q/T$. More explicitly, an energy $T_R S$ is not available to do useful work, where T_R is the temperature of the coldest accessible reservoir or heat sink external to the system. For further discussion, see *Exergy*.

Statistical mechanics demonstrates that entropy is governed by probability, thus allowing for a decrease in disorder even in an isolated system. Although this is possible, such an event has a small probability of occurring, making it unlikely.[30]

3.4   Applications

3.4.1   The fundamental thermodynamic relation

Main article: Fundamental thermodynamic relation

The entropy of a system depends on its internal energy and the external parameters, such as the volume. In the thermodynamic limit this fact leads to an equation relating the change in the internal energy to changes in the entropy and the external parameters. This relation is known as the fundamental thermodynamic relation. If the volume is the only external parameter, this relation is:

$$dU = TdS - PdV$$

Since the internal energy is fixed when one specifies the entropy and the volume, this relation is valid even if the change from one state of thermal equilibrium to another with infinitesimally larger entropy and volume happens in a non-quasistatic way (so during this change the system may be very far out of thermal equilibrium and then the entropy, pressure and temperature may not exist).

The fundamental thermodynamic relation implies many thermodynamic identities that are valid in general, independent of the microscopic details of the system. Important examples are the Maxwell relations and the relations between heat capacities.

3.4.2   Entropy in chemical thermodynamics

Thermodynamic entropy is central in chemical thermodynamics, enabling changes to be quantified and the outcome of reactions predicted. The second law of thermodynamics states that entropy in an isolated system – the combination of a subsystem under study and its surroundings – increases during all spontaneous chemical and physical processes. The Clausius equation of $\delta q_{rev}/T = \Delta S$ introduces the measurement of entropy change, ΔS. Entropy change describes the direction and quantifies the magnitude of simple changes such as heat transfer between systems – always from hotter to cooler spontaneously.

The thermodynamic entropy therefore has the dimension of energy divided by temperature, and the unit joule per kelvin (J/K) in the International System of Units (SI).

Thermodynamic entropy is an extensive property, meaning that it scales with the size or extent of a system. In many processes it is useful to specify the entropy as an intensive property independent of the size, as a specific entropy characteristic of the type of system studied. Specific entropy may be expressed relative to a unit of mass, typically the kilogram (unit: $Jkg^{-1}K^{-1}$). Alternatively, in chemistry, it is also referred to one mole of substance, in which case it is called the *molar entropy* with a unit of $Jmol^{-1}K^{-1}$.

Thus, when one mole of substance at about 0K is warmed by its surroundings to 298K, the sum of the incremental values of q_{rev}/T constitute each element's or compound's standard molar entropy, an indicator of the amount of energy stored by a substance at 298K.[31][32] Entropy change also measures the mixing of substances as a summation of their relative quantities in the final mixture.[33]

Entropy is equally essential in predicting the extent and direction of complex chemical reactions. For such applications, ΔS must be incorporated in an expression that includes both the system and its surroundings, $\Delta S_{universe} = \Delta S_{surroundings} + \Delta S_{system}$. This expression becomes, via some steps, the Gibbs free energy equation for reactants and products in the system: ΔG [the Gibbs free energy change of the system] $= \Delta H$ [the enthalpy change] $-T \Delta S$ [the entropy change].[31]

3.4.3   Entropy balance equation for open systems

In chemical engineering, the principles of thermodynamics are commonly applied to "open systems", i.e. those in which heat, work, and mass flow across the system boundary. Flows of both heat ($\dot{Q}$) and work, i.e. $\dot{W}_S$ (shaft work) and $P(dV/dt)$ (pressure-volume work), across the system boundaries, in general cause changes in the entropy of the system. Transfer as heat entails entropy transfer $\dot{Q}/T$, where T is the absolute thermodynamic temperature of the system at the point of the heat flow. If there are mass flows across the system boundaries, they will also influence the total entropy of the system. This account, in terms of heat and work, is valid only for cases in which the work and heat transfers are by paths physically distinct from the paths of entry and exit of matter from the system.[34][35]

To derive a generalized entropy balanced equation, we start with the general balance equation for the change in any extensive quantity Θ in a thermodynamic system, a quantity that may be either conserved, such as energy, or non-conserved, such as entropy. The basic generic balance expression states that $d\Theta/dt$, i.e. the rate of change of Θ in the system, equals the rate at which Θ enters the system at the boundaries, minus the rate at which Θ leaves the system across the system boundaries, plus the rate at which Θ is generated within the system. For an open thermodynamic system in which heat and work are transferred by paths separate from the paths for transfer of matter, using this generic balance equation, with respect to the rate of change with time of the extensive quantity entropy S, the entropy balance equation is:[36]

$$\frac{dS}{dt} = \sum_{k=1}^{K} \dot{M}_k \hat{S}_k + \frac{\dot{Q}}{T} + \dot{S}_{gen}$$

where

$\sum_{k=1}^{K} \dot{M}_k \hat{S}_k$ = the net rate of entropy flow due to the flows of mass into and out of the system (where $\hat{S}$ = entropy per unit mass).

$\frac{\dot{Q}}{T}$ = the rate of entropy flow due to the flow of heat across the system boundary.

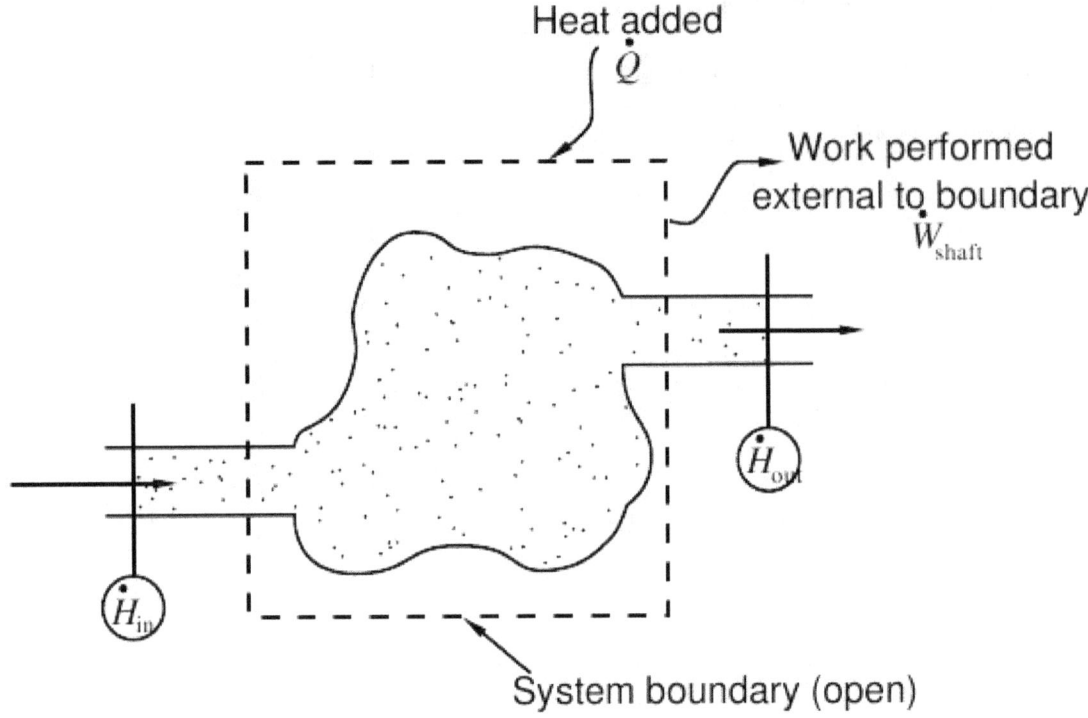

During steady-state continuous operation, an entropy balance applied to an open system accounts for system entropy changes related to heat flow and mass flow across the system boundary.

$\dot{S}_{gen}$ = the rate of entropy production within the system. This entropy production arises from processes within the system, including chemical reactions, internal matter diffusion, internal heat transfer, and frictional effects such as viscosity occurring within the system from mechanical work transfer to or from the system.

Note, also, that if there are multiple heat flows, the term $\dot{Q}/T$ will be replaced by $\sum \dot{Q}_j/T_j$, where $\dot{Q}_j$ is the heat flow and T_j is the temperature at the *jth* heat flow port into the system.

3.5 Entropy change formulas for simple processes

For certain simple transformations in systems of constant composition, the entropy changes are given by simple formulas.[37]

3.5.1 Isothermal expansion or compression of an ideal gas

For the expansion (or compression) of an ideal gas from an initial volume V_0 and pressure P_0 to a final volume V and pressure P at any constant temperature, the change in entropy is given by:

$$\Delta S = nR \ln \frac{V}{V_0} = -nR \ln \frac{P}{P_0}.$$

Here n is the number of moles of gas and R is the ideal gas constant. These equations also apply for expansion into a finite vacuum or a throttling process, where the temperature, internal energy and enthalpy for an ideal gas remain constant.

3.5.2 Cooling and heating

For heating (or cooling) of any system (gas, liquid or solid) at constant pressure from an initial temperature T_0 to a final temperature T, the entropy change is

$$\Delta S = nC_P \ln \frac{T}{T_0}$$

provided that the constant-pressure molar heat capacity (or specific heat) CP is constant and that no phase transition occurs in this temperature interval.

Similarly at constant volume, the entropy change is

$$\Delta S = nC_v \ln \frac{T}{T_0}$$

where the constant-volume heat capacity C_v is constant and there is no phase change.

At low temperatures near absolute zero, heat capacities of solids quickly drop off to near zero, so the assumption of constant heat capacity does not apply.[38]

Since entropy is a state function, the entropy change of any process in which temperature and volume both vary is the same as for a path divided into two steps - heating at constant volume and expansion at constant temperature. For an ideal gas, the total entropy change is[39]

$$\Delta S = nC_v \ln \frac{T}{T_0} + nR \ln \frac{V}{V_0}$$

Similarly if the temperature and pressure of an ideal gas both vary,

$$\Delta S = nC_P \ln \frac{T}{T_0} - nR \ln \frac{P}{P_0}$$

3.5.3 Phase transitions

Reversible phase transitions occur at constant temperature and pressure. The reversible heat is the enthalpy change for the transition, and the entropy change is the enthalpy change divided by the thermodynamic temperature. For fusion (melting) of a solid to a liquid at the melting point T_m, the entropy of fusion is

$$\Delta S_{fus} = \frac{\Delta H_{fus}}{T_m}.$$

Similarly, for vaporization of a liquid to a gas at the boiling point T_b, the entropy of vaporization is

$$\Delta S_{vap} = \frac{\Delta H_{vap}}{T_b}.$$

3.6 Approaches to understanding entropy

As a fundamental aspect of thermodynamics and physics, several different approaches to entropy beyond that of Clausius and Boltzmann are valid.

3.6.1 Standard textbook definitions

The following is a list of additional definitions of entropy from a collection of textbooks:

- a measure of energy dispersal at a specific temperature.

- a measure of disorder in the universe or of the availability of the energy in a system to do work.[40]

- a measure of a system's thermal energy per unit temperature that is unavailable for doing useful work.[41]

In Boltzmann's definition, entropy is a measure of the number of possible microscopic states (or microstates) of a system in thermodynamic equilibrium. Consistent with the Boltzmann definition, the second law of thermodynamics needs to be re-worded as such that entropy increases over time, though the underlying principle remains the same.

3.6.2 Order and disorder

Main article: Entropy (order and disorder)

Entropy has often been loosely associated with the amount of order or disorder, or of chaos, in a thermodynamic system. The traditional qualitative description of entropy is that it refers to changes in the status quo of the system and is a measure of "molecular disorder" and the amount of wasted energy in a dynamical energy transformation from one state or form to another. In this direction, several recent authors have derived exact entropy formulas to account for and measure disorder and order in atomic and molecular assemblies.[42][43][44] One of the simpler entropy order/disorder formulas is that derived in 1984 by thermodynamic physicist Peter Landsberg, based on a combination of thermodynamics and information theory arguments. He argues that when constraints operate on a system, such that it is prevented from entering one or more of its possible or permitted states, as contrasted with its forbidden states, the measure of the total amount of "disorder" in the system is given by:[43][44]

$$\text{Disorder} = \frac{C_D}{C_I}.$$

Similarly, the total amount of "order" in the system is given by:

$$\text{Order} = 1 - \frac{C_O}{C_I}.$$

In which C_D is the "disorder" capacity of the system, which is the entropy of the parts contained in the permitted ensemble, C_I is the "information" capacity of the system, an expression similar to Shannon's channel capacity, and C_O is the "order" capacity of the system.[42]

3.6.3 Energy dispersal

Main article: Entropy (energy dispersal)

The concept of entropy can be described qualitatively as a measure of energy dispersal at a specific temperature.[45] Similar terms have been in use from early in the history of classical thermodynamics, and with the development of statistical thermodynamics and quantum theory, entropy changes have been described in terms of the mixing or "spreading" of the total energy of each constituent of a system over its particular quantized energy levels.

Ambiguities in the terms *disorder* and *chaos*, which usually have meanings directly opposed to equilibrium, contribute to widespread confusion and hamper comprehension of entropy for most students.[46] As the second law of thermodynamics shows, in an isolated system internal portions at different temperatures will tend to adjust to a single uniform temperature and thus produce equilibrium. A recently developed educational approach avoids ambiguous terms and describes such spreading out of energy as dispersal, which leads to loss of the differentials required for work even though the total energy remains constant in accordance with the first law of thermodynamics[47] (compare discussion in next section). Physical chemist Peter Atkins, for example, who previously wrote of dispersal leading to a disordered state, now writes that "spontaneous changes are always accompanied by a dispersal of energy".[48]

3.6.4 Relating entropy to energy *usefulness*

Following on from the above, it is possible (in a thermal context) to regard entropy as an indicator or measure of the *effectiveness* or *usefulness* of a particular quantity of energy.[49] This is because energy supplied at a high temperature (i.e. with low entropy) tends to be more useful than the same amount of energy available at room temperature. Mixing a hot parcel of a fluid with a cold one produces a parcel of intermediate temperature, in which the overall increase in entropy represents a "loss" which can never be replaced.

Thus, the fact that the entropy of the universe is steadily increasing, means that its total energy is becoming less useful: eventually, this will lead to the "heat death of the Universe".

3.6.5 Entropy and adiabatic accessibility

A definition of entropy based entirely on the relation of adiabatic accessibility between equilibrium states was given by E.H.Lieb and J. Yngvason in 1999.[50] This approach has several predecessors, including the pioneering work of Constantin Carathéodory from 1909 [51] and the monograph by R. Giles from 1964.[52] In the setting of Lieb and Yngvason one starts by picking, for a unit amount of the substance under consideration, two reference states X_0 and X_1 such that the latter is adiabatically accessible from the former but not vice versa. Defining the entropies of the reference states to be 0 and 1 respectively the entropy of a state X is defined as the largest number λ such that X is adiabatically accessible from a composite state consisting of an amount λ in the state X_1 and a complementary amount, $(1 - \lambda)$, in the state X_0. A simple but important result within this setting is that entropy is uniquely determined, apart from a choice of unit and an additive constant for each chemical element, by the following properties: It is monotonic with respect to the relation of adiabatic accessibility, additive on composite systems, and extensive under scaling.

3.6.6 Entropy in quantum mechanics

Main article: von Neumann entropy

In quantum statistical mechanics, the concept of entropy was developed by John von Neumann and is generally referred to as "von Neumann entropy",

$$S = -k_B \text{Tr}(\rho \log \rho)$$

where ρ is the density matrix and Tr is the trace operator.

This upholds the correspondence principle, because in the classical limit, when the phases between the basis states used for the classical probabilities are purely random, this expression is equivalent to the familiar classical definition of entropy,

$$S = -k_B \sum_i p_i \log p_i$$

i.e. in such a basis the density matrix is diagonal.

Von Neumann established a rigorous mathematical framework for quantum mechanics with his work *Mathematische Grundlagen der Quantenmechanik*. He provided in this work a theory of measurement, where the usual notion of wave function collapse is described as an irreversible process (the so-called von Neumann or projective measurement). Using this concept, in conjunction with the density matrix he extended the classical concept of entropy into the quantum domain.

3.6.7   Information theory

I thought of calling it "information", but the word was overly used, so I decided to call it "uncertainty". [...] Von Neumann told me, "You should call it entropy, for two reasons. In the first place your uncertainty function has been used in statistical mechanics under that name, so it already has a name. In the second place, and more important, nobody knows what entropy really is, so in a debate you will always have the advantage."

Conversation between Claude Shannon and John von Neumann regarding what name to give to the attenuation in phone-line signals[53]

Main articles: Entropy (information theory), Entropy in thermodynamics and information theory and Entropic uncertainty

When viewed in terms of information theory, the entropy state function is simply the amount of information (in the Shannon sense) that would be needed to specify the full microstate of the system. This is left unspecified by the macroscopic description.

In information theory, *entropy* is the measure of the amount of information that is missing before reception and is sometimes referred to as *Shannon entropy*.[54] Shannon entropy is a broad and general concept which finds applications in information theory as well as thermodynamics. It was originally devised by Claude Shannon in 1948 to study the amount of information in a transmitted message. The definition of the information entropy is, however, quite general, and is expressed in terms of a discrete set of probabilities p_i *so that*

$$H(X) = -\sum_{i=1}^{n} p(x_i) \log p(x_i).$$

In the case of transmitted messages, these probabilities were the probabilities that a particular message was actually transmitted, and the entropy of the message system was a measure of the average amount of information in a message. For the case of equal probabilities (i.e. each message is equally probable), the Shannon entropy (in bits) is just the number of yes/no questions needed to determine the content of the message.[18]

The question of the link between information entropy and thermodynamic entropy is a debated topic. While most authors argue that there is a link between the two,[55][56][57][58][59] a few argue that they have nothing to do with each other.[18]

The expressions for the two entropies are similar. If W is the number of microstates that can yield a given macrostate, and each microstate has the same *A priori* probability, then that probability is $p = 1/W$. The Shannon entropy (in nats) will be:

$$H = -\sum_{i=1}^{W} p \log(p) = \log(W)$$

and if entropy is measured in units of k per nat, then the entropy is given[60] by:

$$H = k \log(W)$$

which is the famous Boltzmann entropy formula when k is Boltzmann's constant, which may be interpreted as the thermodynamic entropy per nat. There are many ways of demonstrating the equivalence of "information entropy" and "physics entropy", that is, the equivalence of "Shannon entropy" and "Boltzmann entropy". Nevertheless, some authors argue for dropping the word entropy for the H function of information theory and using Shannon's other term "uncertainty" instead.[61]

3.7 Interdisciplinary applications of entropy

Although the concept of entropy was originally a thermodynamic construct, it has been adapted in other fields of study, including information theory, psychodynamics, thermoeconomics/ecological economics, and evolution.[42][62][63][64] For instance, an entropic argument has been recently proposed for explaining the preference of cave spiders in choosing a suitable area for laying their eggs. [65]

3.7.1 Thermodynamic and statistical mechanics concepts

- Entropy unit – a non-S.I. unit of thermodynamic entropy, usually denoted "e.u." and equal to one calorie per Kelvin per mole, or 4.184 Joules per Kelvin per mole.[66]

- Gibbs entropy – the usual statistical mechanical entropy of a thermodynamic system.

- Boltzmann entropy – a type of Gibbs entropy, which neglects internal statistical correlations in the overall particle distribution.

- Tsallis entropy – a generalization of the standard Boltzmann-Gibbs entropy.

- Standard molar entropy – is the entropy content of one mole of substance, under conditions of standard temperature and pressure.

- Residual entropy – the entropy present after a substance is cooled arbitrarily close to absolute zero.

- Entropy of mixing – the change in the entropy when two different chemical substances or components are mixed.

- Loop entropy – is the entropy lost upon bringing together two residues of a polymer within a prescribed distance.

- Conformational entropy – is the entropy associated with the physical arrangement of a polymer chain that assumes a compact or globular state in solution.

- Entropic force – a microscopic force or reaction tendency related to system organization changes, molecular frictional considerations, and statistical variations.

- Free entropy – an entropic thermodynamic potential analogous to the free energy.

- Entropic explosion – an explosion in which the reactants undergo a large change in volume without releasing a large amount of heat.

- Entropy change – a change in entropy dS between two equilibrium states is given by the heat transferred $dQrev$ divided by the absolute temperature T of the system in this interval.

- Sackur-Tetrode entropy – the entropy of a monatomic classical ideal gas determined via quantum considerations.

3.7.2 The arrow of time

Main article: Entropy (arrow of time)

Entropy is the only quantity in the physical sciences that seems to imply a particular direction of progress, sometimes called an arrow of time. As time progresses, the second law of thermodynamics states that the entropy of an isolated system never decreases. Hence, from this perspective, entropy measurement is thought of as a kind of clock.

3.7.3   Cosmology

Main article: Heat death of the universe

Since a finite universe is an isolated system, the Second Law of Thermodynamics states that its total entropy is constantly increasing. It has been speculated, since the 19th century, that the universe is fated to a heat death in which all the energy ends up as a homogeneous distribution of thermal energy, so that no more work can be extracted from any source.

If the universe can be considered to have generally increasing entropy, then – as Sir Roger Penrose has pointed out – gravity plays an important role in the increase because gravity causes dispersed matter to accumulate into stars, which collapse eventually into black holes. The entropy of a black hole is proportional to the surface area of the black hole's event horizon.[67] Jacob Bekenstein and Stephen Hawking have shown that black holes have the maximum possible entropy of any object of equal size. This makes them likely end points of all entropy-increasing processes, if they are totally effective matter and energy traps. However, the escape of energy from black holes might be possible due to quantum activity, see Hawking radiation. Hawking has recently changed his stance on some details, in a paper which largely redefined the event horizons of black holes.[68]

The role of entropy in cosmology remains a controversial subject since the time of Ludwig Boltzmann. Recent work has cast some doubt on the heat death hypothesis and the applicability of any simple thermodynamic model to the universe in general. Although entropy does increase in the model of an expanding universe, the maximum possible entropy rises much more rapidly, moving the universe further from the heat death with time, not closer.[69] [70] [71] This results in an "entropy gap" pushing the system further away from the posited heat death equilibrium.[72] Other complicating factors, such as the energy density of the vacuum and macroscopic quantum effects, are difficult to reconcile with thermodynamical models, making any predictions of large-scale thermodynamics extremely difficult.[73]

The entropy gap is widely believed to have been originally opened up by the early rapid exponential expansion of the universe.

3.8   See also

- Autocatalytic reactions and order creation

- Brownian ratchet

- Clausius–Duhem inequality

- Configuration entropy

- Departure function

- Enthalpy

- Entropy (information theory)

- Entropy (computing)

- Entropy and life

- Entropy (order and disorder)

- Entropy rate

- Geometrical frustration

- Laws of thermodynamics

- Multiplicity function

- Orders of magnitude (entropy)

- Stirling's formula

- Thermodynamic databases for pure substances

- Thermodynamic potential

3.9   Notes

[1] A machine in this context includes engineered devices as well as biological organisms.

3.10   References

[1] "Carnot, Sadi (1796–1832)". Wolfram Research. 2007. Retrieved 2010-02-24.

[2] McCulloch, Richard, S. (1876). *Treatise on the Mechanical Theory of Heat and its Applications to the Steam-Engine, etc.* D. Van Nostrand.

[3] Clausius, Rudolf (1850). *On the Motive Power of Heat, and on the Laws which can be deduced from it for the Theory of Heat.* Poggendorff's *Annalen der Physick,* LXXIX (Dover Reprint). ISBN 0-486-59065-8.

[4] *The scientific papers of J. Willard Gibbs in Two Volumes* **1**. Longmans, Green, and Co. 1906. p. 11. Retrieved 2011-02-26.

[5] J. A. McGovern, 2.5 Entropy at the Wayback Machine (archived September 23, 2012)

[6] Irreversibility, Entropy Changes, and "Lost Work" Thermodynamics and Propulsion, Z. S. Spakovszky, 2002

[7] What is entropy? Thermodynamics of Chemical Equilibrium by S. Lower, 2007

[8] B. H. Lavenda, "A New Perspective on Thermodynamics" Springer, 2009, Sec. 2.3.4.

[9] S. Carnot, "Reflexions on the Motive Power of Fire", translated and annotated by R. Fox, Manchester University Press, 1986, p. 26; C. Truesdell, "The Tragicomical History of Thermodynamics, Springer, 1980, pp. 78–85

[10] J. Clerk-Maxwell, "Theory of Heat", 10th ed. Longmans, Green and Co., 1891, pp. 155–158.

[11] R. Clausius, "The Mechanical Theory of Heat", translated by T. Archer Hirst, van Voorst, 1867, p. 28

[12] Clausius, Rudolf (1865). *Ueber verschiedene für die Anwendung bequeme Formen der Hauptgleichungen der mechanischen Wärmetheorie: vorgetragen in der naturforsch. Gesellschaft den 24. April 1865.* p. 46.

[13] Atkins, Peter; Julio De Paula (2006). *Physical Chemistry.* 8th ed. Oxford University Press. p. 79. ISBN 0-19-870072-5.

[14] Engel, Thomas; Philip Reid (2006). *Physical Chemistry.* Pearson Benjamin Cummings. p. 86. ISBN 0-8053-3842-X.

[15] *McGraw-Hill Concise Encyclopedia of Chemistry,* 2004

[16] Sethna, J. *Statistical Mechanics* Oxford University Press 2006 p. 78

[17] Barnes & Noble's *Essential Dictionary of Science,* 2004

[18] Frigg, R. and Werndl, C. "Entropy – A Guide for the Perplexed". In Probabilities in Physics; Beisbart C. and Hartmann, S. Eds; Oxford University Press, Oxford, 2010

[19] Schroeder, Daniel V. *An Introduction to Thermal Physics.* Addison Wesley Longman, 1999, p. 57

[20] "EntropyOrderParametersComplexity.pdf www.physics.cornell.edu" (PDF). Retrieved 2012-08-17.

[21] "Jaynes, E. T., "The Gibbs Paradox," In Maximum Entropy and Bayesian Methods; Smith, C. R; Erickson, G. J; Neudorfer, P. O., Eds; Kluwer Academic: Dordrecht, 1992, pp. 1–22" (PDF). Retrieved 2012-08-17.

[22] Sandler S. I., *Chemical and Engineering Thermodynamics, 3rd Ed.* Wiley, New York, 1999 p. 91

[23] McQuarrie D. A., Simon J. D., *Physical Chemistry: A Molecular Approach*, University Science Books, Sausalito 1997 p. 817

[24] Haynie, Donald, T. (2001). *Biological Thermodynamics*. Cambridge University Press. ISBN 0-521-79165-0.

[25] *Oxford Dictionary of Science*, 2005

[26] de Rosnay, Joel (1979). *The Macroscope – a New World View (written by an M.I.T.-trained biochemist)*. Harper & Row, Publishers. ISBN 0-06-011029-5.

[27] J. A. McGovern, Heat Capacities at the Wayback Machine (archived August 19, 2012)

[28] Ben-Naim, Arieh, On the So-Called Gibbs Paradox, and on the Real Paradox, Entropy, 9, pp. 132–136, 2007 Link

[29] Daintith, John (2005). *Oxford Dictionary of Physics*. Oxford University Press. ISBN 0-19-280628-9.

[30] ""Entropy production theorems and some consequences." Physical Review E; Saha, Arnab; Lahiri, Sourabh; Jayannavar, A. M; The American Physical Society: 14 July 2009, pp. 1–10". Link.aps.org. Retrieved 2012-08-17.

[31] Moore, J. W.; C. L. Stanistski; P. C. Jurs (2005). *Chemistry, The Molecular Science*. Brooks Cole. ISBN 0-534-42201-2.

[32] Jungermann, A.H. (2006). "Entropy and the Shelf Model: A Quantum Physical Approach to a Physical Property". *Journal of Chemical Education* **83** (11): 1686–1694. Bibcode:2006JChEd..83.1686J. doi:10.1021/ed083p1686.

[33] Levine, I. N. (2002). *Physical Chemistry, 5th ed.* McGraw-Hill. ISBN 0-07-231808-2.

[34] Born, M. (1949). *Natural Philosophy of Cause and Chance*, Oxford University Press, London, pp. 44, 146–147.

[35] Haase, R. (1971). Survey of Fundamental Laws, chapter 1 of *Thermodynamics*, pages 1–97 of volume 1, ISBN 0122456017, ed. W. Jost, of *Physical Chemistry, An Advanced Treatise*, ed. H. Eyring, D. Henderson, W. Jost, Academic Press, New York, p. 35.

[36] Sandler, Stanley, I. (1989). *Chemical and Engineering Thermodynamics*. John Wiley & Sons. ISBN 0-471-83050-X.

[37] "GRC.nasa.gov". GRC.nasa.gov. 2000-03-27. Retrieved 2012-08-17.

[38] The Third Law Chemistry 433, Stefan Franzen, ncsu.edu

[39] "GRC.nasa.gov". GRC.nasa.gov. 2008-07-11. Retrieved 2012-08-17.

[40] Gribbin's *Q Is for Quantum: An Encyclopedia of Particle Physics*, Free Press ISBN 0-684-85578-X, 2000

[41] "Entropy" *Encyclopædia Britannica*

[42] Brooks, Daniel, R.; Wiley, E.O. (1988). *Evolution as Entropy– Towards a Unified Theory of Biology*. University of Chicago Press. ISBN 0-226-07574-5.

[43] Landsberg, P.T. (1984). "Is Equilibrium always an Entropy Maximum?". *J. Stat. Physics* **35**: 159–169. Bibcode:1984JSP....35..159L. doi:10.1007/bf01017372.

[44] Landsberg, P.T. (1984). "Can Entropy and "Order" Increase Together?". *Physics Letters* **102A**: 171–173. doi:10.1016/0375-9601(84)90934-4.

[45] Frank L. Lambert, A Student's Approach to the Second Law and Entropy

[46] Carson, E. M. and J. R. Watson (Department of Educational and Professional Studies, Kings College, London), *Undergraduate students' understandings of entropy and Gibbs Free energy*, University Chemistry Education – 2002 Papers, Royal Society of Chemistry

[47] Frank L. Lambert, JCE 2002 (79) 187 [Feb] Disorder – A Cracked Crutch for Supporting Entropy Discussions

[48] Atkins, Peter (1984). *The Second Law*. Scientific American Library. ISBN 0-7167-5004-X.

[49] Sandra Saary (Head of Science, Latifa Girls' School, Dubai) (23 February 1993). "Book Review of "A Science Miscellany"". *Khaleej Times* (Galadari Press, UAE): XI.

[50] Elliott H. Lieb, Jakob Yngvason: *The Physics and Mathematics of the Second Law of Thermodynamics*, Phys. Rep. 310, pp. 1–96 (1999)

[51] Constantin Carathéodory: *Untersuchungen über die Grundlagen der Thermodynamik*, Math. Ann., 67, pp. 355–386, 1909

[52] Robin Giles: *Mathematical Foundations* of Thermodynamics", Pergamon, Oxford 1964

[53] M. Tribus, E.C. McIrvine, Energy and information, Scientific American, 224 (September 1971), pp. 178–184

[54] Balian, Roger (2004). "Entropy, a Protean concept". In Dalibard, Jean. *Poincaré Seminar 2003: Bose-Einstein condensation - entropy*. Basel: Birkhäuser. pp. 119–144. ISBN 9783764371166.

[55] Brillouin, Leon (1956). *Science and Information Theory*. ISBN 0-486-43918-6.

[56] Georgescu-Roegen, Nicholas (1971). *The Entropy Law and the Economic Process*. Harvard University Press. ISBN 0-674-25781-2.

[57] Chen, Jing (2005). *The Physical Foundation of Economics – an Analytical Thermodynamic Theory*. World Scientific. ISBN 981-256-323-7.

[58] Kalinin, M.I.; Kononogov, S.A. (2005). "Boltzmann's constant". *Measurement Techniques* 48: 632–636. doi:10.1007/s11018-005-0195-9.

[59] Ben-Naim A. (2008), *Entropy Demystified* (World Scientific).

[60] "Edwin T. Jaynes – Bibliography". Bayes.wustl.edu. 1998-03-02. Retrieved 2009-12-06.

[61] Schneider, Tom, DELILA system (Deoxyribonucleic acid Library Language), (Information Theory Analysis of binding sites), Laboratory of Mathematical Biology, National Cancer Institute, FCRDC Bldg. 469. Rm 144, P.O. Box. B Frederick, MD 21702-1201, USA

[62] Avery, John (2003). *Information Theory and Evolution*. World Scientific. ISBN 981-238-399-9.

[63] Yockey, Hubert, P. (2005). *Information Theory, Evolution, and the Origin of Life*. Cambridge University Press. ISBN 0-521-80293-8.

[64] Chiavazzo, Eliodoro; Fasano, Matteo; Asinari, Pietro. "Inference of analytical thermodynamic models for biological networks". *Physica A: Statistical Mechanics and its Applications* 392: 1122–1132. Bibcode:2013PhyA..392.1122C. doi:10.1016/j.physa.2012.11.030.

[65] Chiavazzo, Eliodoro; Isaia, Marco; Mammola, Stefano; Lepore, Emiliano; Ventola, Luigi; Asinari, Pietro; Pugno, Nicola Maria. "Cave spiders choose optimal environmental factors with respect to the generated entropy when laying their cocoon". *Scientific Reports* 5: 7611. Bibcode:2015NatSR...5E7611C. doi:10.1038/srep07611.

[66] IUPAC, *Compendium of Chemical Terminology*, 2nd ed. (the "Gold Book") (1997). Online corrected version: (2006–) "Entropy unit".

[67] von Baeyer, Christian, H. (2003). *Information–the New Language of Science*. Harvard University Press. ISBN 0-674-01387-5.Srednicki M (August 1993). "Entropy and area". *Phys. Rev. Lett.* 71 (5): 666–669. arXiv:hep-th/9303048. Bibcode:1993Ph doi:10.1103/PhysRevLett.71.666.PMID10055336.Callaway DJE(April1996). "Surface tension,hydrophobicity,and blackholes: The entropic connection".*Phys.Rev.E*53(4):3738–3744.arXiv:cond-mat/9601111.Bibcode:1996PhRvE..53.3738C.doi:10.1103/PhysRevE.53.3738.PMID9964684.

[68] Buchan, Lizzy. "Black holes do not exist, says Stephen Hawking". Cambridge News. Retrieved 27 January 2014.

[69] Layzer, David (1988). *Growth of Order in the Universe*. MIT Press.

[70] Chaisson, Eric J. (2001). *Cosmic Evolution: The Rise of Complexity in Nature*. Harvard University Press. ISBN 0-674-00342-X.

[71] Lineweaver, Charles H.; Davies, Paul C. W.; Ruse, Michael, eds. (2013). *Complexity and the Arrow of Time*. Cambridge University Press. ISBN 978-1-107-02725-1.

[72] Stenger, Victor J. (2007). *God: The Failed Hypothesis*. Prometheus Books. ISBN 1-59102-481-1.

[73] Benjamin Gal-Or (1981, 1983, 1987). *Cosmology, Physics and Philosophy*. Springer Verlag. ISBN 0-387-96526-2. Check date values in: |date= (help)

3.11 Further reading

- Atkins, Peter; Julio De Paula (2006). *Physical Chemistry, 8th ed*. Oxford University Press. ISBN 0-19-870072-5.

- Baierlein, Ralph (2003). *Thermal Physics*. Cambridge University Press. ISBN 0-521-65838-1.

- Ben-Naim, Arieh (2007). *Entropy Demystified*. World Scientific. ISBN 981-270-055-2.

- Callen, Herbert, B (2001). *Thermodynamics and an Introduction to Thermostatistics, 2nd Ed*. John Wiley and Sons. ISBN 0-471-86256-8.

- Chang, Raymond (1998). *Chemistry, 6th Ed*. New York: McGraw Hill. ISBN 0-07-115221-0.

- Cutnell, John, D.; Johnson, Kenneth, J. (1998). *Physics, 4th ed*. John Wiley and Sons, Inc. ISBN 0-471-19113-2.

- Dugdale, J. S. (1996). *Entropy and its Physical Meaning* (2nd ed.). Taylor and Francis (UK); CRC (US). ISBN 0-7484-0569-0.

- Fermi, Enrico (1937). *Thermodynamics*. Prentice Hall. ISBN 0-486-60361-X.

- Goldstein, Martin; Inge, F (1993). *The Refrigerator and the Universe*. Harvard University Press. ISBN 0-674-75325-9.

- Gyftopoulos, E.P.; G.P. Beretta (1991, 2005, 2010). *Thermodynamics. Foundations and Applications*. Dover. ISBN 0-486-43932-1. Check date values in: |date= (help)

- Haddad, Wassim M.; Chellaboina, VijaySekhar; Nersesov, Sergey G. (2005). *Thermodynamics – A Dynamical Systems Approach*. Princeton University Press. ISBN 0-691-12327-6.

- Kroemer, Herbert; Charles Kittel (1980). *Thermal Physics* (2nd ed.). W. H. Freeman Company. ISBN 0-7167-1088-9.

- Lambert, Frank L.; entropysite.oxy.edu

- Penrose, Roger (2005). *The Road to Reality: A Complete Guide to the Laws of the Universe*. New York: A. A. Knopf. ISBN 0-679-45443-8.

- Reif, F. (1965). *Fundamentals of statistical and thermal physics*. McGraw-Hill. ISBN 0-07-051800-9.

- Schroeder, Daniel V. (2000). *Introduction to Thermal Physics*. New York: Addison Wesley Longman. ISBN 0-201-38027-7.

- Serway, Raymond, A. (1992). *Physics for Scientists and Engineers*. Saunders Golden Subburst Series. ISBN 0-03-096026-6.

- Spirax-Sarco Limited, Entropy – A Basic Understanding A primer on entropy tables for steam engineering

- vonBaeyer; Hans Christian (1998). *Maxwell's Demon: Why Warmth Disperses and Time Passes*. Random House. ISBN 0-679-43342-2.

- Entropy for beginners – a wikibook

- An Intuitive Guide to the Concept of Entropy Arising in Various Sectors of Science – a wikibook

3.12 External links

- Entropy and the Second Law of Thermodynamics - an A-level physics lecture with detailed derivation of entropy based on Carnot cycle

- Khan Academy: entropy lectures, part of Chemistry playlist

  - Proof: S (or Entropy) is a valid state variable
  - Thermodynamic Entropy Definition Clarification
  - Reconciling Thermodynamic and State Definitions of Entropy
  - Entropy Intuition
  - More on Entropy

- The Second Law of Thermodynamics and Entropy - Yale OYC lecture, part of Fundamentals of Physics I (PHYS 200)

- Entropy and the Clausius inequality MIT OCW lecture, part of 5.60 Thermodynamics & Kinetics, Spring 2008

- The Discovery of Entropy by Adam Shulman. Hour-long video, January 2013.

- Moriarty, Philip; Merrifield, Michael (2009). "S Entropy". *Sixty Symbols*. Brady Haran for the University of Nottingham.

- Entropy Scholarpedia

Chapter 4

Entropy and life

Research concerning the relationship between the thermodynamic quantity **entropy** and the evolution of **life** began around the turn of the 20th century. In 1910, American historian Henry Adams printed and distributed to university libraries and history professors the small volume *A Letter to American Teachers of History* proposing a theory of history based on the second law of thermodynamics and on the principle of entropy.[1][2] The 1944 book *What is Life?* by Nobel-laureate physicist Erwin Schrödinger stimulated research in the field. In his book, Schrödinger originally stated that life feeds on negative entropy, or negentropy as it is sometimes called, but in a later edition corrected himself in response to complaints and stated the true source is free energy. More recent work has restricted the discussion to Gibbs free energy because biological processes on Earth normally occur at a constant temperature and pressure, such as in the atmosphere or at the bottom of an ocean, but not across both over short periods of time for individual organisms.

4.1   Origin

In 1863, Rudolf Clausius published his noted memoir "On the Concentration of Rays of Heat and Light, and on the Limits of its Action" wherein he outlined a preliminary relationship, as based on his own work and that of William Thomson, between his newly developed concept of entropy and life. Building on this, one of the first to speculate on a possible thermodynamic perspective of evolution was the Austrian physicist Ludwig Boltzmann. In 1875, building on the works of Clausius and Kelvin, Boltzmann reasoned:

> The general struggle for existence of animate beings is not a struggle for raw materials – these, for organisms, are air, water and soil, all abundantly available – nor for energy which exists in plenty in any body in the form of heat, but a *struggle for [negative] entropy*, which becomes available through the transition of energy from the *hot sun* to the *cold earth*.[3]

4.2   Early views

In 1876, American civil engineer Richard Sears McCulloh, in his *Treatise on the Mechanical Theory of Heat and its Application to the Steam-Engine*, which was an early thermodynamics textbook, states, after speaking about the laws of the physical world, that "there are none that are established on a firmer basis than the two general propositions of Joule and Carnot; which constitute the fundamental laws of our subject." McCulloch then goes on to show that these two laws may be combined in a single expression as follows:

$$S = \int \frac{dQ}{\tau}$$

where

S = entropy

dQ = a differential amount of heat passed into a thermodynamic system

τ = absolute temperature

McCulloch then declares that the applications of these two laws, i.e. what are currently known as the first law of thermodynamics and the second law of thermodynamics, are innumerable. He then states:

> When we reflect how generally physical phenomena are connected with thermal changes and relations, it at once becomes obvious that there are few, if any, branches of natural science which are not more or less dependent upon the great truths under consideration. Nor should it, therefore, be a matter of surprise that already, in the short space of time, not yet one generation, elapsed since the mechanical theory of heat has been freely adopted, whole branches of physical science have been revolutionized by it.[4]:p. 267

McCulloch then gives a few examples of what he calls the "more interesting examples" of the application of these laws in extent and utility. The first example he gives, is physiology wherein he states that "the body of an animal, not less than a steamer, or a locomotive, is truly a heat engine, and the consumption of food in the one is precisely analogous to the burning of fuel in the other; in both, the chemical process is the same: that called combustion." He then incorporates a discussion of Lavoisier's theory of respiration with cycles of digestion and excretion, perspiration, but then contradicts Lavoisier with recent findings, such as internal heat generated by friction, according to the new theory of heat, which, according to McCulloch, states that the "heat of the body generally and uniformly is diffused instead of being concentrated in the chest". McCulloch then gives an example of the second law, where he states that friction, especially in the smaller blooded-vessels, must develop heat. Without doubt, animal heat is thus in part produced. He then asks: "but whence the expenditure of energy causing that friction, and which must be itself accounted for?"

To answer this question he turns to the mechanical theory of heat and goes on to loosely outline how the heart is what he calls a "force-pump", which receives blood and sends it to every part of the body, as discovered by William Harvey, that "acts like the piston of an engine and is dependent upon and consequently due to the cycle of nutrition and excretion which sustains physical or organic life." It is likely, here, that McCulloch was modeling parts of this argument on that of the famous Carnot cycle. In conclusion, he summarizes his first and second law argument as such:

> Everything physical being subject to the law of conservation of energy, it follows that no physiological action can take place except with expenditure of energy derived from food; also, that an animal performing mechanical work must from the same quantity of food generate less heat than one abstaining from exertion, the difference being precisely the heat equivalent of that of work.[4]:p. 270

4.3 Negative entropy

Later, building on this premise, in the famous 1944 book *What is Life?*, Nobel-laureate physicist Erwin Schrödinger theorizes that life, contrary to the general tendency dictated by the Second law of thermodynamics, decreases or maintains its entropy by feeding on negative entropy.[5] In his note to Chapter 6 of *What is Life?*, however, Schrödinger remarks on his usage of the term negative entropy:

> Let me say first, that if I had been catering for them [physicists] alone I should have let the discussion turn on *free energy* instead. It is the more familiar notion in this context. But this highly technical term seemed linguistically too near to *energy* for making the average reader alive to the contrast between the two things.

This is what is argued to differentiate life from other forms of matter organization. In this direction, although life's dynamics may be argued to go against the tendency of second law, which states that the entropy of an isolated system tends to increase, it does not in any way conflict or invalidate this law, because the principle that entropy can only increase or remain constant applies only to a closed system which is adiabatically isolated, meaning no heat can enter or leave. Whenever a system can exchange either heat or matter with its environment, an entropy decrease of that system is entirely compatible with the second law.[6] The problem of organization in living systems increasing despite the second law is known as the Schrödinger paradox.[7]

In 1964, James Lovelock was among a group of scientists who were requested by NASA to make a theoretical life detection system to look for life on Mars during the upcoming space mission. When thinking about this problem, Lovelock wondered "how can we be sure that Martian life, if any, will reveal itself to tests based on Earth's lifestyle?"[8] To Lovelock, the basic question was "What is life, and how should it be recognized?" When speaking about this issue with some of his colleagues at the Jet Propulsion Laboratory, he was asked what he would do to look for life on Mars. To this, Lovelock replied:

> I'd look for an entropy reduction, since this must be a general characteristic of life.[8]

Thus, according to Lovelock, to find signs of life, one must look for a "reduction or a reversal of entropy."

4.4   Gibbs free energy and biological evolution

In recent years, the thermodynamic interpretation of evolution in relation to entropy has begun to utilize the concept of the Gibbs free energy, rather than entropy.[9] This is because biological processes on earth take place at roughly constant temperature and pressure, a situation in which the Gibbs free energy is an especially useful way to express the second law of thermodynamics. The Gibbs free energy is given by:

$$\Delta G \equiv \Delta H - T\Delta S$$

The minimization of the Gibbs free energy is a form of the principle of minimum energy, which follows from the entropy maximization principle for closed systems. Moreover, the Gibbs free energy equation, in modified form, can be utilized for open systems when chemical potential terms are included in the energy balance equation. In a popular 1982 textbook, *Principles of Biochemistry* by noted American biochemist Albert Lehninger, it is argued that the order produced within cells as they grow and divide is more than compensated for by the disorder they create in their surroundings in the course of growth and division. In short, according to Lehninger, "living organisms preserve their internal order by taking from their surroundings free energy, in the form of nutrients or sunlight, and returning to their surroundings an equal amount of energy as heat and entropy."[10]

Similarly, according to the chemist John Avery, from his recent 2003 book *Information Theory and Evolution*, we find a presentation in which the phenomenon of life, including its origin and evolution, as well as human cultural evolution, has its basis in the background of thermodynamics, statistical mechanics, and information theory. The (apparent) paradox between the second law of thermodynamics and the high degree of order and complexity produced by living systems, according to Avery, has its resolution "in the information content of the Gibbs free energy that enters the biosphere from outside sources."[11] The process of natural selection responsible for such local increase in order may be mathematically derived directly from the expression of the second law equation for connected non-equilibrium open systems.[12]

4.5   Entropy and the origin of life

The second law of thermodynamics applied on the origin of life is a far more complicated issue than the further development of life, since there is no "standard model" of how the first biological lifeforms emerged; only a number of competing hypotheses. The problem is discussed within the area of abiogenesis, implying gradual pre-Darwinian *chemical evolution*. In 1924, Alexander Oparin suggested that sufficient energy was provided in a *primordial soup*. The Belgian scientist Ilya Prigogine was awarded with a Nobel prize in 1977 for an analysis in this area. A related topic is the probability that life would emerge, which has been discussed in several studies, for example by Russell Doolittle.[13]

4.6 Entropy and the search of life elsewhere in the Universe

In 2013 Azua-Bustos and Vega argued that disregarding the type of lifeform that could be envisioned both on Earth and elsewhere in the Universe, all should share in common the attribute of being entities that decrease their internal entropy at the expense of free energy obtained from its surroundings. As entropy allows the quantification of the degree of disorder in a system, any envisioned lifeform must have a higher degree of order than its supporting environment. These authors showed that by using fractal mathematics analysis alone, they could readily quantify the degree of structural complexity difference (and thus entropy) of living processes as distinct entities separate from its similar abiotic surroundings. This approach may allow the future detection of unknown forms of life both in the Solar System and on recently discovered exoplanets based on nothing more than entropy differentials of complementary datasets (morphology, coloration, temperature, pH, isotopic composition, etc). Detecting 'life as we don't know it' by fractal analysis

4.7 Other terms

For nearly a century and a half, beginning with Clausius' 1863 memoir "On the Concentration of Rays of Heat and Light, and on the Limits of its Action", much writing and research has been devoted to the relationship between thermodynamic entropy and the evolution of life. The argument that life feeds on negative entropy or negentropy was asserted by physicist Erwin Schrödinger in a 1944 book *What is Life?*. He posed, "How does the living organism avoid decay?" The obvious answer is: "By eating, drinking, breathing and (in the case of plants) assimilating." Recent writings have used the concept of Gibbs free energy to elaborate on this issue.[14] While energy from nutrients is necessary to sustain an organism's order, there is also the Schrödinger prescience: "An organism's astonishing gift of concentrating a stream of order on itself and thus escaping the decay into atomic chaos – of drinking orderliness from a suitable environment – seems to be connected with the presence of the aperiodic solids..." We now know that the 'aperiodic' crystal is DNA and that the irregular arrangement is a form of information. "The DNA in the cell nucleus contains the master copy of the software, in duplicate. This software seems to control by "specifying an algorithm, or set of instructions, for creating and maintaining the entire organism containing the cell."[15] DNA and other macromolecules determine an organism's life cycle: birth, growth, maturity, decline, and death. Nutrition is necessary but not sufficient to account for growth in size as genetics is the governing factor. At some point, organisms normally decline and die even while remaining in environments that contain sufficient nutrients to sustain life. The controlling factor must be internal and not nutrients or sunlight acting as causal exogenous variables. Organisms inherit the ability to create unique and complex biological structures; it is unlikely for those capabilities to be reinvented or be taught each generation. Therefore DNA must be operative as the prime cause in this characteristic as well. Applying Boltzmann's perspective of the second law, the change of state from a more probable, less ordered and high entropy arrangement to one of less probability, more order, and lower entropy seen in biological ordering calls for a function like that known of DNA. DNA's apparent information processing function provides a resolution of the paradox posed by life and the entropy requirement of the second law.[16]

In 1982, American biochemist Albert Lehninger argued that the "order" produced within cells as they grow and divide is more than compensated for by the "disorder" they create in their surroundings in the course of growth and division. "Living organisms preserve their internal order by taking from their surroundings free energy, in the form of nutrients or sunlight, and returning to their surroundings an equal amount of energy as heat and entropy."[17]

Evolution-related concepts:

- **Negentropy** – a shorthand colloquial phrase for negative entropy.[18]

- **Ectropy** – a measure of the tendency of a dynamical system to do useful work and grow more organized.[19]

- **Extropy** – a metaphorical term defining the extent of a living or organizational system's intelligence, functional order, vitality, energy, life, experience, and capacity and drive for improvement and growth.

- **Ecological entropy** – a measure of biodiversity in the study of biological ecology.

In a study titled "Natural selection for least action" published in the *Proceedings of The Royal Society A.*, Ville Kaila and Arto Annila of the University of Helsinki describe how the second law of thermodynamics can be written as an equation of motion to describe evolution, showing how natural selection and the principle of least action can be connected by expressing natural selection in terms of chemical thermodynamics. In this view, evolution explores possible paths to level differences in energy densities and so increase entropy most rapidly. Thus, an organism serves as an energy transfer mechanism, and beneficial mutations allow successive organisms to transfer more energy within their environment.[20]

4.8   Objections

Since entropy is defined for equilibrium systems,[21] objections to the extension of the second law and entropy to biological systems, especially as it pertains to its use to support or discredit the theory of evolution, have been stated.[22] Live systems and indeed much of the systems and processes in the universe operate far from equilibrium, whereas the second law succinctly states that isolated systems evolve toward thermodynamic equilibrium — the state of maximum entropy.

On the other hand, (1) live systems cannot persist in isolation and (2) the second principle of thermodynamics does not require that free energy be transformed into entropy along the shortest path: live organisms absorb energy from sun-light or from energy-rich chemical compounds and finally return part of such energy to the environment as entropy (heat and low free-energy compounds such as water and CO_2).

4.9   See also

- Complex systems

- Dissipative system

- Entropy (order and disorder)

4.10   References

[1] Adams, Henry. (1986). History of the United States of America During the Administration of Thomas Jefferson (pg. 1299). Library of America.

[2] Adams, Henry. (1910). A Letter to American Teachers of History. Google Books, Scanned PDF. Washington.

[3] Boltzmann, Ludwig (1974). *The second law of thermodynamics (Theoretical physics and philosophical problems)*. Springer-Verlag New York, LLC. ISBN 978-90-277-0250-0.

[4] McCulloch, Richard Sears (1876). *Treatise on the mechanical theory of heat and its applications to the steam-engine, etc.* New York: D. Van Nostrand.

[5] Schrödinger, Erwin (1944). *What is Life – the Physical Aspect of the Living Cell*. Cambridge University Press. ISBN 0-521-42708-8.

[6] The common justification for this argument, for example, according to renowned chemical engineer Kenneth Denbigh, from his 1955 book *The Principles of Chemical Equilibrium*, is that "living organisms are open to their environment and can build up at the expense of foodstuffs which they take in and degrade."

[7] Schneider, Eric D.; Sagan, Dorion (2005). *Into the Cool: Energy Flow Thermodynamics and Life*. Chicago, United States: The University of Chicago Press. p. 15.

[8] Lovelock, James (1979). *GAIA – A New Look at Life on Earth*. Oxford University Press. ISBN 0-19-286218-9.

[9] See, for example, Moroz, 2011

[10] Lehninger, Albert (1993). *Principles of Biochemistry, 2nd Ed*. Worth Publishers. ISBN 0-87901-711-2.

[11] Avery, John (2003). *Information Theory and Evolution*. World Scientific. ISBN 981-238-399-9.

[12] Kaila, V. R. and Annila, A. (8 November 2008). "Natural selection for least action". *Proceedings of the Royal Society A* **464** (2099): 3055–3070. Bibcode:2008RSPSA.464.3055K. doi:10.1098/rspa.2008.0178.

[13] Russell Doolittle, "The Probability and Origin of Life" in *Scientists Confront Creationism* (1984) Ed. Laurie R. Godfrey, p. 85

[14] Higgs, P. G., & Pudritz, R. E. (2009). "A thermodynamic basis for prebiotic amino acid synthesis and the nature of the first genetic code" Accepted for publication in Astrobiology

[15] Nelson, P. (2004). Biological Physics, Energy, Information, Life. W.H. Freeman and Company. ISBN 0-7167-4372-8

[16] Peterson, Jacob, Understanding the Thermodynamics of Biological Order, The American Biology Teacher, 74, Number 1, January 2012, pp. 22-24

[17] Lehninger, Albert (1993). *Principles of Biochemistry, 2nd Ed*. Worth Publishers. ISBN 0-87901-711-2.

[18] Schrödinger, Erwin (1944). *What is Life – the Physical Aspect of the Living Cell*. Cambridge University Press. ISBN 0-521-42708-8.

[19] Haddad, Wassim M.; Chellaboina, VijaySekhar; Nersesov, Sergey G. (2005). *Thermodynamics – A Dynamical Systems Approach*. Princeton University Press. ISBN 0-691-12327-6.

[20] Lisa Zyga (11 August 2008). "Evolution as Described by the Second Law of Thermodynamics". Physorg.com. Retrieved 2008-08-14.

[21] Callen, Herbert B (1985). Thermodynamics and an Introduction to Statistical Thermodynamics. John Wiley and Sons.

[22] Ben-Naim, Arieh (2012). Entropy and the Second Law. World Scientific Publishing.

4.11   Further reading

- Schneider, E. and Sagan, D. (2005). *Into the Cool: Energy Flow, Thermodynamics, and Life*. University of Chicago Press, Chicago. ISBN 9780226739366

- La Cerra, P. (2003). "The First Law of Psychology is the Second Law of Thermodynamics: The Energetic Evolutionary Model of the Mind and the Generation of Human Psychological Phenomena". *Human Nature Review*, Volume 3: 440–447. Full text.

- Moroz, A. (2011). *The Common Extremalities in Biology and Physics*. Elsevier Insights, NY. ISBN 978-0-12-385187-1

4.12   External links

- Thermodynamic Evolution of the Universe

Chapter 5

Entropy (classical thermodynamics)

Entropy is a property of thermodynamical systems invented by Rudolf Clausius who named it from the Greek word τροπή, "transformation". Later Ludwig Boltzmann described the entropy as a measure of the number of possible microscopic configurations Ω of the individual atoms and molecules of the system (microstates) which comply with the macroscopic state (macrostate) of the system. Boltzmann then went on to show that $k\ln\Omega$ was equal to the thermodynamic entropy. The factor k has since been known as Boltzmann's constant.

5.1   Introduction

In a thermodynamic system, in a horizontal plane, pressure differences, density differences, and temperature differences all tend to equalize over time. For example, consider a room containing a glass of melting ice as one system. The difference in temperature between the warm room and the cold glass of ice and water is equalized as heat from the room is transferred to the cooler ice and water mixture. Over time the temperature of the glass and its contents and the temperature of the room achieve balance. The entropy of the room has decreased. However, the entropy of the glass of ice and water has increased more than the entropy of the room has decreased. In an isolated system, such as the room and ice water taken together, the dispersal of energy from warmer to cooler regions always results in a net increase in entropy. Thus, when the system of the room and ice water system has reached temperature equilibrium, the entropy change from the initial state is at its maximum. The entropy of the thermodynamic system is a measure of how far the equalization has progressed.

There are many irreversible processes that result in an increase of the entropy. See: Entropy production. One of them is mixing of two or more different substances. The mixing is accompanied by the entropy of mixing. If the substances originally are at the same temperature and pressure, there will be no net exchange of heat or work in many important cases, such as mixing of ideal gases. The entropy increase will be entirely due to the mixing of the different substances.[1]

From a *macroscopic perspective*, in classical thermodynamics, the entropy is a state function of a thermodynamic system: that is, a property depending only on the current state of the system, independent of how that state came to be achieved. Entropy is a key ingredient of the Second law of thermodynamics, which has important consequences e.g. for the performance of heat engines, refrigerators, and heat pumps.

5.2   Definition

According to the Clausius equality, for a closed homogeneous system, in which only reversible processes take place,

$$\oint \frac{\delta Q}{T} = 0.$$

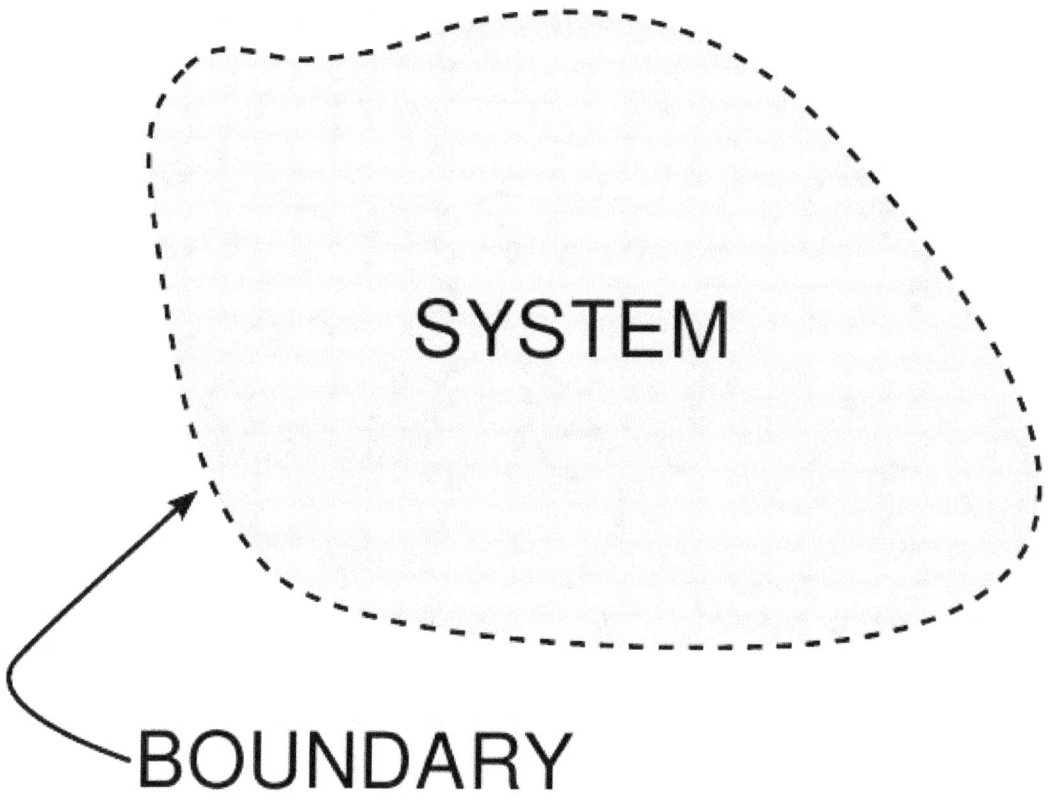

Figure 1. A thermodynamic model system

With T being the uniform temperature of the closed system and delta Q the incremental reversible transfer of heat energy into that system.

That means the line integral $\int_L \frac{\delta Q}{T}$ is path independent.

So we can define a state function S, called entropy, which satisfies

$$dS = \frac{\delta Q}{T}.$$

5.3 Entropy measurement

For simplicity, we examine a uniform closed system, whose thermodynamic state is determined by its temperature T and pressure P. A change in entropy can be written as

$$dS = \left(\frac{\partial S}{\partial T} \right)_P dT + \left(\frac{\partial S}{\partial P} \right)_T dP.$$

The first contribution depends on the heat capacity at constant pressure CP through

$$\left(\frac{\partial S}{\partial T}\right)_P = \frac{C_P}{T}.$$

This is the result of the definition of the heat capacity by $\delta Q = C_P dT$ and $TdS = \delta Q$. For rewriting the second term we use one of the Maxwell relations

$$\left(\frac{\partial S}{\partial P}\right)_T = -\left(\frac{\partial V}{\partial T}\right)_P$$

and the definition of the volumetric thermal-expansion coefficient

$$\alpha_V = \frac{1}{V}\left(\frac{\partial V}{\partial T}\right)_P$$

so that

$$dS = \frac{C_P}{T}dT - \alpha_V V dP.$$

With this expression the entropy S at arbitrary P and T can be related to the entropy S_0 at some reference state at P_0 and T_0 according to

$$S(P,T) = S(P_0, T_0) + \int_{T_0}^{T} \frac{C_P(P_0, T')}{T'} dT' - \int_{P_0}^{P} \alpha_V(P', T)V(P', T)dP'.$$

In classical thermodynamics the entropy of the reference state can be put equal to zero at any convenient temperature and pressure. E.g., for pure substances, one can take the entropy of the solid at the melting point at 1 bar equal to zero. From a more fundamental point of view, the third law of thermodynamics suggests that there is a preference to take $S = 0$ at $T = 0$ (absolute zero) for perfectly ordered materials such as crystals.

In order to determine $S(P,T)$ we followed a specific path in the P-T diagram: first we integrated over T at constant pressure P_0, so that $dP=0$, and in the second integral we integrated over P at constant temperature T, so that $dT=0$. As the entropy is a function of state the result is independent of the path.

The above relation shows that the determination of the entropy requires knowledge of the heat capacity and the equation of state (which is the relation between P, V, and T of the substance involved). Normally these are complicated functions and numerical integration is needed. In simple cases it is possible to get analytical expressions for the entropy. E.g., in the case of an ideal gas, the heat capacity is constant and the ideal-gas law $PV = nRT$ gives that $\alpha_V V = V/T = nR/p$, with n the number of moles and R the molar ideal-gas constant. So, the molar entropy of an ideal gas is given by

$$S_m(P,T) = S_m(P_0, T_0) + C_P \ln \frac{T}{T_0} - R \ln \frac{P}{P_0}.$$

In this expression CP now is the *molar* heat capacity.

The entropy of inhomogeneous systems is the sum of the entropies of the various subsystems. The laws of thermodynamics hold rigorously for inhomogeneous systems even though they may be far from internal equilibrium. The only condition is that the thermodynamic parameters of the composing subsystems are (reasonably) well-defined.

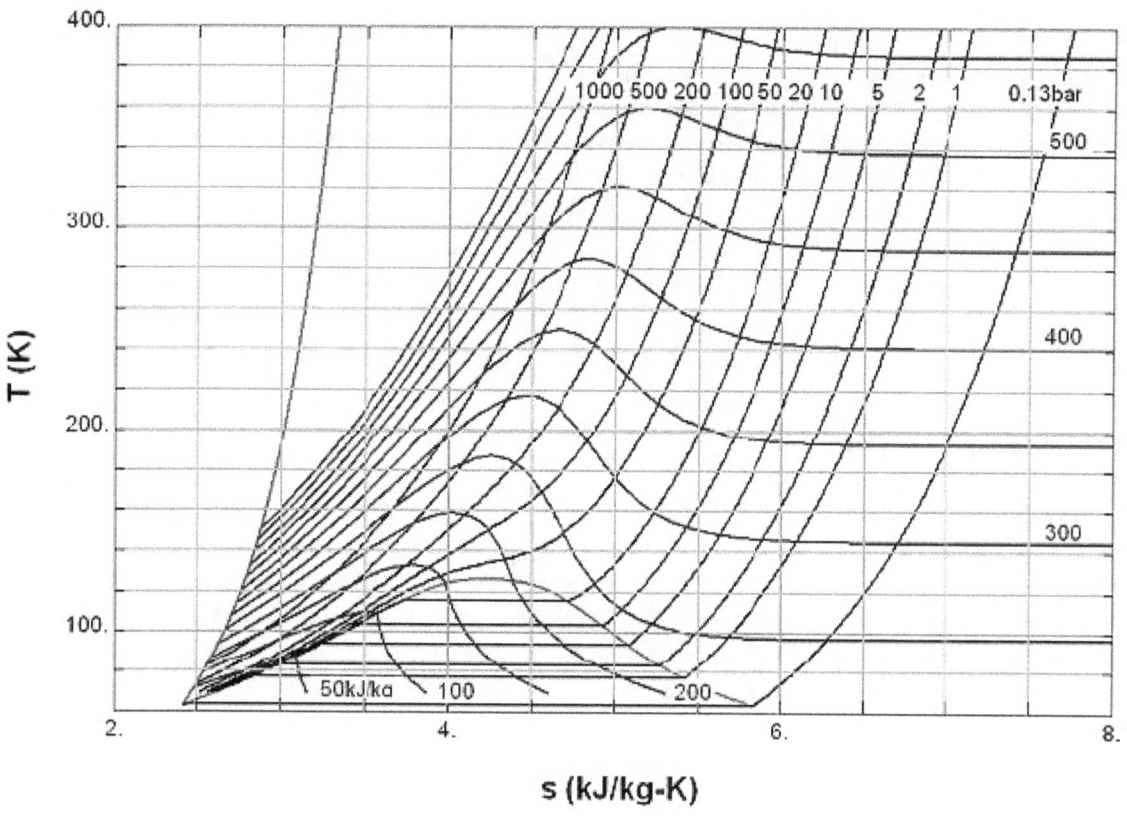

Fig.2 Temperature–entropy diagram of nitrogen. The red curve at the left is the melting curve. The red dome represents the two-phase region with the low-entropy side the saturated liquid and the high-entropy side the saturated gas. The black curves give the Ts relation along isobars. The pressures are indicated in bar. The blue curves are isenthalps (curves of constant enthalpy). The values are indicated in blue in kJ/kg.

5.4 Temperature-entropy diagrams

Nowadays the entropy values of important substances can be obtained via commercial software in tabular form or as diagrams. One of the most common diagrams is the temperature-entropy diagram (Ts-diagram). An example is Fig.2 which is the Ts-diagram of nitrogen.[2] It gives the melting curve and saturated liquid and vapor values together with isobars and isenthalps.

5.5 Entropy change in irreversible transformations

See also: exergy
See also: entropy production

We now consider inhomogeneous systems in which internal transformations (processes) can take place. If we calculate the entropy S_1 before and S_2 after such an internal process the Second Law of Thermodynamics demands that $S_2 \geq S_1$ where the equality sign holds if the process is reversible. The difference $S_i = S_2 - S_1$ is the entropy production due to the irreversible process. The Second law demands that the entropy of an isolated system cannot decrease.

Suppose a system is thermally and mechanically isolated from the environment (isolated system). For example, consider an insulating rigid box divided by a movable partition into two volumes, each filled with gas. If the pressure of one gas is higher, it will expand by moving the partition, thus performing work on the other gas. Also, if the gases are at different temperatures, heat can flow from one gas to the other provided the partition allows heat conduction. Our above result indicates that the entropy of the system *as a whole* will increase during these processes. There exists a maximum amount of entropy the system may possess under the circumstances. This entropy corresponds to a state of *stable equilibrium*, since a transformation to any other equilibrium state would cause the entropy to decrease, which is forbidden. Once the system reaches this maximum-entropy state, no part of the system can perform work on any other part. It is in this sense that entropy is a measure of the energy in a system that cannot be used to do work.

An irreversible process degrades the performance of a thermodynamic system, designed to do work or produce cooling, and results in entropy production. The entropy generation during a reversible process is zero. Thus entropy production is a measure of the irreversibility and may be used to compare engineering processes and machines.

5.6 Thermal machines

Clausius' identification of S as a significant quantity was motivated by the study of reversible and irreversible thermodynamic transformations. A heat engine is a thermodynamic system that can undergo a sequence of transformations which ultimately return it to its original state. Such a sequence is called a cyclic process, or simply a *cycle*. During some transformations, the engine may exchange energy with its environment. The net result of a cycle is

1. mechanical work done by the system (which can be positive or negative, the latter meaning that work is done *on* the engine),

2. heat transferred from one part of the environment to another. In the steady state, by the conservation of energy, the net energy lost by the environment is equal to the work done by the engine.

If every transformation in the cycle is reversible, the cycle is reversible, and it can be run in reverse, so that the heat transfers occur in the opposite directions and the amount of work done switches sign.

5.6.1 Heat engines

Consider a heat engine working between two temperatures TH and T_a. With T_a we have ambient temperature in mind, but, in principle it may also be some other low temperature. The heat engine is in thermal contact with two heat reservoirs which are supposed to have a very large heat capacity so that their temperatures do not change significantly if heat QH is removed from the hot reservoir and Q_a is added to the lower reservoir. Under normal operation $TH > T_a$ and QH, Q_a, and W are all positive.

As our thermodynamical system we take a big system which includes the engine and the two reservoirs. It is indicated in Fig.3 by the dotted rectangle. It is inhomogeneous, closed (no exchange of matter with its surroundings), and adiabatic (no exchange of heat *with its surroundings*). It is not isolated since per cycle a certain amount of work W is produced by the system given by the First law of thermodynamics

$$W = Q_H - Q_a.$$

We used the fact that the engine itself is periodic, so its internal energy has not changed after one cycle. The same is true for its entropy, so the entropy increase $S_2 - S_1$ of our system after one cycle is given by the reduction of entropy of the hot source and the increase of the cold sink. The entropy increase of the total system $S_2 - S_1$ is equal to the entropy production S_i due to irreversible processes in the engine so

$$S_i = -\frac{Q_H}{T_H} + \frac{Q_a}{T_a}.$$

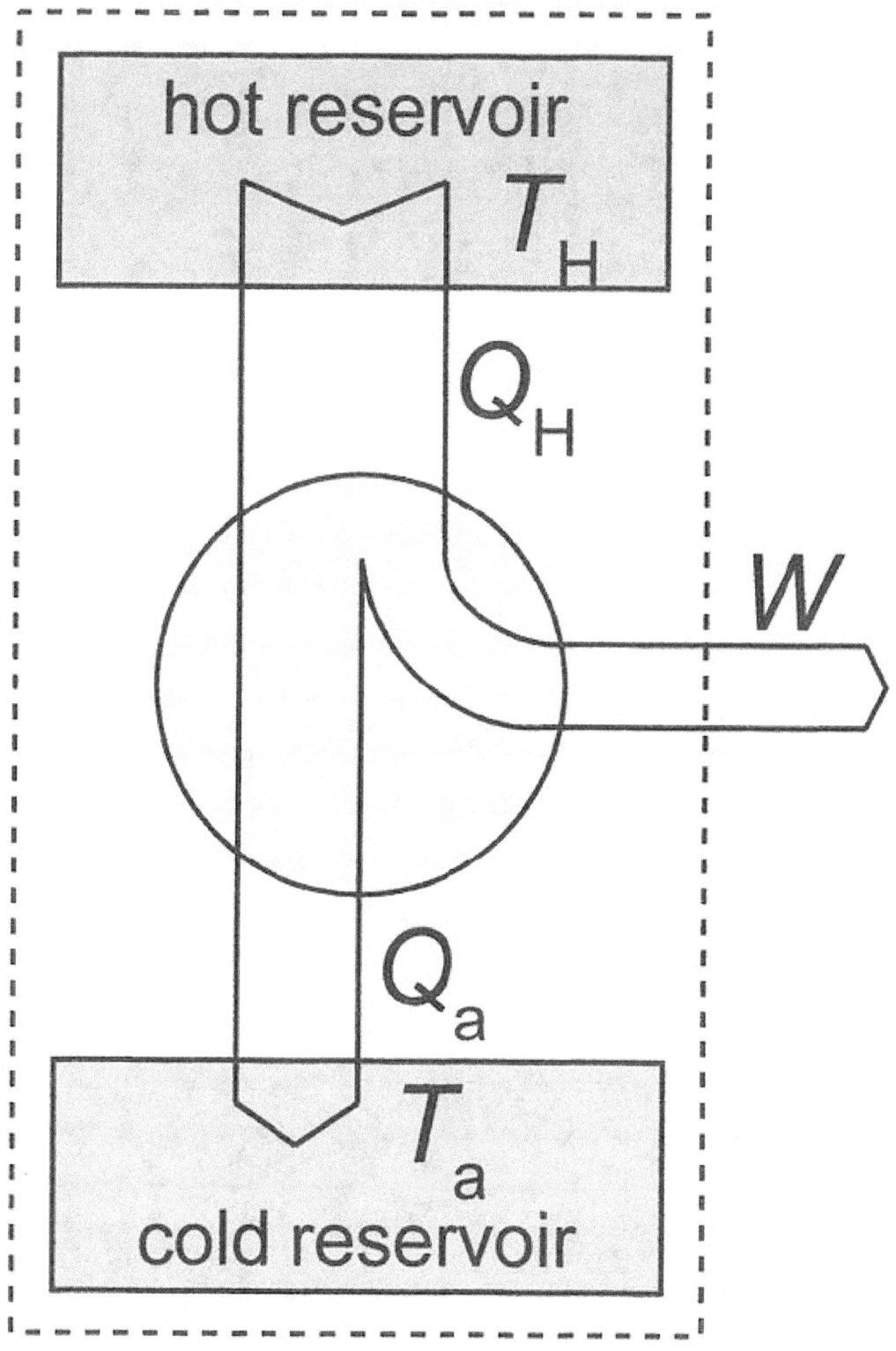

Figure 3: **Heat engine diagram**. The system, discussed in the text, is indicated by the dotted rectangle. It contains the two reservoirs and the heat engine. The arrows define the positive directions of the flows of heat and work.

The Second law demands that $S_i \geq 0$. Eliminating Q_a from the two relations gives

$$W = \left(1 - \frac{T_a}{T_H}\right) Q_H - T_a S_i.$$

The first term is the maximum possible work for a heat engine, given by a reversible engine, as one operating along a Carnot cycle. Finally

$$W = W_{max} - T_a S_i.$$

This equation tells us that the production of work is reduced by the generation of entropy. The term $T_a S_i$ gives the *lost work*, or dissipated energy, by the machine.

Correspondingly, the amount of heat, discarded to the cold sink, is increased by the entropy generation

$$Q_a = \frac{T_a}{T_H} Q_H + T_a S_i = Q_{a,min} + T_a S_i.$$

These important relations can also be obtained without the inclusion of the heat reservoirs. See the Article on entropy production.

5.6.2   Refrigerators

The same principle can be applied to a refrigerator working between a low temperature TL and ambient temperature. The schematic drawing is exactly the same as Fig.3 with TH replaced by TL, QH by QL, and the sign of W reversed. In this case the entropy production is

$$S_i = \frac{Q_a}{T_a} - \frac{Q_L}{T_L}$$

and the work needed to extract heat QL from the cold source is

$$W = Q_L(T_a/T_L - 1) + T_a S_i.$$

The first term is the minimum required work, which corresponds to a reversible refrigerator, so we have

$$W = W_{min} + T_a S_i$$

i.e., the refrigerator compressor has to perform extra work to compensate for the dissipated energy due to irreversible processes which lead to entropy production.

5.7   See also

- Entropy
- Enthalpy
- Entropy production

- Fundamental thermodynamic relation

- Thermodynamic free energy

- History of entropy

- Entropy (statistical views)

5.8 References

[1] See, e.g., Notes for a "Conversation About Entropy" for a brief discussion of *both* thermodynamic and "configurational" ("positional") entropy in chemistry.

[2] Figure composed with data obtained with RefProp, NIST Standard Reference Database 23

5.9 Further reading

- E.A. Guggenheim *Thermodynamics, an advanced treatment for chemists and physicists* North-Holland Publishing Company, Amsterdam, 1959.

- C. Kittel and H. Kroemer *Thermal Physics* W.H. Freeman and Company, New York, 1980.

- Goldstein, Martin, and Inge F., 1993. *The Refrigerator and the Universe*. Harvard Univ. Press. A gentle introduction at a lower level than this entry.

Chapter 6

Entropy (statistical thermodynamics)

For other uses, see Statistical thermodynamics.

In classical statistical mechanics, the entropy function earlier introduced by Clausius is interpreted as **statistical entropy** using probability theory. The statistical entropy perspective was introduced in 1870 with the work of the Austrian physicist Ludwig Boltzmann.

6.1   Gibbs Entropy Formula

The macroscopic state of the system is defined by a distribution on the microstates that are accessible to a system in the course of its thermal fluctuations. So the entropy is defined for two different levels of description of the given system. At one of these levels, the entropy is given by the Gibbs entropy formula, named after J. Willard Gibbs. For a classical system (i.e., a collection of classical particles) with a discrete set of microstates, if E_i is the energy of microstate i, and p_i is the probability that it occurs during the system's fluctuations, then the entropy of the system is

$$S = -k_B \sum_i p_i \ln p_i$$

Entropy changes for systems in a canonical state

A system with a well-defined temperature, i.e., one in thermal equilibrium with a thermal reservoir, has a probability of being in a microstate i given by Boltzmann's distribution.

Changes in the entropy caused by changes in the external constraints are then given by:

$$dS = -k_B \sum_i dp_i \ln p_i$$

$$= -k_B \sum_i dp_i (-E_i/k_B T - \ln Z)$$

$$= \sum_i E_i dp_i / T$$

$$= \sum_i [d(E_i p_i) - (dE_i)p_i]/T$$

where we have twice used the conservation of probability, $\sum dp_i = 0$.

Now, $\sum i\, d(Ei\, pi)$ is the expectation value of the change in the total energy of the system.

If the changes are sufficiently slow, so that the system remains in the same microscopic state, but the state slowly (and reversibly) changes, then $\sum i\, (dEi)\, pi$ is the expectation value of the work done on the system through this reversible process, dw_{rev}.

But from the first law of thermodynamics, $\delta E = \delta w + \delta q$. Therefore,

$$dS = \frac{\delta \langle q_{rev} \rangle}{T}$$

In the thermodynamic limit, the fluctuation of the macroscopic quantities from their average values becomes negligible; so this reproduces the definition of entropy from classical thermodynamics, given above.

The quantity k_B is a physical constant known as Boltzmann's constant, which, like the entropy, has units of heat capacity. The logarithm is dimensionless.

This definition remains meaningful even when the system is far away from equilibrium. Other definitions assume that the system is in thermal equilibrium, either as an isolated system, or as a system in exchange with its surroundings. The set of microstates (with probability distribution) on which the sum is done is called a statistical ensemble. Each type of statistical ensemble (micro-canonical, canonical, grand-canonical, etc.) describes a different configuration of the system's exchanges with the outside, varying from a completely isolated system to a system that can exchange one or more quantities with a reservoir, like energy, volume or molecules. In every ensemble, the equilibrium configuration of the system is dictated by the maximization of the entropy of the union of the system and its reservoir, according to the second law of thermodynamics (see the statistical mechanics article).

Neglecting correlations (or, more generally, statistical dependencies) between the states of individual particles will lead to an incorrect probability distribution on the microstates and thence to an overestimate of the entropy.[1] Such correlations occur in any system with nontrivially interacting particles, that is, in all systems more complex than an ideal gas.

This S is almost universally called simply the *entropy*. It can also be called the *statistical entropy* or the *thermodynamic entropy* without changing the meaning. Note the above expression of the statistical entropy is a discretized version of Shannon entropy. The von Neumann entropy formula is an extension of the Gibbs entropy formula to the quantum mechanical case.

It has been shown[1] that the Gibb's Entropy is equal to the classical "heat engine" entropy characterized by $dS = \frac{\delta Q}{T}$

6.2 Boltzmann's principle

Main article: Boltzmann's entropy formula

In Boltzmann's definition, entropy is a measure of the number of possible microscopic states (or **microstates**) of a system in thermodynamic equilibrium, consistent with its macroscopic thermodynamic properties (or **macrostate**). To understand what microstates and macrostates are, consider the example of a gas in a container. At a microscopic level, the gas consists of a vast number of freely moving atoms, which occasionally collide with one another and with the walls of the container. The microstate of the system is a description of the positions and momenta of all the atoms. In principle, all the physical properties of the system are determined by its microstate. However, because the number of atoms is so large, the details of the motion of individual atoms is mostly irrelevant to the behavior of the system as a whole. Provided the system is in thermodynamic equilibrium, the system can be adequately described by a handful of macroscopic quantities, called "thermodynamic variables": the total energy E, volume V, pressure P, temperature T, and so forth. The macrostate of the system is a description of its thermodynamic variables.

There are three important points to note. Firstly, to specify any one microstate, we need to write down an impractically long list of numbers, whereas specifying a macrostate requires only a few numbers (E, V, etc.). However, and this is the second point, the usual thermodynamic equations only describe the macrostate of a system adequately when this system is in equilibrium; non-equilibrium situations can generally *not* be described by a small number of variables. As a simple example, consider adding a drop of food coloring to a glass of water. The food coloring diffuses in a complicated matter, which is in practice very difficult to precisely predict. However, after sufficient time has passed the system will reach a uniform color, which is much less complicated to describe. Actually, the macroscopic state of the system will be described by a small number of variables only if the system is at global thermodynamic equilibrium. Thirdly, more than one microstate can correspond to a single macrostate. In fact, for any given macrostate, there will be a huge number of microstates that are consistent with the given values of E, V, etc.

We are now ready to provide a definition of entropy. The entropy S is defined as

$$S = k_B \ln \Omega$$

where

  kB is Boltzmann's constant and

  Ω is the number of microstates consistent with the given macrostate.

The statistical entropy reduces to Boltzmann's entropy when all the accessible microstates of the system are equally likely. It is also the configuration corresponding to the maximum of a system's entropy for a given set of accessible microstates, in other words the macroscopic configuration in which the lack of information is maximal. As such, according to the second law of thermodynamics, it is the equilibrium configuration of an isolated system. Boltzmann's entropy is the expression of entropy at thermodynamic equilibrium in the micro-canonical ensemble.

This postulate, which is known as Boltzmann's principle, may be regarded as the foundation of statistical mechanics, which describes thermodynamic systems using the statistical behaviour of its constituents. It turns out that S is itself a thermodynamic property, just like E or V. Therefore, it acts as a link between the microscopic world and the macroscopic. One important property of S follows readily from the definition: since Ω is a natural number (1,2,3,....), S is either *zero* or *positive* ($\ln(1) = 0$, $\ln \Omega \geq 0$.)

6.2.1   Ensembles

The various ensembles used in statistical thermodynamics are linked to the entropy by the following relations:

$$S = k_B \ln \Omega_{mic} = k_B(\ln Z_{can} + \beta \bar{E}) = k_B(\ln \mathcal{Z}_{gr} + \beta(\bar{E} - \mu \bar{N}))$$

Ω_{mic} is the microcanonical partition function
Z_{can} is the canonical partition function
$\mathcal{Z}_{gr}$ is the grand canonical partition function

6.3   Lack of knowledge and the second law of thermodynamics

We can view Ω as a measure of our lack of knowledge about a system. As an illustration of this idea, consider a set of 100 coins, each of which is either heads up or tails up. The macrostates are specified by the total number of heads and tails, whereas the microstates are specified by the facings of each individual coin. For the macrostates of 100 heads or 100 tails, there is exactly one possible configuration, so our knowledge of the system is complete. At the opposite extreme, the macrostate which gives us the least knowledge about the system consists of 50 heads and 50 tails in any order, for which there are 100,891,344,545,564,193,334,812,497,256 (100 choose 50) $\approx 10^{29}$ possible microstates.

Even when a system is entirely isolated from external influences, its microstate is constantly changing. For instance, the particles in a gas are constantly moving, and thus occupy a different position at each moment of time; their momenta are also constantly changing as they collide with each other or with the container walls. Suppose we prepare the system in an artificially highly ordered equilibrium state. For instance, imagine dividing a container with a partition and placing a gas on one side of the partition, with a vacuum on the other side. If we remove the partition and watch the subsequent behavior of the gas, we will find that its microstate evolves according to some chaotic and unpredictable pattern, and that on average these microstates will correspond to a more disordered macrostate than before. It is *possible*, but *extremely unlikely*, for the gas molecules to bounce off one another in such a way that they remain in one half of the container. It is overwhelmingly probable for the gas to spread out to fill the container evenly, which is the new equilibrium macrostate of the system.

This is an example illustrating the Second Law of Thermodynamics:

> *the total entropy of any isolated thermodynamic system tends to increase over time, approaching a maximum value.*

Since its discovery, this idea has been the focus of a great deal of thought, some of it confused. A chief point of confusion is the fact that the Second Law applies only to *isolated* systems. For example, the Earth is not an isolated system because it is constantly receiving energy in the form of sunlight. In contrast, the universe may be considered an isolated system, so that its total entropy is constantly increasing.

6.4 Counting of microstates

In classical statistical mechanics, the number of microstates is actually uncountably infinite, since the properties of classical systems are continuous. For example, a microstate of a classical ideal gas is specified by the positions and momenta of all the atoms, which range continuously over the real numbers. If we want to define Ω, we have to come up with a method of grouping the microstates together to obtain a countable set. This procedure is known as coarse graining. In the case of the ideal gas, we count two states of an atom as the "same" state if their positions and momenta are within δx and δp of each other. Since the values of δx and δp can be chosen arbitrarily, the entropy is not uniquely defined. It is defined only up to an additive constant. (As we will see, the thermodynamic definition of entropy is also defined only up to a constant.)

This ambiguity can be resolved with quantum mechanics. The quantum state of a system can be expressed as a superposition of "basis" states, which can be chosen to be energy eigenstates (i.e. eigenstates of the quantum Hamiltonian). Usually, the quantum states are discrete, even though there may be an infinite number of them. For a system with some specified energy E, one takes Ω to be the number of energy eigenstates within a macroscopically small energy range between E and $E + \delta E$. In the thermodynamical limit, the specific entropy becomes independent on the choice of δE.

An important result, known as Nernst's theorem or the third law of thermodynamics, states that the entropy of a system at zero absolute temperature is a well-defined constant. This is because a system at zero temperature exists in its lowest-energy state, or ground state, so that its entropy is determined by the degeneracy of the ground state. Many systems, such as crystal lattices, have a unique ground state, and (since $\ln(1) = 0$) this means that they have zero entropy at absolute zero. Other systems have more than one state with the same, lowest energy, and have a non-vanishing "zero-point entropy". For instance, ordinary ice has a zero-point entropy of 3.41 J/(mol·K), because its underlying crystal structure possesses multiple configurations with the same energy (a phenomenon known as geometrical frustration).

The third law of thermodynamics states that the entropy of a perfect crystal at absolute zero, or 0 kelvin is zero. This means that in a perfect crystal, at 0 kelvin, nearly all molecular motion should cease in order to achieve $\Delta S=0$. A perfect crystal is one in which the internal lattice structure is the same at all times; in other words, it is fixed and non-moving, and does not have rotational or vibrational energy. This means that there is only one way in which this order can be attained: when every particle of the structure is in its proper place.

However, the oscillator equation for predicting quantized vibrational levels shows that even when the vibrational quantum number is 0, the molecule still has vibrational energy. This means that no matter how cold the temperature gets, the lattice will always vibrate. This is in keeping with the Heisenberg uncertainty principle, which states that both the position and the momentum of a particle cannot be known precisely, at a given time:

$$E_\nu = h\nu_0(n + \tfrac{1}{2})$$

where h is Planck's constant, ν_0 is the characteristic frequency of the vibration, and n is the vibrational quantum number. Note that even when $n = 0$ (the zero-point energy), E_n does not equal 0.

6.5   See also

- Boltzmann constant

- Configuration entropy

- Conformational entropy

- Enthalpy

- Entropy

- Entropy (classical thermodynamics)

- Entropy (energy dispersal)

- Entropy of mixing

- Entropy (order and disorder)

- Entropy (information theory)

- History of entropy

- Information theory

- Thermodynamic free energy

6.6   References

[1] E.T. Jaynes: Gibbs vs Boltzmann Entropies; American Journal of Physics, 391, 1965

- Boltzmann, Ludwig (1896, 1898). Vorlesungen über Gastheorie : 2 Volumes - Leipzig 1895/98 UB: O 5262-6. English version: Lectures on gas theory. Translated by Stephen G. Brush (1964) Berkeley: University of California Press; (1995) New York: Dover ISBN 0-486-68455-5

Chapter 7

Entropy (information theory)

$$S = \log_2 2^N = N = 2$$

2 shannons of entropy: Information entropy is the log-base-2 of the number of possible outcomes; with two coins there are four outcomes, and the entropy is two bits.

In information theory, **entropy** (more specifically, **Shannon entropy**) is the expected value (average) of the information contained in each message received. 'Messages' don't have to be text; in this context a 'message' is simply any flow of information. The entropy of the message is its amount of uncertainty; it increases when the message is closer to random, and decreases when it is less random. The idea here is that the less likely an event is, the more information it provides when it occurs. This seems backwards at first: it seems like messages which have more structure would contain more information, but this is not true. For example, the message 'aaaaaaaaaa' (which appears to be very structured and not random at all [although in fact it *could* result from a random process]) contains much less information than the message 'alphabet' (which is somewhat structured, but more random) or even the message 'axraefy6h' (which is very random). In information theory, 'information' doesn't necessarily mean *useful* information; it simply describes the amount of randomness of the message, so in the example above the first message has the least information and the last message has the most information, even though in everyday terms we would say that the middle message, 'alphabet', contains more information than a stream of random letters. Therefore, we would say in information theory that the first message has low entropy, the second has higher entropy, and the third has the highest entropy.

In a more technical sense, there are reasons (explained below) to define information as the negative of the logarithm of the probability distribution. The probability distribution of the events, coupled with the information amount of every event, forms a random variable whose average (also termed expected value) is the average amount of information, a.k.a. entropy, generated by this distribution. Units of entropy are the shannon, nat, or hartley, depending on the base of the logarithm used to define it, though the shannon is commonly referred to as a bit.

The logarithm of the probability distribution is useful as a measure of entropy because it is additive for independent sources. For instance, the entropy of a coin toss is 1 shannon, whereas of m tosses it is m shannons. Generally, you need $\log_2(n)$ bits to represent a variable that can take one of n values if n is a power of 2. If these values are equiprobable, the entropy (in shannons) is equal to the number of bits. Equality between number of bits and shannons holds only while all outcomes are equally probable. If one of the events is more probable than others, observation of that event is less informative. Conversely, observing rarer events compensate by providing more information when observed. Since observation of less probable events occurs more rarely, the net effect is that the entropy (thought of as the average information) received from non-uniformly distributed data is less than $\log_2(n)$. Entropy is zero when one outcome is certain. Shannon entropy quantifies all these considerations exactly when a probability distribution of the source is known. The meaning of the events observed (a.k.a. the meaning of *messages*) do not matter in the definition of entropy. Entropy only takes into account the probability of observing a specific event, so the information it encapsulates is information about the underlying probability distribution, not the meaning of the events themselves.

Generally, *entropy* refers to disorder or uncertainty. Shannon entropy was introduced by Claude E. Shannon in his 1948 paper "A Mathematical Theory of Communication".[1] Shannon entropy provides an absolute limit on the best possible average length of lossless encoding or compression of an information source. Rényi entropy generalizes Shannon entropy.

7.1   Introduction

Entropy is a measure of *unpredictability* of *information content*. To get an informal, intuitive understanding of the connection between these three English terms, consider the example of a poll on some political issue. Usually, such polls happen because the outcome of the poll isn't already known. In other words, the outcome of the poll is relatively *unpredictable*, and actually performing the poll and learning the results gives some new *information*; these are just different ways of saying that the *entropy* of the poll results is large. Now, consider the case that the same poll is performed a second time shortly after the first poll. Since the result of the first poll is already known, the outcome of the second poll can be predicted well and the results should not contain much new information; in this case the entropy of the second poll result relative to the first is small.

Now consider the example of a coin toss. When the coin is fair, that is, when the probability of heads is the same as the probability of tails, then the entropy of the coin toss is as high as it could be. This is because there is no way to predict the outcome of the coin toss ahead of time—the best we can do is predict that the coin will come up heads, and our prediction will be correct with probability 1/2. Such a coin toss has one bit of entropy since there are two possible outcomes that occur with equal probability, and learning the actual outcome contains one bit of information. Contrarily, a coin toss with a coin that has two heads and no tails has zero entropy since the coin will always come up heads, and the outcome can be predicted perfectly.

English text has fairly low entropy. In other words, it is fairly predictable. Even if we don't know exactly what is going to come next, we can be fairly certain that, for example, there will be many more e's than z's, that the combination 'qu' will be much more common than any other combination with a 'q' in it, and that the combination 'th' will be more common than 'z', 'q', or 'qu'. After the first few letters one can often guess the rest of the word. English text has between 0.6 and 1.3 bits of entropy for each character of message.[2][3]

The Chinese version of Wikipedia points out that Chinese characters have a much higher entropy than English. Each character of Chinese has about -log2(1/2500)=11.3bits which is almost three times higher than English. However, the discussion could be much more sophisticated than this simple calculation because in English the usage of words, not only characters, and redundancy factors could be considered.

If a compression scheme is lossless—that is, you can always recover the entire original message by decompressing—then a compressed message has the same quantity of information as the original, but communicated in fewer characters. That is, it has more information, or a higher entropy, per character. This means a compressed message has less redundancy. Roughly speaking, Shannon's source coding theorem says that a lossless compression scheme cannot compress messages, on average, to have *more* than one bit of information per bit of message, but that any value *less* than one bit of information per bit of message can be attained by employing a suitable coding scheme. The entropy of a message per bit multiplied by the length of that message is a measure of how much total information the message contains.

Shannon's theorem also implies that no lossless compression scheme can shorten *all* messages. If some messages come out shorter, at least one must come out longer due to the pigeonhole principle. In practical use, this is generally not a problem, because we are usually only interested in compressing certain types of messages, for example English documents as opposed to gibberish text, or digital photographs rather than noise, and it is unimportant if a compression algorithm makes some unlikely or uninteresting sequences larger. However, the problem can still arise even in everyday use when applying a compression algorithm to already compressed data: for example, making a ZIP file of music that is already in the FLAC audio format is unlikely to achieve much extra saving in space.

7.2 Definition

Named after Boltzmann's H-theorem, Shannon defined the entropy H (Greek letter Eta) of a discrete random variable X with possible values $\{x_1, ..., xn\}$ and probability mass function $P(X)$ as:

$$H(X) = E[I(X)] = E[-\ln(P(X))].$$

Here E is the expected value operator, and I is the information content of X.[4][5] $I(X)$ is itself a random variable.

The entropy can explicitly be written as

$$H(X) = \sum_i P(x_i) I(x_i) = -\sum_i P(x_i) \log_b P(x_i),$$

where b is the base of the logarithm used. Common values of b are 2, Euler's number e, and 10, and the unit of entropy is shannon for $b = 2$, nat for $b = e$, and hartley for $b = 10$.[6] When $b = 2$, the units of entropy are also commonly referred to as bits.

In the case of $p(xi) = 0$ for some i, the value of the corresponding summand $0 \log b(0)$ is taken to be 0, which is consistent with the limit:

$$\lim_{p \to 0+} p \log(p) = 0.$$

When the distribution is continuous rather than discrete, the sum is replaced with an integral as

$$H(X) = \int P(x) I(x) \, dx = -\int P(x) \log_b P(x) \, dx,$$

where $P(x)$ represents a probability density function.

One may also define the conditional entropy of two events X and Y taking values xi and yj respectively, as

$$H(X|Y) = \sum_{i,j} p(x_i, y_j) \log \frac{p(y_j)}{p(x_i, y_j)}$$

where $p(xi, yj)$ is the probability that $X = xi$ and $Y = yj$. This quantity should be understood as the amount of randomness in the random variable X given that you know the value of Y.

7.3   Example

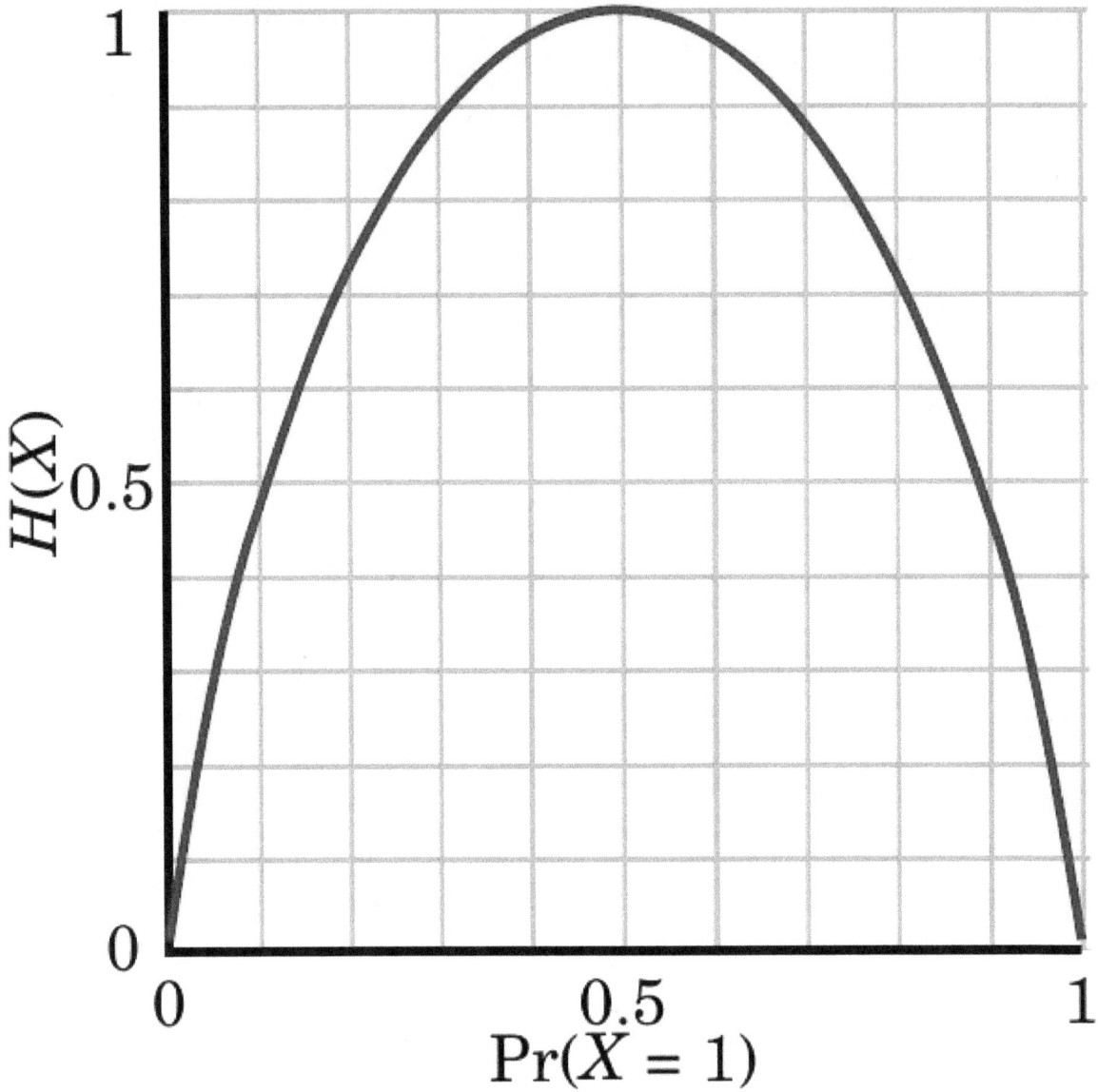

Entropy H(X) (i.e. the expected surprisal) of a coin flip, measured in shannons, graphed versus the fairness of the coin Pr(X = 1), where X = 1 represents a result of heads.

Note that the maximum of the graph depends on the distribution. Here, the entropy is at most 1 shannon, and to communicate the outcome of a fair coin flip (2 possible values) will require an average of at most 1 bit. The result of a fair die (6 possible values) would require on average $\log_2 6$ bits.

Main article: Binary entropy function
Main article: Bernoulli process

Consider tossing a coin with known, not necessarily fair, probabilities of coming up heads or tails; this is known as the Bernoulli process.

The entropy of the unknown result of the next toss of the coin is maximized if the coin is fair (that is, if heads and tails both have equal probability 1/2). This is the situation of maximum uncertainty as it is most difficult to predict the outcome of the next toss; the result of each toss of the coin delivers one full bit of information.

However, if we know the coin is not fair, but comes up heads or tails with probabilities p and q, where $p \neq q$, then there is less uncertainty. Every time it is tossed, one side is more likely to come up than the other. The reduced uncertainty is quantified in a lower entropy: on average each toss of the coin delivers less than one full bit of information.

The extreme case is that of a double-headed coin that never comes up tails, or a double-tailed coin that never results in a head. Then there is no uncertainty. The entropy is zero: each toss of the coin delivers no new information as the outcome of each coin toss is always certain. In this respect, entropy can be normalized by dividing it by information length. This ratio is called metric entropy and is a measure of the randomness of the information.

7.4 Rationale

To understand the meaning of $\sum pi \log(1/pi)$, at first, try to define an information function, I, in terms of an event i with probability pi. How much information is acquired due to the observation of event i? Shannon's solution follows from the fundamental properties of information:[7]

1. $I(p) \geq 0$ – information is a non-negative quantity

2. $I(1) = 0$ – events that always occur do not communicate information

3. $I(p_1 \, p_2) = I(p_1) + I(p_2)$ – information due to independent events is additive

The last is a crucial property. It states that joint probability communicates as much information as two individual events separately. Particularly, if the first event can yield one of n equiprobable outcomes and another has one of m equiprobable outcomes then there are mn possible outcomes of the joint event. This means that if $\log_2(n)$ bits are needed to encode the first value and $\log_2(m)$ to encode the second, one needs $\log_2(mn) = \log_2(m) + \log_2(n)$ to encode both. Shannon discovered that the proper choice of function to quantify information, preserving this additivity, is logarithmic, i.e.,

$$I(p) = \log(1/p)$$

The base of the logarithm can be any fixed real number greater than 1. The different units of information (bits for $\log_2$, trits for $\log_3$, nats for the natural logarithm ln and so on) are just constant multiples of each other. (In contrast, the entropy would be negative if the base of the logarithm were less than 1.) For instance, in case of a fair coin toss, heads provides $\log_2(2) = 1$ bit of information, which is approximately 0.693 nats or 0.631 trits. Because of additivity, n tosses provide n bits of information, which is approximately $0.693n$ nats or $0.631n$ trits.

Now, suppose we have a distribution where event i can happen with probability pi. Suppose we have sampled it N times and outcome i was, accordingly, seen $ni = N \, pi$ times. The total amount of information we have received is

$$\sum_i n_i I(p_i) = \sum N p_i \log(1/p_i)$$

The average amount of information that we receive with every event is therefore

$$\sum_i p_i \log \frac{1}{p_i}.$$

7.5 Aspects

7.5.1   Relationship to thermodynamic entropy

Main article: Entropy in thermodynamics and information theory

The inspiration for adopting the word *entropy* in information theory came from the close resemblance between Shannon's formula and very similar known formulae from statistical mechanics.

In statistical thermodynamics the most general formula for the thermodynamic entropy S of a thermodynamic system is the Gibbs entropy,

$$S = -k_B \sum p_i \ln p_i$$

where kB is the Boltzmann constant, and pi is the probability of a microstate. The Gibbs entropy was defined by J. Willard Gibbs in 1878 after earlier work by Boltzmann (1872).[8]

The Gibbs entropy translates over almost unchanged into the world of quantum physics to give the von Neumann entropy, introduced by John von Neumann in 1927,

$$S = -k_B \operatorname{Tr}(\rho \ln \rho)$$

where ρ is the density matrix of the quantum mechanical system and Tr is the trace.

At an everyday practical level the links between information entropy and thermodynamic entropy are not evident. Physicists and chemists are apt to be more interested in *changes* in entropy as a system spontaneously evolves away from its initial conditions, in accordance with the second law of thermodynamics, rather than an unchanging probability distribution. And, as the minuteness of Boltzmann's constant kB indicates, the changes in S/kB for even tiny amounts of substances in chemical and physical processes represent amounts of entropy that are extremely large compared to anything in data compression or signal processing. Furthermore, in classical thermodynamics the entropy is defined in terms of macroscopic measurements and makes no reference to any probability distribution, which is central to the definition of information entropy.

At a multidisciplinary level, however, connections can be made between thermodynamic and informational entropy, although it took many years in the development of the theories of statistical mechanics and information theory to make the relationship fully apparent. In fact, in the view of Jaynes (1957), thermodynamic entropy, as explained by statistical mechanics, should be seen as an *application* of Shannon's information theory: the thermodynamic entropy is interpreted as being proportional to the amount of further Shannon information needed to define the detailed microscopic state of the system, that remains uncommunicated by a description solely in terms of the macroscopic variables of classical thermodynamics, with the constant of proportionality being just the Boltzmann constant. For example, adding heat to a system increases its thermodynamic entropy because it increases the number of possible microscopic states of the system that are consistent with the measurable values of its macroscopic variables, thus making any complete state description longer. (See article: *maximum entropy thermodynamics*). Maxwell's demon can (hypothetically) reduce the thermodynamic entropy of a system by using information about the states of individual molecules; but, as Landauer (from 1961) and co-workers have shown, to function the demon himself must increase thermodynamic entropy in the process, by at least the amount of Shannon information he proposes to first acquire and store; and so the total thermodynamic entropy does not decrease (which resolves the paradox). Landauer's principle imposes a lower bound on the amount of heat a computer must generate to process a given amount of information, though modern computers are far less efficient.

7.5.2   Entropy as information content

Main article: Shannon's source coding theorem

Entropy is defined in the context of a probabilistic model. Independent fair coin flips have an entropy of 1 bit per flip. A source that always generates a long string of B's has an entropy of 0, since the next character will always be a 'B'.

The entropy rate of a data source means the average number of bits per symbol needed to encode it. Shannon's experiments with human predictors show an information rate between 0.6 and 1.3 bits per character in English;[9] the PPM compression algorithm can achieve a compression ratio of 1.5 bits per character in English text.

From the preceding example, note the following points:

1. The amount of entropy is not always an integer number of bits.

2. Many data bits may not convey information. For example, data structures often store information redundantly, or have identical sections regardless of the information in the data structure.

Shannon's definition of entropy, when applied to an information source, can determine the minimum channel capacity required to reliably transmit the source as encoded binary digits (see caveat below in italics). The formula can be derived by calculating the mathematical expectation of the *amount of information* contained in a digit from the information source. *See also* Shannon-Hartley theorem.

Shannon's entropy measures the information contained in a message as opposed to the portion of the message that is determined (or predictable). *Examples of the latter include redundancy in language structure or statistical properties relating to the occurrence frequencies of letter or word pairs, triplets etc.* See Markov chain.

7.5.3 Entropy as a measure of diversity

Main article: Diversity index

Entropy is one of several ways to measure diversity. Specifically, Shannon entropy is the logarithm of ^1D, the true diversity index with parameter equal to 1.

7.5.4 Data compression

Main article: Data compression

Entropy effectively bounds the performance of the strongest lossless compression possible, which can be realized in theory by using the typical set or in practice using Huffman, Lempel–Ziv or arithmetic coding. The performance of existing data compression algorithms is often used as a rough estimate of the entropy of a block of data.[10][11] See also Kolmogorov complexity. In practice, compression algorithms deliberately include some judicious redundancy in the form of checksums to protect against errors.

7.5.5 World's technological capacity to store and communicate information

A 2011 study in *Science* estimates the world's technological capacity to store and communicate optimally compressed information normalized on the most effective compression algorithms available in the year 2007, therefore estimating the entropy of the technologically available sources.[12]

The authors estimate humankind technological capacity to store information (fully entropically compressed) in 1986 and again in 2007. They break the information into three categories—to store information on a medium, to receive information through a one-way broadcast networks, or to exchange information through two-way telecommunication networks.[12]

7.5.6 Limitations of entropy as information content

There are a number of entropy-related concepts that mathematically quantify information content in some way:

- the **self-information** of an individual message or symbol taken from a given probability distribution,

- the **entropy** of a given probability distribution of messages or symbols, and

- the **entropy rate** of a stochastic process.

(The "rate of self-information" can also be defined for a particular sequence of messages or symbols generated by a given stochastic process: this will always be equal to the entropy rate in the case of a stationary process.) Other quantities of information are also used to compare or relate different sources of information.

It is important not to confuse the above concepts. Often it is only clear from context which one is meant. For example, when someone says that the "entropy" of the English language is about 1 bit per character, they are actually modeling the English language as a stochastic process and talking about its entropy *rate*. Shannon himself used the term in this way.[3]

Although entropy is often used as a characterization of the information content of a data source, this information content is not absolute: it depends crucially on the probabilistic model. A source that always generates the same symbol has an entropy rate of 0, but the definition of what a symbol is depends on the alphabet. Consider a source that produces the string ABABABABAB... in which A is always followed by B and vice versa. If the probabilistic model considers individual letters as independent, the entropy rate of the sequence is 1 bit per character. But if the sequence is considered as "AB AB AB AB AB..." with symbols as two-character blocks, then the entropy rate is 0 bits per character.

However, if we use very large blocks, then the estimate of per-character entropy rate may become artificially low. This is because in reality, the probability distribution of the sequence is not knowable exactly; it is only an estimate. For example, suppose one considers the text of every book ever published as a sequence, with each symbol being the text of a complete book. If there are N published books, and each book is only published once, the estimate of the probability of each book is $1/N$, and the entropy (in bits) is $-\log_2(1/N) = \log_2(N)$. As a practical code, this corresponds to assigning each book a unique identifier and using it in place of the text of the book whenever one wants to refer to the book. This is enormously useful for talking about books, but it is not so useful for characterizing the information content of an individual book, or of language in general: it is not possible to reconstruct the book from its identifier without knowing the probability distribution, that is, the complete text of all the books. The key idea is that the complexity of the probabilistic model must be considered. Kolmogorov complexity is a theoretical generalization of this idea that allows the consideration of the information content of a sequence independent of any particular probability model; it considers the shortest program for a universal computer that outputs the sequence. A code that achieves the entropy rate of a sequence for a given model, plus the codebook (i.e. the probabilistic model), is one such program, but it may not be the shortest.

For example, the Fibonacci sequence is 1, 1, 2, 3, 5, 8, 13, ... Treating the sequence as a message and each number as a symbol, there are almost as many symbols as there are characters in the message, giving an entropy of approximately $\log_2(n)$. So the first 128 symbols of the Fibonacci sequence has an entropy of approximately 7 bits/symbol. However, the sequence can be expressed using a formula [F(n) = F($n-1$) + F($n-2$) for $n=\{3,4,5,....\}$, F(1)=1, F(2)=1] and this formula has a much lower entropy and applies to any length of the Fibonacci sequence.

7.5.7 Limitations of entropy as a measure of unpredictability

In cryptanalysis, entropy is often roughly used as a measure of the unpredictability of a cryptographic key. For example, a 128-bit key that is randomly generated has 128 bits of entropy. It takes (on average) 2^{128-1} guesses to break by brute force. If the key's first digit is 0, and the others random, then the entropy is 127 bits, and it takes (on average) 2^{127-1} guesses.

However, entropy fails to capture the number of guesses required if the possible keys are not of equal probability.[13][14] If the key is half the time "password" and half the time a true random 128-bit key, then the entropy is approximately 65 bits. Yet half the time the key may be guessed on the first try, if your first guess is "password", and on average, it takes around 2^{126} guesses (not 2^{65-1}) to break this password.

Similarly, consider a 1000000-digit binary one-time pad. If the pad has 1000000 bits of entropy, it is perfect. If the pad has 999999 bits of entropy, evenly distributed (each individual bit of the pad having 0.999999 bits of entropy) it may still be considered very good. But if the pad has 999999 bits of entropy, where the first digit is fixed and the remaining 999999 digits are perfectly random, then the first digit of the ciphertext will not be encrypted at all.

7.5.8   Data as a Markov process

A common way to define entropy for text is based on the Markov model of text. For an order-0 source (each character is selected independent of the last characters), the binary entropy is:

$$H(\mathcal{S}) = -\sum p_i \log_2 p_i.$$

where pi is the probability of i. For a first-order Markov source (one in which the probability of selecting a character is dependent only on the immediately preceding character), the **entropy rate** is:

$$H(\mathcal{S}) = -\sum_i p_i \sum_j p_i(j) \log_2 p_i(j).$$

where i is a **state** (certain preceding characters) and $p_i(j)$ is the probability of j given i as the previous character.

For a second order Markov source, the entropy rate is

$$H(\mathcal{S}) = -\sum_i p_i \sum_j p_i(j) \sum_k p_{i,j}(k) \log_2 p_{i,j}(k).$$

7.5.9   *b*-ary entropy

In general the ***b*-ary entropy** of a source $\mathcal{S} = (S,P)$ with source alphabet $S = \{a_1, ..., an\}$ and discrete probability distribution $P = \{p_1, ..., pn\}$ where pi is the probability of ai (say $pi = p(ai)$) is defined by:

$$H_b(\mathcal{S}) = -\sum_{i=1}^{n} p_i \log_b p_i.$$

Note: the b in "b-ary entropy" is the number of different symbols of the *ideal alphabet* used as a standard yardstick to measure source alphabets. In information theory, two symbols are necessary and sufficient for an alphabet to encode information. Therefore, the default is to let $b = 2$ ("binary entropy"). Thus, the entropy of the source alphabet, with its given empiric probability distribution, is a number equal to the number (possibly fractional) of symbols of the "ideal alphabet", with an optimal probability distribution, necessary to encode for each symbol of the source alphabet. Also note that "optimal probability distribution" here means a uniform distribution: a source alphabet with n symbols has the highest possible entropy (for an alphabet with n symbols) when the probability distribution of the alphabet is uniform. This optimal entropy turns out to be $\log b(n)$.

7.6   Efficiency

A source alphabet with non-uniform distribution will have less entropy than if those symbols had uniform distribution (i.e. the "optimized alphabet"). This deficiency in entropy can be expressed as a ratio called efficiency:

$$\eta(X) = -\sum_{i=1}^{n} \frac{p(x_i) \log_b(p(x_i))}{\log_b(n)}$$

Efficiency has utility in quantifying the effective use of a communications channel. This formulation is also referred to as the normalized entropy, as the entropy is divided by the maximum entropy $\log_b(n)$.

7.7   Characterization

Shannon entropy is characterized by a small number of criteria, listed below. Any definition of entropy satisfying these assumptions has the form

$$-K \sum_{i=1}^{n} p_i \log(p_i)$$

where K is a constant corresponding to a choice of measurement units.

In the following, $pi = \Pr(X = xi)$ and $H_n(p_1, \ldots, p_n) = H(X)$.

7.7.1   Continuity

The measure should be continuous, so that changing the values of the probabilities by a very small amount should only change the entropy by a small amount.

7.7.2   Symmetry

The measure should be unchanged if the outcomes xi are re-ordered.

$$H_n(p_1, p_2, \ldots) = H_n(p_2, p_1, \ldots)$$

7.7.3   Maximum

The measure should be maximal if all the outcomes are equally likely (uncertainty is highest when all possible events are equiprobable).

$$H_n(p_1, \ldots, p_n) \leq H_n\left(\frac{1}{n}, \ldots, \frac{1}{n}\right) = \log_b(n).$$

For equiprobable events the entropy should increase with the number of outcomes.

$$H_n\left(\underbrace{\frac{1}{n}, \ldots, \frac{1}{n}}_{n}\right) = \log_b(n) < \log_b(n+1) = H_{n+1}\left(\underbrace{\frac{1}{n+1}, \ldots, \frac{1}{n+1}}_{n+1}\right).$$

7.7.4   Additivity

The amount of entropy should be independent of how the process is regarded as being divided into parts.

This last functional relationship characterizes the entropy of a system with sub-systems. It demands that the entropy of a system can be calculated from the entropies of its sub-systems if the interactions between the sub-systems are known.

Given an ensemble of n uniformly distributed elements that are divided into k boxes (sub-systems) with b_1, ..., bk elements each, the entropy of the whole ensemble should be equal to the sum of the entropy of the system of boxes and the individual entropies of the boxes, each weighted with the probability of being in that particular box.

For positive integers bi where $b_1 + \ldots + bk = n$,

$$H_n \left(\frac{1}{n}, \ldots, \frac{1}{n} \right) = H_k \left(\frac{b_1}{n}, \ldots, \frac{b_k}{n} \right) + \sum_{i=1}^{k} \frac{b_i}{n} H_{b_i} \left(\frac{1}{b_i}, \ldots, \frac{1}{b_i} \right).$$

Choosing $k = n$, $b_1 = \ldots = bn = 1$ this implies that the entropy of a certain outcome is zero: $H_1(1) = 0$. This implies that the efficiency of a source alphabet with n symbols can be defined simply as being equal to its n-ary entropy. See also Redundancy (information theory).

7.8   Further properties

The Shannon entropy satisfies the following properties, for some of which it is useful to interpret entropy as the amount of information learned (or uncertainty eliminated) by revealing the value of a random variable X:

- Adding or removing an event with probability zero does not contribute to the entropy:

$$H_{n+1}(p_1, \ldots, p_n, 0) = H_n(p_1, \ldots, p_n)$$

- It can be confirmed using the Jensen inequality that

$$H(X) = E\left[\log_b\left(\frac{1}{p(X)}\right)\right] \leq \log_b\left(E\left[\frac{1}{p(X)}\right]\right) = \log_b(n)$$

This maximal entropy of $\log b(n)$ is effectively attained by a source alphabet having a uniform probability distribution: uncertainty is maximal when all possible events are equiprobable.

- The entropy or the amount of information revealed by evaluating (X,Y) (that is, evaluating X and Y simultaneously) is equal to the information revealed by conducting two consecutive experiments: first evaluating the value of Y, then revealing the value of X given that you know the value of Y. This may be written as

$$H(X,Y) = H(X|Y) + H(Y) = H(Y|X) + H(X).$$

- If $Y = f(X)$ where f is deterministic, then $H(f(X)|X) = 0$. Applying the previous formula to $H(X, f(X))$ yields

$$H(X) + H(f(X)|X) = H(f(X)) + H(X|f(X)),$$

so $H(f(X)) \leq H(X)$, thus the entropy of a variable can only decrease when the latter is passed through a deterministic function.

- If X and Y are two independent experiments, then knowing the value of Y doesn't influence our knowledge of the value of X (since the two don't influence each other by independence):

$$H(X|Y) = H(X).$$

- The entropy of two simultaneous events is no more than the sum of the entropies of each individual event, and are equal if the two events are independent. More specifically, if X and Y are two random variables on the same probability space, and (X, Y) denotes their Cartesian product, then

$$\mathrm{H}(X, Y) \leq \mathrm{H}(X) + \mathrm{H}(Y).$$

Proving this mathematically follows easily from the previous two properties of entropy.

7.9   Extending discrete entropy to the continuous case

7.9.1   Differential entropy

Main article: Differential entropy

The Shannon entropy is restricted to random variables taking discrete values. The corresponding formula for a continuous random variable with probability density function $f(x)$ with finite or infinite support $\mathbb{X}$ on the real line is defined by analogy, using the above form of the entropy as an expectation:

$$h[f] = \mathrm{E}[-\ln(f(x))] = -\int_{\mathbb{X}} f(x) \ln(f(x)) \, dx.$$

This formula is usually referred to as the **continuous entropy**, or differential entropy. A precursor of the continuous entropy $h[f]$ is the expression for the functional H in the H-theorem of Boltzmann.

Although the analogy between both functions is suggestive, the following question must be set: is the differential entropy a valid extension of the Shannon discrete entropy? Differential entropy lacks a number of properties that the Shannon discrete entropy has – it can even be negative – and thus corrections have been suggested, notably limiting density of discrete points.

To answer this question, we must establish a connection between the two functions:

We wish to obtain a generally finite measure as the bin size goes to zero. In the discrete case, the bin size is the (implicit) width of each of the n (finite or infinite) bins whose probabilities are denoted by pn. As we generalize to the continuous domain, we must make this width explicit.

To do this, start with a continuous function f discretized into bins of size Δ. By the mean-value theorem there exists a value xi in each bin such that

$$f(x_i)\Delta = \int_{i\Delta}^{(i+1)\Delta} f(x) \, dx$$

and thus the integral of the function f can be approximated (in the Riemannian sense) by

$$\int_{-\infty}^{\infty} f(x) \, dx = \lim_{\Delta \to 0} \sum_{i=-\infty}^{\infty} f(x_i)\Delta$$

where this limit and "bin size goes to zero" are equivalent.

We will denote

$$H^\Delta := - \sum_{i=-\infty}^{\infty} f(x_i)\Delta \log\left(f(x_i)\Delta\right)$$

and expanding the logarithm, we have

$$H^\Delta = - \sum_{i=-\infty}^{\infty} f(x_i)\Delta \log(f(x_i)) - \sum_{i=-\infty}^{\infty} f(x_i)\Delta \log(\Delta).$$

As $\Delta \to 0$, we have

$$\sum_{i=-\infty}^{\infty} f(x_i)\Delta \to \int_{-\infty}^{\infty} f(x)\,dx = 1$$

$$\sum_{i=-\infty}^{\infty} f(x_i)\Delta \log(f(x_i)) \to \int_{-\infty}^{\infty} f(x)\log f(x)\,dx.$$

But note that $\log(\Delta) \to -\infty$ as $\Delta \to 0$, therefore we need a special definition of the differential or continuous entropy:

$$h[f] = \lim_{\Delta \to 0} \left(H^\Delta + \log \Delta\right) = - \int_{-\infty}^{\infty} f(x)\log f(x)\,dx,$$

which is, as said before, referred to as the **differential entropy**. This means that the differential entropy *is not* a limit of the Shannon entropy for $n \to \infty$. Rather, it differs from the limit of the Shannon entropy by an infinite offset.

It turns out as a result that, unlike the Shannon entropy, the differential entropy is *not* in general a good measure of uncertainty or information. For example, the differential entropy can be negative; also it is not invariant under continuous co-ordinate transformations.

7.9.2 Relative entropy

Main article: Generalized relative entropy

Another useful measure of entropy that works equally well in the discrete and the continuous case is the **relative entropy** of a distribution. It is defined as the Kullback–Leibler divergence from the distribution to a reference measure m as follows. Assume that a probability distribution p is absolutely continuous with respect to a measure m, i.e. is of the form $p(dx) = f(x)m(dx)$ for some non-negative m-integrable function f with m-integral 1, then the relative entropy can be defined as

$$D_{\mathrm{KL}}(p\|m) = \int \log(f(x))p(dx) = \int f(x)\log(f(x))m(dx).$$

In this form the relative entropy generalises (up to change in sign) both the discrete entropy, where the measure m is the counting measure, and the differential entropy, where the measure m is the Lebesgue measure. If the measure m is itself a probability distribution, the relative entropy is non-negative, and zero if $p = m$ as measures. It is defined for any measure space, hence coordinate independent and invariant under co-ordinate reparameterizations if one properly takes into account the transformation of the measure m. The relative entropy, and implicitly entropy and differential entropy, do depend on the "reference" measure m.

7.10   Use in combinatorics

Entropy has become a useful quantity in combinatorics.

7.10.1   Loomis-Whitney inequality

A simple example of this is an alternate proof of the Loomis-Whitney inequality: for every subset $A \subseteq \mathbf{Z}^d$, we have

$$|A|^{d-1} \leq \prod_{i=1}^{d} |P_i(A)|$$

where Pi is the orthogonal projection in the ith coordinate:

$$P_i(A) = \{(x_1, ..., x_{i-1}, x_{i+1}, ..., x_d) : (x_1, ..., x_d) \in A\}.$$

The proof follows as a simple corollary of Shearer's inequality: if X_1, ..., Xd are random variables and S_1, ..., Sn are subsets of $\{1, ..., d\}$ such that every integer between I and d lies in exactly r of these subsets, then

$$H[(X_1, ..., X_d)] \leq \frac{1}{r} \sum_{i=1}^{n} H[(X_j)_{j \in S_i}]$$

where $(X_j)_{j \in S_i}$ is the Cartesian product of random variables Xj with indexes j in Si (so the dimension of this vector is equal to the size of Si).

We sketch how Loomis-Whitney follows from this: Indeed, let X be a uniformly distributed random variable with values in A and so that each point in A occurs with equal probability. Then (by the further properties of entropy mentioned above) $H(X) = \log|A|$, where $|A|$ denotes the cardinality of A. Let $Si = \{1, 2, ..., i-1, i+1, ..., d\}$. The range of $(X_j)_{j \in S_i}$ is contained in $Pi(A)$ and hence $H[(X_j)_{j \in S_i}] \leq \log|P_i(A)|$. Now use this to bound the right side of Shearer's inequality and exponentiate the opposite sides of the resulting inequality you obtain.

7.10.2   Approximation to binomial coefficient

For integers $0 < k < n$ let $q = k/n$. Then

$$\frac{2^{nH(q)}}{n+1} \leq \binom{n}{k} \leq 2^{nH(q)}.$$

where

$$H(q) = -q \log_2(q) - (1-q) \log_2(1-q). \text{[15]}$$

Here is a sketch proof. Note that $\binom{n}{k} q^{qn}(1-q)^{n-nq}$ is one term of the expression

$$\sum_{i=0}^{n} \binom{n}{i} q^i (1-q)^{n-i} = (q + (1-q))^n = 1.$$

Rearranging gives the upper bound. For the lower bound one first shows, using some algebra, that it is the largest term in the summation. But then,

$$\binom{n}{k} q^{qn}(1-q)^{n-nq} \geq \frac{1}{n+1}$$

since there are $n+1$ terms in the summation. Rearranging gives the lower bound.

A nice interpretation of this is that the number of binary strings of length n with exactly k many 1's is approximately $2^{nH(k/n)}$. [16]

7.11 See also

- Conditional entropy

- Cross entropy – is a measure of the average number of bits needed to identify an event from a set of possibilities between two probability distributions

- Diversity index – alternative approaches to quantifying diversity in a probability distribution

- Entropy (arrow of time)

- Entropy encoding – a coding scheme that assigns codes to symbols so as to match code lengths with the probabilities of the symbols.

- Entropy estimation

- Entropy power inequality

- Entropy rate

- Fisher information

- Hamming distance

- History of entropy

- History of information theory

- Information geometry

- Joint entropy – is the measure how much entropy is contained in a joint system of two random variables.

- Kolmogorov-Sinai entropy in dynamical systems

- Levenshtein distance

- Mutual information

- Negentropy

- Perplexity

- Qualitative variation – other measures of statistical dispersion for nominal distributions

- Quantum relative entropy – a measure of distinguishability between two quantum states.

- Rényi entropy – a generalisation of Shannon entropy; it is one of a family of functionals for quantifying the diversity, uncertainty or randomness of a system.

- Randomness

- Shannon index

- Theil index

- Typoglycemia

7.12   References

[1] Shannon, Claude E. (July–October 1948). "A Mathematical Theory of Communication". *Bell System Technical Journal* **27** (3): 379–423. doi:10.1002/j.1538-7305.1948.tb01338.x. (PDF)

[2] Schneier, B: *Applied Cryptography*, Second edition, page 234. John Wiley and Sons.

[3] Shannon, C. E. (January 1951). "Prediction and Entropy of Printed English" (PDF). *Bell System Technical Journal* **30** (1): 50–64. doi:10.1002/j.1538-7305.1951.tb01366.x. Retrieved 30 March 2014.

[4] Borda, Monica (2011). *Fundamentals in Information Theory and Coding*. Springer. p. 11. ISBN 978-3-642-20346-6.

[5] Han, Te Sun & Kobayashi, Kingo (2002). *Mathematics of Information and Coding*. American Mathematical Society. pp. 19–20. ISBN 978-0-8218-4256-0.

[6] Schneider, T.D, Information theory primer with an appendix on logarithms, National Cancer Institute, 14 April 2007.

[7] Carter, Tom (March 2014). *An introduction to information theory and entropy* (PDF). Santa Fe. Retrieved Aug 2014.

[8] Compare: Boltzmann, Ludwig (1896, 1898). Vorlesungen über Gastheorie : 2 Volumes – Leipzig 1895/98 UB: O 5262-6. English version: Lectures on gas theory. Translated by Stephen G. Brush (1964) Berkeley: University of California Press; (1995) New York: Dover ISBN 0-486-68455-5

[9] Mark Nelson (24 August 2006). "The Hutter Prize". Retrieved 2008-11-27.

[10] T. Schürmann and P. Grassberger, Entropy Estimation of Symbol Sequences, *CHAOS*, Vol. 6, No. 3 (1996) 414–427

[11] T. Schürmann, Bias Analysis in Entropy Estimation J. Phys. A: Math. Gen. 37 (2004) L295-L301.

[12] "The World's Technological Capacity to Store, Communicate, and Compute Information", Martin Hilbert and Priscila López (2011), Science (journal), 332(6025), 60–65; free access to the article through here: martinhilbert.net/WorldInfoCapacity.html

[13] Massey, James (1994). "Guessing and Entropy" (PDF). *Proc. IEEE International Symposium on Information Theory*. Retrieved December 31, 2013.

[14] Malone, David; Sullivan, Wayne (2005). "Guesswork is not a Substitute for Entropy" (PDF). *Proceedings of the Information Technology & Telecommunications Conference*. Retrieved December 31, 2013.

[15] Aoki, New Approaches to Macroeconomic Modeling. page 43.

[16] Probability and Computing, M. Mitzenmacher and E. Upfal, Cambridge University Press

This article incorporates material from Shannon's entropy on PlanetMath, which is licensed under the Creative Commons Attribution/Share-Alike License.

7.13   Further reading

7.13.1   Textbooks on information theory

- Arndt, C. (2004), *Information Measures: Information and its Description in Science and Engineering*, Springer, ISBN 978-3-540-40855-0

- Cover, T. M., Thomas, J. A. (2006), *Elements of information theory*, 2nd Edition. Wiley-Interscience. ISBN 0-471-24195-4.

- Gray, R. M. (2011), *Entropy and Information Theory*, Springer.

- Martin, Nathaniel F.G. & England, James W. (2011). *Mathematical Theory of Entropy*. Cambridge University Press. ISBN 978-0-521-17738-2.

- Shannon, C.E., Weaver, W. (1949) *The Mathematical Theory of Communication*, Univ of Illinois Press. ISBN 0-252-72548-4

- Stone, J. V. (2014), Chapter 1 of *Information Theory: A Tutorial Introduction*, University of Sheffield, England. ISBN 978-0956372857.

7.14 External links

- Hazewinkel, Michiel, ed. (2001), "Entropy", *Encyclopedia of Mathematics*, Springer, ISBN 978-1-55608-010-4

- Introduction to entropy and information on Principia Cybernetica Web

- *Entropy* an interdisciplinary journal on all aspect of the entropy concept. Open access.

- Description of information entropy from "Tools for Thought" by Howard Rheingold

- A java applet representing Shannon's Experiment to Calculate the Entropy of English

- Slides on information gain and entropy

- *An Intuitive Guide to the Concept of Entropy Arising in Various Sectors of Science* – a wikibook on the interpretation of the concept of entropy.

- Calculator for Shannon entropy estimation and interpretation

- A Light Discussion and Derivation of Entropy

- Network Event Detection With Entropy Measures, Dr. Raimund Eimann, University of Auckland, PDF; 5993 kB – a PhD thesis demonstrating how entropy measures may be used in network anomaly detection.

Chapter 8

Entropy (order and disorder)

Boltzmann's molecules (1896) shown at a "rest position" in a solid

In thermodynamics, **entropy** is commonly associated with the amount of order, disorder, or chaos in a thermodynamic system. This stems from Rudolf Clausius' 1862 assertion that any thermodynamic process always "admits to being reduced to the alteration in some way or another of the *arrangement* of the constituent parts of the working body" and that internal work associated with these alterations is quantified energetically by a measure of "entropy" change, according to the following differential expression:[1]

$$\int \frac{\delta Q}{T} \geq 0$$

In the years to follow, Ludwig Boltzmann translated these "alterations" into that of a probabilistic view of order and disorder in gas phase molecular systems.

In recent years, in chemistry textbooks there has been a shift away from using the terms "order" and "disorder" to that of the concept of energy dispersion to describe entropy, among other theories. In the 2002 encyclopedia Encarta, for example, *entropy* is defined as a thermodynamic property which serves as a measure of how close a system is to equilibrium, as well as a measure of the disorder in the system. [2] In the context of entropy, *"perfect internal disorder"* is synonymous with "equilibrium", but since that definition is so far different from the usual definition implied in normal speech, the use of the term in science has caused a great deal of confusion and misunderstanding.

Locally, the entropy can be lowered by external action. This applies to machines, such as a refrigerator, where the entropy in the cold chamber is being reduced, and to living organisms. This local decrease in entropy is, however, only possible at the expense of an entropy increase in the surroundings.

8.1 History

This "molecular ordering" entropy perspective traces its origins to molecular movement interpretations developed by Rudolf Clausius in the 1850s, particularly with his 1862 visual conception of molecular disgregation. Similarly, in 1859, after reading a paper on the diffusion of molecules by Clausius, Scottish physicist James Clerk Maxwell formulated the Maxwell distribution of molecular velocities, which gave the proportion of molecules having a certain velocity in a specific range. This was the first-ever statistical law in physics.[3]

In 1864, Ludwig Boltzmann, a young student in Vienna, came across Maxwell's paper and was so inspired by it that he spent much of his long and distinguished life developing the subject further. Later, Boltzmann, in efforts to develop a kinetic theory for the behavior of a gas, applied the laws of probability to Maxwell's and Clausius' molecular interpretation of entropy so to begin to interpret entropy in terms of order and disorder. Similarly, in 1882 Hermann von Helmholtz used the word "Unordnung" (disorder) to describe entropy.[4]

8.2 Overview

To highlight the fact that order and disorder are commonly understood to be measured in terms of entropy, below are current science encyclopedia and science dictionary definitions of entropy:

- A measure of the unavailability of a system's energy to do work; also a measure of disorder; the higher the entropy the greater the disorder.[5]

- A measure of disorder; the higher the entropy the greater the disorder.[6]

- In thermodynamics, a parameter representing the state of disorder of a system at the atomic, ionic, or molecular level; the greater the disorder the higher the entropy.[7]

- A measure of disorder in the universe or of the availability of the energy in a system to do work.[8]

Entropy and disorder also have associations with equilibrium.[9] Technically, *entropy*, from this perspective, is defined as a thermodynamic property which serves as a measure of how close a system is to equilibrium — that is, to perfect internal

disorder.[2] Likewise, the value of the entropy of a distribution of atoms and molecules in a thermodynamic system is a measure of the disorder in the arrangements of its particles.[10] In a stretched out piece of rubber, for example, the arrangement of the molecules of its structure has an "ordered" distribution and has zero entropy, while the "disordered" kinky distribution of the atoms and molecules in the rubber in the non-stretched state has positive entropy. Similarly, in a gas, the **order** is perfect and the measure of entropy of the system has its lowest value when all the molecules are in one place, whereas when more points are occupied the gas is all the more disorderly and the measure of the entropy of the system has its largest value.[10]

In systems ecology, as another example, the entropy of a collection of items comprising a system is defined as a measure of their disorder or equivalently the relative likelihood of the instantaneous configuration of the items.[11] Moreover, according to theoretical ecologist and chemical engineer Robert Ulanowicz, "that entropy might provide a quantification of the heretofore subjective notion of disorder has spawned innumerable scientific and philosophical narratives."[11][12] In particular, many biologists have taken to speaking in terms of the entropy of an organism, or about its antonym negentropy, as a measure of the structural order within an organism.[11]

The mathematical basis with respect to the association entropy has with order and disorder began, essentially, with the famous Boltzmann formula, $S = k \ln W$, which relates entropy S to the number of possible states W in which a system can be found.[13] As an example, consider a box that is divided into two sections. What is the probability that a certain number, or all of the particles, will be found in one section versus the other when the particles are randomly allocated to different places within the box? If you only have one particle, then that system of one particle can subsist in two states, one side of the box versus the other. If you have more than one particle, or define states as being further locational subdivisions of the box, the entropy is lower because the number of states is greater. The relationship between entropy, order, and disorder in the Boltzmann equation is so clear among physicists that according to the views of thermodynamic ecologists Sven Jorgensen and Yuri Svirezhev, "it is obvious that entropy is a measure of order or, most likely, disorder in the system."[13] In this direction, the second law of thermodynamics, as famously enunciated by Rudolf Clausius in 1865, states that:

Thus, if entropy is associated with disorder and if the entropy of the universe is headed towards maximal entropy, then many are often puzzled as to the nature of the "ordering" process and operation of evolution in relation to Clausius' most famous version of the second law, which states that the universe is headed towards maximal "disorder". In the recent 2003 book *SYNC – the Emerging Science of Spontaneous Order* by Steven Strogatz, for example, we find "Scientists have often been baffled by the existence of spontaneous order in the universe. The laws of thermodynamics seem to dictate the opposite, that nature should inexorably degenerate toward a state of greater disorder, greater entropy. Yet all around us we see magnificent structures—galaxies, cells, ecosystems, human beings—that have all somehow managed to assemble themselves."[14]

The common argument used to explain this is that, locally, entropy can be lowered by external action, e.g. solar heating action, and that this applies to machines, such as a refrigerator, where the entropy in the cold chamber is being reduced, to growing crystals, and to living organisms.[2] This local increase in order is, however, only possible at the expense of an entropy increase in the surroundings; here more disorder must be created.[2][15] The conditioner of this statement suffices that living systems are open systems in which both heat, mass, and or work may transfer into or out of the system. Unlike temperature, the putative entropy of a living system would drastically change if the organism were thermodynamically isolated. If an organism was in this type of "isolated" situation, its entropy would increase markedly as the once-living components of the organism decayed to an unrecognizable mass.[11]

8.3   Phase change

Owing to these early developments, the typical example of entropy change ΔS is that associated with phase change. In solids, for example, which are typically ordered on the molecular scale, usually have smaller entropy than liquids, and liquids have smaller entropy than gases and colder gases have smaller entropy than hotter gases. Moreover, according to the third law of thermodynamics, at absolute zero temperature, crystalline structures are approximated to have perfect "order" and zero entropy. This correlation occurs because the numbers of different microscopic quantum energy states available to an ordered system are usually much smaller than the number of states available to a system that appears to be disordered.

From his famous 1896 *Lectures on Gas Theory*, Boltzmann diagrams the structure of a solid body, as shown above, by postulating that each molecule in the body has a "rest position". According to Boltzmann, if it approaches a neighbor molecule it is repelled by it, but if it moves farther away there is an attraction. This, of course was a revolutionary perspective in its time; many, during these years, did not believe in the existence of either atoms or molecules (see: history of the molecule).[16] According to these early views, and others such as those developed by William Thomson, if energy in the form of heat is added to a solid, so to make it into a liquid or a gas, a common depiction is that the ordering of the atoms and molecules becomes more random and chaotic with an increase in temperature:

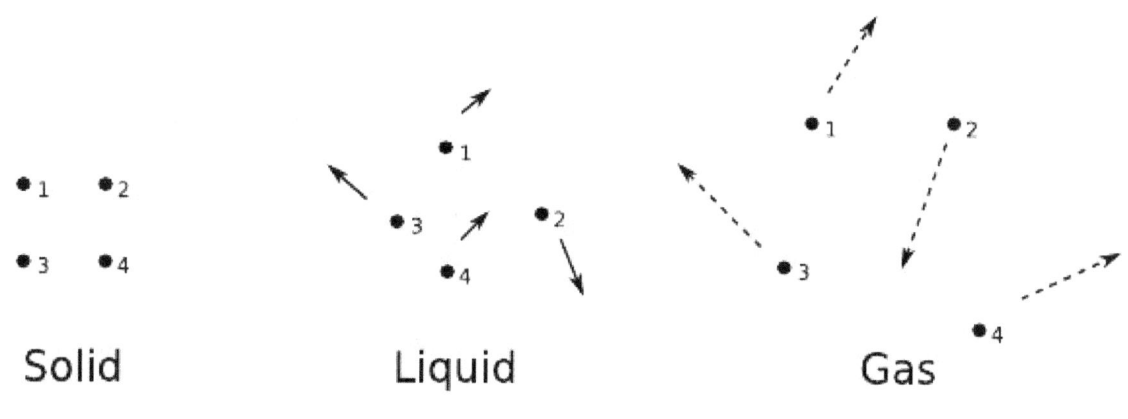

Thus, according to Boltzmann, owing to increases in thermal motion, whenever heat is added to a working substance, the rest position of molecules will be pushed apart, the body will expand, and this will create more *molar-disordered* distributions and arrangements of molecules. These disordered arrangements, subsequently, correlate, via probability arguments, to an increase in the measure of entropy.[17]

8.4 Adiabatic demagnetization

In the quest for ultra-cold temperatures, a temperature lowering technique called adiabatic demagnetization is used, where atomic entropy considerations are utilized which can be described in order-disorder terms.[18] In this process, a sample of solid such as chrome-alum salt, whose molecules are equivalent to tiny magnets, is inside an insulated enclosure cooled to a low temperature, typically 2 or 4 kelvins, with a strong magnetic field being applied to the container using a powerful external magnet, so that the tiny molecular magnets are aligned forming a well-ordered "initial" state at that low temperature. This magnetic alignment means that the magnetic energy of each molecule is minimal.[19] The external magnetic field is then reduced, a removal that is considered to be closely reversible. Following this reduction, the atomic magnets then assume random less-ordered orientations, owing to thermal agitations, in the "final" state:

The "disorder" and hence the entropy associated with the change in the atomic alignments has clearly increased.[18] In terms of energy flow, the movement from a magnetically aligned state requires energy from the thermal motion of the molecules, converting thermal energy into magnetic energy.[19] Yet, according to the second law of thermodynamics, because no heat can enter or leave the container, due to its adiabatic insulation, the system should exhibit no change in entropy, i.e. $\Delta S = 0$. The increase in disorder, however, associated with the randomizing directions of the atomic magnets represents an entropy *increase*? To compensate for this, the disorder (entropy) associated with the temperature of the specimen must *decrease* by the same amount.[18] The temperature thus falls as a result of this process of thermal energy being converted into magnetic energy. If the magnetic field is then increased, the temperature rises and the magnetic salt has to be cooled again using a cold material such as liquid helium.[19]

8.5 Difficulties with the term "disorder"

In recent years the long-standing use of term "disorder" to discuss entropy has met with some criticism.[20][21][22]

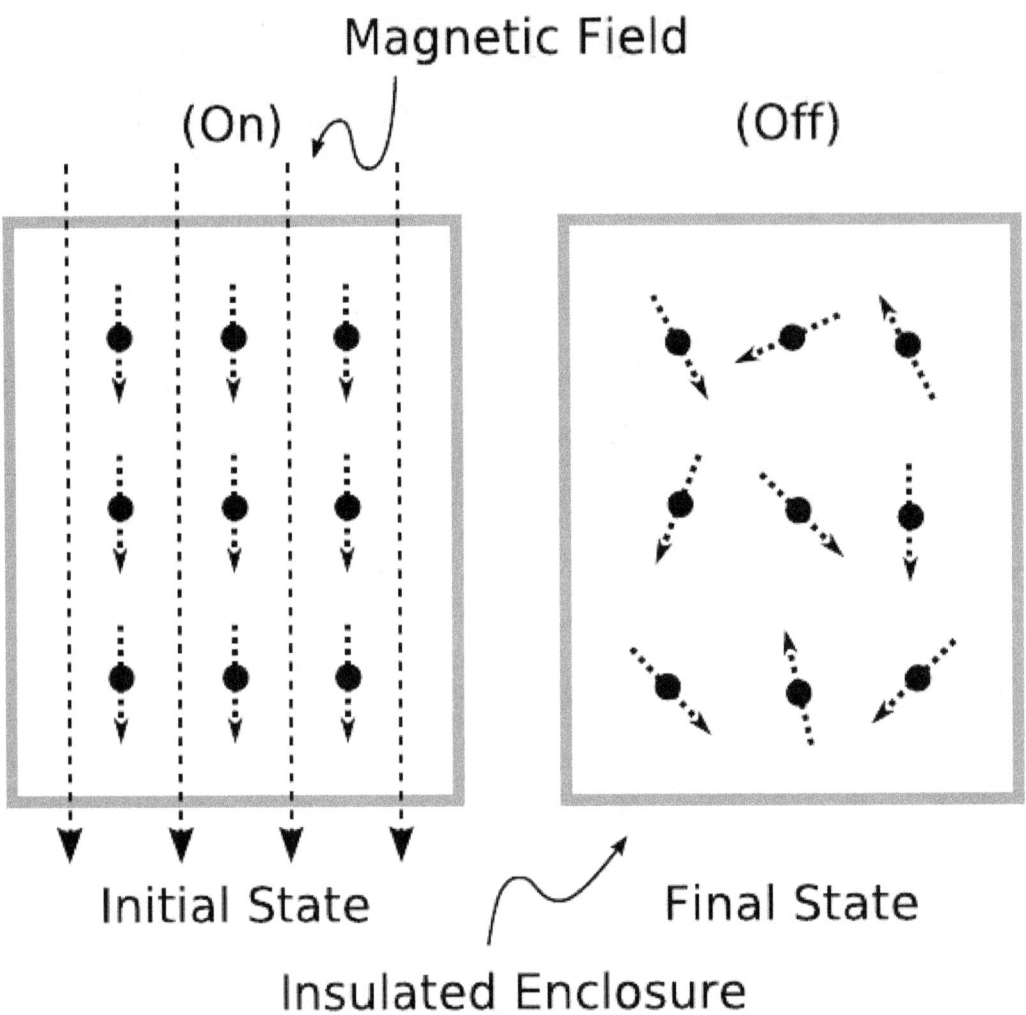

Entropy "order"/"disorder" considerations in the process of adiabatic demagnetization

When considered at a *microscopic* level, the term disorder may quite correctly suggest an increased range of accessible possibilities; but this may result in confusion because, at the *macroscopic* level of everyday perception, more ordered things seem more disordered, and more disordered things seem more ordered. For example, mixing water and oil counterintuitively creates more order from a thermodynamics perspective, because of the way water molecules and oil molecules interact. Equally, one can imagine on a beach in the summer if everyone arranges their towels in a "disorderly" fashion, people will struggle more to move and rearrange themselves (therefore more ordered from a thermodynamics perspective), whilst a more "ordered" towel arrangement means people are more free to move about (therefore more disordered from a thermodynamics perspective). [23] It has to be stressed, therefore, that "disorder", as used in a thermodynamic sense, relates to a full microscopic description of the system, rather than its apparent macroscopic properties. Many popular chemistry textbooks in recent editions increasingly have tended to instead present entropy through the idea of degrees of freedom and energy dispersal, which is a dominant contribution to entropy in most everyday situations. The textbook examples of a messy (disordered) and tidy (ordered) bedroom for describing entropy do not provide particularly good analogies, because (being a textbook) they're both still images, meaning they have an entropy of 0, because everything is fixed, and so there are no degrees of freedom. A better comparison to the tidy bedroom would be a bedroom where the socks are free to fly around the room randomly, rather than being confined to the sock drawer. Thermodynamics, unlike mothers of teenage boys, doesn't recognise whether the socks are in a specific place in the sock drawer, or a specific place on the bedroom floor, both represent highly ordered states.

8.6 See also

- Entropy

- History of entropy

- Entropy of mixing

- Entropy (information theory)

- Entropy (computing)

- Entropy (energy dispersal)

- Second law of thermodynamics

- Entropy (statistical thermodynamics)

- Entropy (classical thermodynamics)

8.7 References

[1] *Mechanical Theory of Heat* – Nine Memoirs on the development of concept of "Entropy" by Rudolf Clausius [1850–1865]

[2] Microsoft Encarta 2006. © 1993–2005 Microsoft Corporation. All rights reserved.

[3] Mahon, Basil (2003). *The Man Who Changed Everything – the Life of James Clerk Maxwell*. Hoboken, NJ: Wiley. ISBN 0-470-86171-1.

[4] Anderson, Greg (2005). *Thermodynamics of Natural Systems*. Cambridge University Press. ISBN 0-521-84772-9.

[5] *Oxford Dictionary of Science*, 2005

[6] *Oxford Dictionary of Chemistry*, 2004

[7] Barnes & Noble's *Essential Dictionary of Science*, 2004

[8] Gribbin's *Encyclopedia of Particle Physics*, 2000

[9] Landsberg, P.T. (1984). "Is Equilibrium always an Entropy Maximum?" J. Stat. Physics 35: 159-69.

[10] Greven, Andreas; Keller, Gerhard; Warnercke, Gerald (2003). *Entropy – Princeton Series in Applied Mathematics*. Princeton University Press. ISBN 0-691-11338-6.

[11] Ulanowicz, Robert, E. (2000). *Growth and Development – Ecosystems Phenomenology*. toExcel Press. ISBN 0-595-00145-9.

[12] Kubat, L.; Zeman, J. (1975). *Entropy and Information in Science and Philosophy*. Elsevier.

[13] Jorgensen, Sven, J.; Svirezhev, Yuri, M.; (2004). *Towards a Thermodynamic Theory for Ecological Systems*. Elsevier. ISBN 0-08-044167-X.

[14] Strogatz, Steven (2003). *the Emerging Science of Spontaneous Order*. Theia. ISBN 0-7868-6844-9.

[15] Brooks, Daniel, R.; Wiley, E.O. (1988). *Entropy as Evolution – Towards a Unified Theory of Biology*. University of Chicago Press. ISBN 0-226-07574-5.

[16] Cercignani, Carlo (1998). *Ludwig Boltzmann: The Man Who Trusted Atoms*. Oxford University Press. ISBN 9780198501541.

[17] Boltzmann, Ludwig (1896). *Lectures on Gas Theory*. Dover (reprint). ISBN 0-486-68455-5.

[18] Halliday, David; Resnick, Robert (1988). *Fundamentals of Physics, Extended 3rd ed*. Wiley. ISBN 0-471-81995-6.

[19] NASA - How does an Adiabatic Demagnetization Refrigerator Work ?

[20] Frank L. Lambert, 2002, "Disorder--A Cracked Crutch for Supporting Entropy Discussions," *Journal of Chemical Education* 79: 187. Updated version at here.

[21] Carson, E. M., and Watson, J. R., (Department of Educational and Professional Studies, Kings College, London), 2002, "Undergraduate students' understandings of entropy and Gibbs Free energy," University Chemistry Education - 2002 Papers, Royal Society of Chemistry.

[22] Sozbilir, Mustafa, PhD studies: Turkey, *A Study of Undergraduates' Understandings of Key Chemical Ideas in Thermodynamics*, Ph.D. Thesis, Department of Educational Studies, The University of York, 2001.

[23] Biel.R. and Mu-Jeong Kho (2009)"The Issue of Energy within a Dialectical Approach to the Regulationist Problematique]." Recherches & Régulation Working Papers, RR Série ID 2009-1, Association Recherche & Régulation: 1-21." (PDF). http://theorie-regulation.org. 2009-11-23. Retrieved 2012-08-09.

8.8   External links

- Lambert, F.L. Entropy Sites — A Guide

- Lambert, F.L. *Shuffled Cards, Messy Desks, and Disorderly Dorm Rooms - Examples of Entropy Increase? Nonsense!* Journal of Chemical Education

Chapter 9

Entropy (energy dispersal)

The description of **entropy as energy dispersal** provides an introductory method of teaching the thermodynamic concept of entropy. In physics and physical chemistry, entropy has commonly been defined as a scalar measure of the disorder of a thermodynamic system. This newer approach sets out a variant approach to entropy, namely as a measure of energy *dispersal* or *distribution* at a specific temperature. Under this approach, changes in entropy can be quantitatively related to the distribution or the spreading out of the energy of a thermodynamic system, divided by its temperature.

The energy dispersal approach to teaching entropy was developed to facilitate teaching entropy to students beginning university chemistry and biology. This new approach also avoids ambiguous terms such as disorder and chaos, which have multiple everyday meanings.

9.1 Problem: entropy as disorder is hard to teach

The term "entropy" has been in use from early in the history of classical thermodynamics, and with the development of statistical thermodynamics and quantum theory, entropy changes have been described in terms of the mixing or "spreading" of the total energy of each constituent of a system over its particular quantized energy levels.

Such descriptions have tended to be used together with commonly used terms such as disorder and chaos which are ambiguous, and whose everyday meaning is the opposite of what they are intended to mean in thermodynamics. Not only does this situation cause confusion, but it also hampers the teaching of thermodynamics. Students were being asked to grasp meanings directly contradicting their normal usage, with equilibrium being equated to "perfect internal disorder" and the mixing of milk in coffee from apparent chaos to uniformity being described as a transition from an ordered state into a disordered state.[1]

The description of entropy as the amount of "mixedupness" or "disorder," as well as the abstract nature of the statistical mechanics grounding this notion, can lead to confusion and considerable difficulty for those beginning the subject.[2][3] Even though courses emphasised microstates and energy levels, most students could not get beyond simplistic notions of randomness or disorder. Many of those who learned by practising calculations did not understand well the intrinsic meanings of equations, and there was a need for qualitative explanations of thermodynamic relationships.[4][5]

9.2 Solution: entropy as energy dispersal

Entropy can be described in terms of "energy dispersal" and the "spreading of energy," while avoiding all mention of "disorder" and "chaos" except when explaining misconceptions. All explanations of where and how energy is dispersing or spreading have been recast in terms of energy dispersal, so as to emphasise the underlying qualitative meaning.[2]

In this approach, the second law of thermodynamics is introduced as "Energy spontaneously disperses from being localized to becoming spread out if it is not hindered from doing so." in the context of common experiences such as a rock falling, a hot frying pan cooling down, iron rusting, air leaving a punctured tyre and ice melting in a warm room. Entropy is then depicted as a sophisticated kind of "before and after" yardstick — measuring how much energy is spread out over time as a result of a process such as heating a system, or how widely spread out the energy is after something happens in comparison with its previous state, in a process such as gas expansion or fluids mixing (at a constant temperature). The equations are explored with reference to the common experiences, with emphasis that in chemistry the energy that entropy measures as dispersing is the internal energy of molecules.

The statistical interpretation is related to quantum mechanics in describing the way that energy is distributed (quantized) amongst molecules on specific energy levels, with all the energy of the macrostate always in only one microstate at one instant. Entropy is described as measuring the energy dispersal for a system by the number of accessible microstates, the number of different arrangements of all its energy at the next instant. Thus, an increase in entropy means a greater number of microstates for the final state than for the initial state, and hence more possible arrangements of a system's total energy at any one instant. Here, the greater 'dispersal of the total energy of a system' means the existence of many possibilities.[6]

Continuous movement and molecular collisions visualised as being like bouncing balls blown by air as used in a lottery can then lead on to showing the possibilities of many Boltzmann distributions and continually changing "distribution of the instant", and on to the idea that when the system changes, dynamic molecules will have a greater number of accessible microstates. In this approach, all everyday spontaneous physical happenings and chemical reactions are depicted as involving some type of energy flows from being localized or concentrated to becoming spread out to a larger space, always to a state with a greater number of microstates.[7]

This approach provides a good basis for understanding the conventional approach, except in very complex cases where the qualitative relation of energy dispersal to entropy change can be so inextricably obscured that it is moot.[7] Thus in situations such as the entropy of mixing when the two or more different substances being mixed are at the same temperature and pressure so there will be no net exchange of heat or work, the entropy increase will be due to the literal spreading out of the motional energy of each substance in the larger combined final volume. Each component's energetic molecules become more separated from one another than they would be in the pure state, when in the pure state they were colliding only with identical adjacent molecules, leading to an increase in its number of accessible microstates.[8]

Variants of the energy dispersal approach have been adopted in number of undergraduate chemistry texts, mainly in the United States. An advanced text, *Physical Chemistry* 8th edition, by Peter Atkins of Oxford University and Julio De Paula, says "The concept of the number of microstates makes quantitative the ill-defined qualitative concepts of 'disorder' and 'the dispersal of matter and energy' that are used widely to introduce the concept of entropy: a more 'disorderly' distribution of energy and matter corresponds to a greater number of microstates associated with the same total energy." --- p. 81[9]

Websites have made the energy dispersal approach accessible not only to all students of chemistry, but also to the lay public seeking a basic intuitive understanding of thermodynamic entropy. For example, here [7] is a page setting out the qualitative simplicity of the notion of entropy.

The energy dispersal approach has been criticised by Arieh Ben-Naim.[10]

9.3   History of energy dispersal

The concept of "energy dispersal" as a description of entropy appeared in William Thomson's (Lord Kelvin) 1852 article "On a Universal Tendency in Nature to the Dissipation of Mechanical Energy."[11] Thomson distinguished between two types or "stores" of mechanical energy: "statical" and "dynamical." He discussed how these two types of energy can change from one form to the other during a thermodynamic transformation. When heat is created by any irreversible process (such as friction), or when heat is diffused by conduction, mechanical energy is dissipated, and it is impossible to restore the initial state.[12][13]

In the mid-1950s, with the development of quantum theory, researchers began speaking about entropy changes in terms of the mixing or "spreading" of the total energy of each constituent of a system over its particular quantized energy levels, such as by the reactants and products of a chemical reaction.[14]

In 1984, the Oxford physical chemist Peter Atkins, in a book *The Second Law*, written for laypersons, presented a non-mathematical interpretation of what he called the "infinitely incomprehensible entropy" in simple terms, describing the Second Law of thermodynamics as "energy tends to disperse". His analogies included an imaginary intelligent being called "Boltzmann's Demon," who runs around reorganizing and dispersing energy, in order to show how the W in Boltzmann's entropy formula relates to energy dispersion. This dispersion is transmitted via atomic vibrations and collisions. Atkins wrote: "each atom carries kinetic energy, and the spreading of the atoms spreads the energy...the Boltzmann equation therefore captures the aspect of dispersal: the dispersal of the entities that are carrying the energy."[15]

Stanley Sandler, in his 1989 *Chemical and Engineering Thermodynamics*, described how given any thermodynamic process, a quantity TS can be interpreted as the amount of mechanical energy that has been converted into thermal energy by viscous dissipation, dispersion, and other system irreversibilities.[16] In 1997, John Wrigglesworth described spatial particle distributions as represented by distributions of energy states. According to the second law of thermodynamics, isolated systems will tend to redistribute the energy of the system into a more probable arrangement or a maximum probability energy distribution, i.e. from that of being concentrated to that of being spread out. By virtue of the First law of thermodynamics, the total energy does not change; instead, the energy tends to disperse from a coherent to a more incoherent state.[17] In his 1999 *Statistical Thermodynamics*, M.C. Gupta defined entropy as a function that measures how energy disperses when a system changes from one state to another.[18] Other authors defining entropy in a way that embodies energy dispersal are Cecie Starr[19] and Andrew Scott.[20]

In a 1996 article, the physicist Harvey S. Leff set out what he called "the spreading and sharing of energy."[21] Another physicist, Daniel F. Styer, published an article in 2000 showing that "entropy as disorder" was inadequate.[22] In an article published in the 2002 *Journal of Chemical Education*, Frank L. Lambert argued that portraying entropy as "disorder" is confusing and should be abandoned. He has gone on to develop detailed resources for chemistry instructors, equating entropy increase as the spontaneous dispersal of energy, namely how much energy is spread out in a process, or how widely dispersed it becomes – at a specific temperature.[2][23]

9.4 References

[1] Microsoft Encarta 2006. © 1993–2005 Microsoft Corporation. All rights reserved.

[2] Frank L. Lambert, 2002, "Disorder--A Cracked Crutch for Supporting Entropy Discussions." *Journal of Chemical Education* 79: 187. Updated version at here.

[3] Frank L. Lambert, "The Second Law of Thermodynamics (6)."

[4] Carson, E. M., and Watson, J. R., (Department of Educational and Professional Studies, Kings College, London), 2002, "Undergraduate students' understandings of entropy and Gibbs Free energy." University Chemistry Education - 2002 Papers, Royal Society of Chemistry.

[5] Sozbilir, Mustafa, PhD studies: Turkey, *A Study of Undergraduates' Understandings of Key Chemical Ideas in Thermodynamics*, Ph.D. Thesis, Department of Educational Studies, The University of York, 2001.

[6] Frank L. Lambert, The Molecular Basis for Understanding Simple Entropy Change

[7] Frank L. Lambert, Entropy is simple, qualitatively

[8] Frank L. Lambert, Notes for a "Conversation About Entropy": a brief discussion of *both* thermodynamic and "configurational" ("positional") entropy in chemistry.

[9] Atkins, Peter; Julio De Paula (2006). *Physical Chemistry , 8th edition.* Oxford University Press. ISBN 0-19-870072-5.

[10] Review of "Entropy and the second law: interpretation and misss-interpretationsss" in Chemistry World

[11] Jensen, William. (2004). "Entropy and Constraint of Motion." *Journal of Chemical Education* (81) 693, May

[12] Thomson, William (1852). "On a Universal Tendency in Nature to the Dissipation of Mechanical Energy." *Proceedings of the Royal Society of Edinburg*, April 19.

[13] Thomson, William (1874). "Kinetic Theory of the Dissipation of Energy", *Nature* IX: 441-44. (April 9).

[14] Denbigh, Kenneth (1981). *The Principles of Chemical Equilibrium, 4th Ed.* Cambridge University Press. ISBN 0-521-28150-4.

[15] Atkins, Peter (1984). *The Second Law*. Scientific American Library. ISBN 0-7167-5004-X.

[16] Sandler, Stanley, I. (1989). *Chemical and Engineering Thermodynamics*. John Wiley & Sons. ISBN 0-471-83050-X.

[17] Wrigglesworth, John (1997). *Energy and Life (Modules in Life Sciences)*. CRC. ISBN 0-7484-0433-3. (see excerpt)

[18] Gupta, M.C. (1999). *Statistical Thermodynamics*. New Age Publishers. ISBN 81-224-1066-9. (see excerpt)

[19] Starr, Cecie; Taggart, R. (1992). *Biology - the Unity and Diversity of Life*. Wadsworth Publishing Co. ISBN 0-534-16566-4.

[20] Scott, Andrew (2001). *101 Key ideas in Chemistry*. Teach Yourself Books. ISBN 0-07-139665-9.

[21] Leff, H. S., 1996, "Thermodynamic entropy: The spreading and sharing of energy," *Am. J. Phys.* 64: 1261-71.

[22] Styer D. F., 2000, *Am. J. Phys.* 68: 1090-96.

[23] "A Student's Approach to the Second Law and Entropy". 2009-07-17. Retrieved 2014-12-12.

9.5   Further reading

- Carson, E. M., and Watson, J. R., (Department of Educational and Professional Studies, Kings College, London), 2002, "Undergraduate students' understandings of entropy and Gibbs Free energy," University Chemistry Education - 2002 Papers, Royal Society of Chemistry.

- Frank L. Lambert, 2002, "Disorder - A Cracked Crutch For Supporting Entropy Discussions," *Journal of Chemical Education* **79**: 187-92. Updated version here.

- Leff, Harvey S., 2007, "Entropy, Its Language and Interpretation," *Foundations of Physics* 37(12): 1744-66.

9.6   Texts using the energy dispersal approach

- Atkins, P. W., *Physical Chemistry for the Life Sciences*. Oxford University Press, ISBN 0-19-928095-9; W. H. Freeman, ISBN 0-7167-8628-1

- Benjamin Gal-Or, "Cosmology, Physics and Philosophy", Springer-Verlag, New York, 1981, 1983, 1987 ISBN 0-387-90581-2

- Bell, J., *et al.*, 2005. *Chemistry: A General Chemistry Project of the American Chemical Society*, 1st ed. W. H. Freeman, 820pp, ISBN 0-7167-3126-6

- Brady, J.E., and F. Senese, 2004. *Chemistry, Matter and Its Changes*, 4th ed. John Wiley, 1256pp, ISBN 0-471-21517-1

- Brown, T. L., H. E. LeMay, and B. E. Bursten, 2006. *Chemistry: The Central Science*, 10th ed. Prentice Hall, 1248pp, ISBN 0-13-109686-9

- Ebbing, D.D., and S. D. Gammon, 2005. *General Chemistry*, 8th ed. Houghton-Mifflin, 1200pp, ISBN 0-618-39941-0

- Ebbing, Gammon, and Ragsdale. *Essentials of General Chemistry*, 2nd ed.

- Hill, Petrucci, McCreary and Perry. *General Chemistry*, 4th ed.

- Kotz, Treichel, and Weaver. *Chemistry and Chemical Reactivity*, 6th ed.

- Moog, Spencer, and Farrell. *Thermodynamics, A Guided Inquiry*.

- Moore, J. W., C. L. Stanistski, P. C. Jurs, 2005. *Chemistry, The Molecular Science*, 2nd ed. Thompson Learning, 1248pp, ISBN 0-534-42201-2

- Olmsted and Williams, *Chemistry*, 4th ed.

- Petrucci, Harwood, and Herring. *General Chemistry*, 9th ed.

- Silberberg, M.S., 2006. *Chemistry, The Molecular Nature of Matter and Change*, 4th ed. McGraw-Hill, 1183pp, ISBN 0-07-255820-2

- Suchocki, J., 2004. *Conceptual Chemistry* 2nd ed. Benjamin Cummings. 706pp, ISBN 0-8053-3228-6

9.7 External links

- welcome to entropysite. A large website, maintained by Frank L. Lambert, with links to work on the energy disperal approach to entropy.

- "The Second Law of Thermodynamics (6)."

Chapter 10

Entropy production

Entropy production determines the performance of thermal machines such as power plants, heat engines, refrigerators, heat pumps, and air conditioners. It also plays a key role in the thermodynamics of irreversible processes.[1]

10.1   Short history

Entropy is produced in irreversible processes. The importance of avoiding irreversible processes (hence reducing the entropy production) was recognized as early as 1824 by Carnot.[2] In 1867 Rudolf Clausius expanded his previous work from 1854[3] on the concept of "unkompensierte Verwandlungen" (uncompensated transformations), which, in our modern nomenclature, would be called the entropy production. In the same article as where he introduced the name entropy,[4] Clausius gives the expression for the entropy production (for a closed system), which he denotes by N, in equation (71) which reads

$$N = S - S_0 - \int \frac{dQ}{T}.$$

Here S is the entropy in the final state and the integral is to be taken from the initial state to the final state. From the context it is clear that $N = 0$ if the process is reversible and $N > 0$ in case of an irreversible process.

10.2   First and second law

The laws of thermodynamics system apply to well-defined systems. Fig.1 is a general representation of a thermodynamic system. We consider systems which, in general, are inhomogeneous. Heat and mass are transferred across the boundaries (nonadiabatic, open systems), and the boundaries are moving (usually through pistons). In our formulation we assume that heat and mass transfer and volume changes take place only separately at well-defined regions of the system boundary. The expression, given here, are not the most general formulations of the first and second law. E.g. kinetic energy and potential energy terms are missing and exchange of matter by diffusion is excluded.

The rate of entropy production, denoted by $\dot{S}_i$, is a key element of the second law of thermodynamics for open inhomogeneous systems which reads

$$\frac{dS}{dt} = \Sigma_k \frac{\dot{Q}_k}{T_k} + \Sigma_k \dot{S}_k + \Sigma_k \dot{S}_{ik} \text{ with } \dot{S}_{ik} \geq 0.$$

Here S is the entropy of the system; T_k is the temperature at which the heat flow $\dot{Q}_k$ enters the system; $\dot{S}_k = \dot{n}_k S_{mk} = \dot{m}_k s_k$ represents the entropy flow into the system at position k, due to matter flowing into the system ($\dot{n}_k$, $\dot{m}_k$ are the

Rudolf Clausius

molar flow and mass flow and S_{mk} and s_k are the molar entropy (i.e. entropy per mole) and specific entropy (i.e. entropy per unit mass) of the matter, flowing into the system, respectively; $\dot{S}_{ik}$ represents the entropy production rates due to internal processes. The index i in $\dot{S}_{ik}$ refers to the fact that the entropy is produced due to irreversible processes. The entropy-production rate of every process in nature is always positive or zero. This is an essential aspect of the second law.

The $\sum$'s indicate the algebraic sum of the respective contributions if there are more heat flows, matter flows, and internal processes.

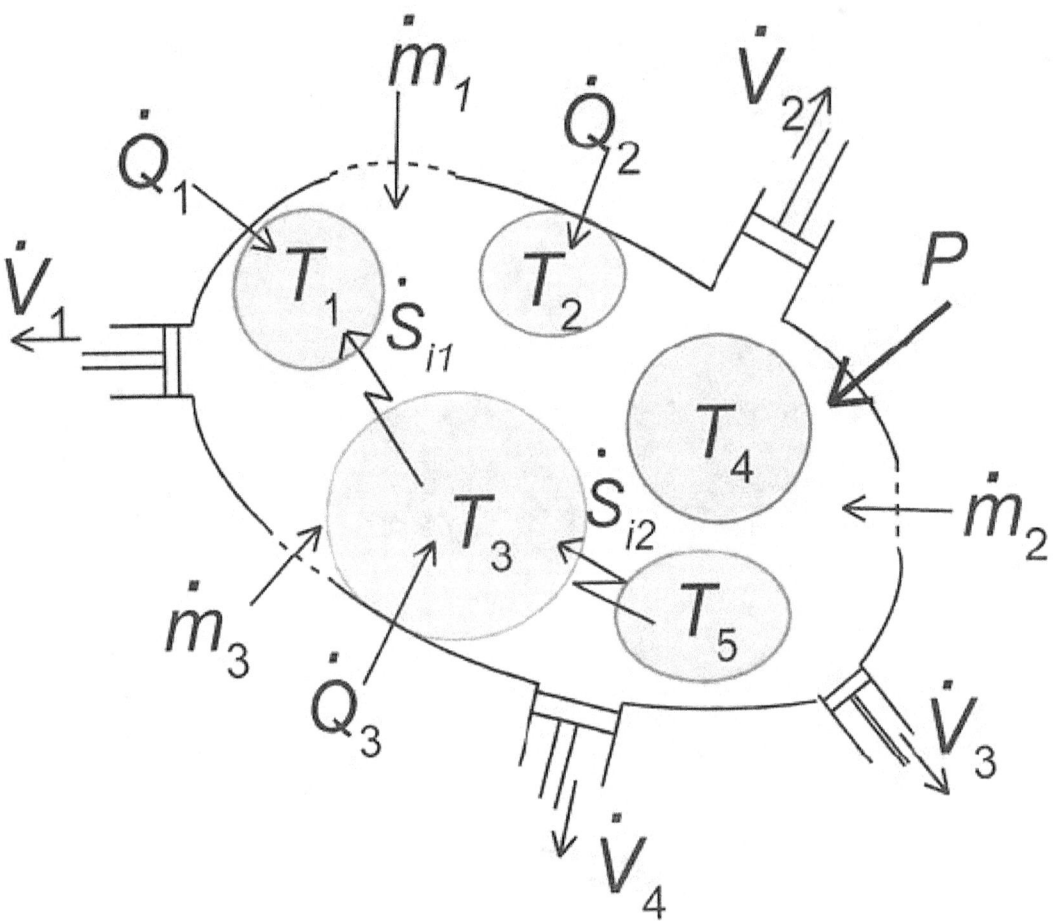

Fig.1 General representation of an inhomogeneous system that consists of a number of subsystems. The interaction of the system with the surroundings is through exchange of heat and other forms of energy, flow of matter, and changes of shape. The internal interactions between the various subsystems are of a similar nature and lead to entropy production.

In order to demonstrate the impact of the second law, and the role of entropy production, it has to be combined with the first law which reads

$$\frac{\mathrm{d}U}{\mathrm{d}t} = \Sigma_k \dot{Q}_k + \Sigma_k \dot{H}_k - \Sigma_k p_k \frac{\mathrm{d}V_k}{\mathrm{d}t} + P.$$

with U the internal energy of the system; $\dot{H}_k = \dot{n}_k H_{mk} = \dot{m}_k h_k$ the enthalpy flows into the system due to the matter that flows into the system (H_{mk} its molar enthalpy, h_k the specific enthalpy (i.e. enthalpy per unit mass)), and $\mathrm{d}V_k/\mathrm{d}t$ are the rates of change of the volume of the system due to a moving boundary at position k while p_k is the pressure behind that boundary; P represents all other forms of power application (such as electrical).

The first and second law have been formulated in terms of time derivatives of U and S rather than in terms of total differentials $\mathrm{d}U$ and $\mathrm{d}S$ where it is tacitly assumed that $\mathrm{d}t > 0$. So, the formulation in terms of time derivatives is more elegant. An even bigger advantage of this formulation is, however, that it is emphasizes that *heat flow* and *power* are the basic thermodynamic properties and that heat and work are derived quantities being the time integrals of the heat flow and the power respectively.

10.3   Examples of irreversible processes

Entropy is produced in irreversible processes. Some important irreversible processes are:

- heat flow through a thermal resistance

- fluid flow through a flow resistance such as in the Joule expansion or the Joule-Thomson effect

- diffusion

- chemical reactions

- Joule heating

- friction between solid surfaces

- fluid viscosity within a system.

The expression for the rate of entropy production in the first two cases will be derived in separate sections.

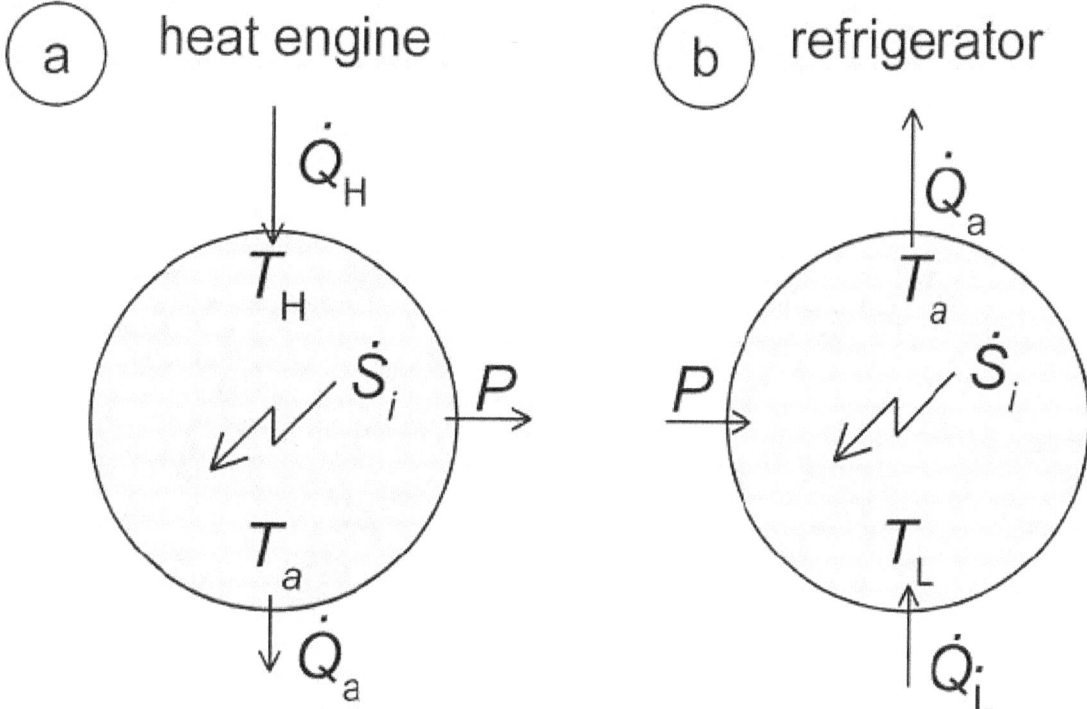

Fig.2 a: Schematic diagram of a heat engine. A heating power $\dot{Q}_H$ enters the engine at the high temperature TH, and $\dot{Q}_a$ is released at ambient temperature Ta. A power P is produced and the entropy production rate is $\dot{S}_i$. b: Schematic diagram of a refrigerator. $\dot{Q}_L$ is the cooling power at the low temperature TL, and $\dot{Q}_a$ is released at ambient temperature. The power P is supplied and $\dot{S}_i$ is the entropy production rate. The arrows define the positive directions of the flows of heat and power in the two cases. They are positive under normal operating conditions.

10.4   Performance of heat engines and refrigerators

Most heat engines and refrigerators are closed cyclic machines.[5] In the steady state the internal energy and the entropy of the machines after one cycle are the same as at the start of the cycle. Hence, on average, $dU/dt = 0$ and $dS/dt = 0$ since

U and S are functions of state. Furthermore they are closed systems ($\dot{n} = 0$) and the volume is fixed ($dV/dt = 0$). This leads to a significant simplification of the first and second law:

$$0 = \Sigma_k \dot{Q}_k + P$$

and

$$0 = \Sigma_k \frac{\dot{Q}_k}{T_k} + \dot{S}_i.$$

The summation is over the (two) places where heat is added or removed.

10.4.1   Engines

For a heat engine (Fig.2a) the first and second law obtain the form

$$0 = \dot{Q}_H - \dot{Q}_a - P$$

and

$$0 = \frac{\dot{Q}_H}{T_H} - \frac{\dot{Q}_a}{T_a} + \dot{S}_i.$$

Here $\dot{Q}_H$ is the heat supplied at the high temperature TH, $\dot{Q}_a$ is the heat removed at ambient temperature T_a, and P is the power delivered by the engine. Eliminating $\dot{Q}_a$ gives

$$P = \frac{T_H - T_a}{T_H}\dot{Q}_H - T_a\dot{S}_i.$$

The efficiency is defined by

$$\eta = \frac{P}{\dot{Q}_H}.$$

If $\dot{S}_i = 0$ the performance of the engine is at its maximum and the efficiency is equal to the Carnot efficiency

$$\eta_C = \frac{T_H - T_a}{T_H}.$$

10.4.2   Refrigerators

For refrigerators (fig.2b) holds

$$0 = \dot{Q}_L - \dot{Q}_a + P$$

and

$$0 = \frac{\dot{Q}_L}{T_L} - \frac{\dot{Q}_a}{T_a} + \dot{S}_i.$$

Here P is the power, supplied to produce the cooling power $\dot{Q}_L$ at the low temperature TL. Eliminating $\dot{Q}_a$ now gives

$$\dot{Q}_L = \frac{T_L}{T_a - T_L}(P - T_a \dot{S}_i).$$

The Coefficient Of Performance of refrigerators is defined by

$$\xi = \frac{\dot{Q}_L}{P}.$$

If $\dot{S}_i = 0$ the performance of the cooler is at its maximum. The COP is then given by the Carnot Coefficient Of Performance

$$\xi_C = \frac{T_L}{T_a - T_L}.$$

10.4.3   Power dissipation

In both cases we find a contribution $T_a \dot{S}_i$ which reduces the system performance. This product of ambient temperature and the (average) entropy production rate $P_{diss} = T_a \dot{S}_i$ is called the dissipated power.

10.5   Equivalence with other formulations

It is interesting investigate how the above mathematical formulation of the second law relates with other well-known formulations of the second law.

We first look at a heat engine, assuming that $\dot{Q}_a = 0$. In other words: the heat flow $\dot{Q}_H$ is completely converted into power. In this case the second law would reduce to

$$0 = \frac{\dot{Q}_H}{T_H} + \dot{S}_i.$$

Since $\dot{Q}_H \geq 0$ and $T_H > 0$ this would result in $\dot{S}_i \leq 0$ which violates the condition that the entropy production is always positive. Hence: *No process is possible in which the sole result is the absorption of heat from a reservoir and its complete conversion into work.* This is the Kelvin statement of the second law.

Now look at the case of the refrigerator and assume that the input power is zero. In other words: heat is transported from a low temperature to a high temperature without doing work on the system. The first law with $P = 0$ would give

$$\dot{Q}_L = \dot{Q}_a$$

and the second law then yields

$$0 = \frac{\dot{Q}_L}{T_L} - \frac{\dot{Q}_L}{T_a} + \dot{S}_i$$

or

$$\dot{S}_i = \dot{Q}_L \left(\frac{1}{T_a} - \frac{1}{T_L} \right).$$

Since $\dot{Q}_L \geq 0$ and $T_a > T_L$ this would result in $\dot{S}_i \leq 0$ which again violates the condition that the entropy production is always positive. Hence: *No process is possible whose sole result is the transfer of heat from a body of lower temperature to a body of higher temperature.* This is the Clausius statement of the second law.

10.6 Expressions for the entropy production

10.6.1 Heat flow

In case of a heat flow $\dot{Q}$ from T_1 to T_2 the rate of entropy production is given by

$$\dot{S}_i = \dot{Q} \left(\frac{1}{T_2} - \frac{1}{T_1} \right).$$

If the heat flow is in a bar with length L, cross-sectional area A, and thermal conductivity κ, and the temperature difference is small

$$\dot{Q} = \kappa \frac{A}{L} (T_1 - T_2)$$

the entropy production rate is

$$\dot{S}_i = \kappa \frac{A}{L} \frac{(T_1 - T_2)^2}{T_1 T_2}.$$

10.6.2 Flow of matter

In case of a volume flow $\dot{V}$ from a pressure p_1 to p_2

$$\dot{S}_i = - \int_1^2 \frac{\dot{V}}{T} dp.$$

For small pressure drops and defining the flow conductance C by $\dot{V} = C(p_1 - p_2)$ we get

$$\dot{S}_i = C \frac{(p_1 - p_2)^2}{T}.$$

The dependences of $\dot{S}$, on $(T_1 - T_2)$ and on $(p_1 - p_2)$ are quadratic. This is typical for expressions of the entropy production rates in general. They guarantee that the entropy production is positive.

10.6.3 Entropy of mixing

In this Section we will calculate the entropy of mixing when two ideal gases diffuse into each other. Consider a volume V_t divided in two volumes V_a and V_b so that $V_t = V_a + V_b$. The volume V_a contains n_a moles of an ideal gas a and V_b contains n_b moles of gas b. The total amount is $n_t = n_a + n_b$. The temperature and pressure in the two volumes is the same. The entropy at the start is given by

$$S_{t1} = S_{a1} + S_{b1}.$$

When the division between the two gases is removed the two gases expand, comparable to a Joule-Thomson expansion. In the final state the temperature is the same as initially but the two gases now both take the volume V_t. The relation of the entropy of n moles an ideal gas is

$$S = nC_V \ln \frac{T}{T_0} + nR \ln \frac{V}{V_0}$$

with CV the molar heat capacity at constant volume and R the molar ideal gas constant. The system is an adiabatic closed system, so the entropy increase during the mixing of the two gases is equal to the entropy production. It is given by

$$S_i = S_{t2} - S_{t1}.$$

As the initial and final temperature are the same the temperature terms plays no role, so we can focus on the volume terms. The result is

$$S_i = n_a R \ln \frac{V_t}{V_a} + n_b R \ln \frac{V_t}{V_b}.$$

Introducing the concentration $x = n_a/n_t = V_a/V_t$ we arrive at the well known expression

$$S_i = -n_t R [x \ln x + (1 - x) \ln(1 - x)].$$

10.6.4 Joule expansion

The Joule expansion is similar to the mixing described above. It takes place in an adiabatic system consisting of a gas and two rigid vessels (a and b) of equal volume, connected by a valve. Initially the valve is closed. Vessel (a) contains the gas under high pressure while the other vessel (b) is empty. When the valve is opened the gas flows from vessel (a) into (b) until the pressures in the two vessels are equal. The volume, taken by the gas, is doubled while the internal energy of the system is constant (adiabatic and no work done). Assuming that the gas is ideal the molar internal energy is given by $U_m = CVT$. As CV is constant, constant U means constant T. The molar entropy of an ideal gas, as function of the molar volume V_m and T, is given by

$$S_m = C_V \ln \frac{T}{T_0} + R \ln \frac{V_m}{V_0}.$$

The system, of the two vessels and the gas, is closed and adiabatic, so the entropy production during the process is equal to the increase of the entropy of the gas. So, doubling the volume with T constant, gives that the entropy production per mole gas is

$$S_{mi} = R \ln 2.$$

10.7   Microscopic interpretation

The Joule expansion gives a nice opportunity to explain the entropy production in statistical mechanical (microscopic) terms. At the expansion the volume, that the gas can occupy, is doubled. That means that, for every molecule there are now two possibilities: it can be placed in container an or in b. If we have one mole of gas the number of molecules is equal to Avogadro's number NA. The increase of the microscopic possibilities is a factor 2 per molecule so in total a factor 2NA. Using the well-known Boltzmann expression for the entropy

$$S_m = k \ln \Omega,$$

with k Boltzmann's constant and Ω the number of microscopic possibilities to realize the macroscopic state, gives

$$S_{mi} = k \ln(2^{N_A}) = k N_A \ln 2 = R \ln 2.$$

So, at an irreversible process, the number of microscopic possibilities to realize the macroscopic state is increased by a certain factor.

10.8   Basic inequalities and stability conditions

In this Section we derive the basic inequalities and stability conditions for closed systems. For closed systems the first law reduces to

$$\frac{dU}{dt} = \dot{Q} - p\frac{dV}{dt} + P.$$

The second law we write as

$$\frac{dS}{dt} - \frac{\dot{Q}}{T} \geq 0.$$

For **adiabatic systems** $\dot{Q} = 0$ so **dS/d$t \geq$ 0**. In other words: the entropy of adiabatic systems can only increase. In equilibrium the entropy is at its maximum. Isolated systems are a special case of adiabatic systems, so this statement is also valid for isolated systems.

Now consider systems with **constant temperature and volume**. In most cases T is the temperature of the surroundings with which the system is in good thermal contact. Since V is constant the first law gives $\dot{Q} = dU/dt - P$. Substitution in the second law, and using that T is constant, gives

$$\frac{d(TS)}{dt} - \frac{dU}{dt} + P \geq 0.$$

With the Helmholtz free energy, defined as

$$F = U - TS,$$

we get

$$\frac{dF}{dt} - P \leq 0.$$

If $P = 0$ this is the mathematical formulation of the general property that the free energy of systems with fixed temperature and volume tends to a minimum. The expression can be integrated from the initial state i to the final state f resulting in

$$W_S \leq F_i - F_f$$

where WS is the work done *by* the system. If the process inside the system is completely reversible the equality sign holds. Hence the maximum work, that can be extrated from the system, is equal to the free energy of the initial state minus the free energy of the final state.

Finally we consider systems with **constant temperature and pressure** and take $P = 0$. As p is constant the first laws gives

$$\frac{\mathrm{d}U}{\mathrm{d}t} = \dot{Q} - \frac{\mathrm{d}(pV)}{\mathrm{d}t}.$$

Combining with in the second law, and using that T is constant, gives

$$\frac{\mathrm{d}(TS)}{\mathrm{d}t} - \frac{\mathrm{d}U}{\mathrm{d}t} - \frac{\mathrm{d}(pV)}{\mathrm{d}t} \geq 0.$$

With the Gibbs free energy, defined as

$$G = U + pV - TS,$$

we get

$$\frac{\mathrm{d}G}{\mathrm{d}t} \leq 0.$$

10.9 Homogeneous systems

In homogeneous systems the temperature and pressure are well-defined and all internal processes are reversible. Hence $\dot{S}_i = 0$. As a result the second law, multiplied by T, reduces to

$$T\frac{\mathrm{d}S}{\mathrm{d}t} = \dot{Q} + \dot{n}TS_m.$$

With $P=0$ the first law becomes

$$\frac{\mathrm{d}U}{\mathrm{d}t} = \dot{Q} + \dot{n}H_m - p\frac{\mathrm{d}V}{\mathrm{d}t}.$$

Eliminating $\dot{Q}$ and multiplying with dt gives

$$\mathrm{d}U = T\mathrm{d}S - p\mathrm{d}V + (H_m - TS_m)\mathrm{d}n.$$

Since

$$H_m - TS_m = G_m = \mu$$

with G_m the molar Gibbs free energy and μ the molar chemical potential we obtain the well-known result

$$dU = TdS - pdV + \mu dn.$$

10.10   See also

- Thermodynamics

- First law of thermodynamics

- Second law of thermodynamics

- Irreversible process

- Non-equilibrium thermodynamics

- High entropy alloys

10.11   References

[1] S.R. de Groot and P. Mazur, Non-equilibrium thermodynamics (North-Holland Publishing Company, Amsterdam-London, 1969)

[2] S. Carnot *Reflexions sur la puissance motrice du feu* Bachelier, Paris, 1824

[3] Clausius, R. (1854). "Ueber eine veränderte Form des zweiten Hauptsatzes der mechanischen Wärmetheoriein". *Annalen der Physik und Chemie* **93** (12): 481–506. Retrieved 25 June 2012.. Clausius, R. (August 1856). "On a Modified Form of the Second Fundamental Theorem in the Mechanical Theory of Heat". *Philos. Mag.* 4 **12** (77): 81–98. Retrieved 25 June 2012.

[4] R. Clausius *Über verschiedene für die Anwendung bequeme Formen der Hauptgleigungen der mechanische Wärmetheorie* in Abhandlungen über die Anwendung bequeme Formen der Haubtgleichungen der mechanischen Wärmetheorie Ann.Phys. [2] 125, 390 (1865). This paper is translated and can be found in: The second law of thermodynamics, Edited by J. Kestin, Dowden, Hutchinson, & Ross, Inc., Stroudsburg, Pennsylvania, pp. 162-193.

[5] A.T.A.M. de Waele, Basic operation of cryocoolers and related thermal machines, Review article, Journal of Low Temperature Physics, Vol.164, pp. 179-236, (2011), DOI: 10.1007/s10909-011-0373-x.

10.12   Further reading

- Belandria, José Iraides (2008) Positive and negative entropy production in thermodynamic systems. Universidad de Los Andes, Venezuela.

- Crooks, G. (1999). "Entropy production fluctuation theorem and the non-equilibrium work relation for free energy differences". *Physical Review E* (Free PDF) **60** (3): 2721. arXiv:cond-mat/9901352. Bibcode:1999PhRvE..60.2721C.doi:10.1103/PhysRevE.60.2721.

- Seifert, Udo (2005). "Entropy Production along a Stochastic Trajectory and an Integral Fluctuation Theorem" (Free PDF). *Physical Review Letters* **95** (4): 040602. arXiv:cond-mat/0503686. Bibcode:2005PhRvL..95d0602S. doi:10.1103/PhysRevLett.95.040602. PMID 16090792.

Chapter 11

Entropy in thermodynamics and information theory

There are close parallels between the mathematical expressions for the thermodynamic entropy, usually denoted by S, of a physical system in the statistical thermodynamics established by Ludwig Boltzmann and J. Willard Gibbs in the 1870s, and the information-theoretic entropy, usually expressed as H, of Claude Shannon and Ralph Hartley developed in the 1940s. Shannon, although not initially aware of this similarity, commented on it upon publicizing information theory in *A Mathematical Theory of Communication*.

This article explores what links there are between the two concepts, and how far they can be regarded as connected.

11.1   Equivalence of form of the defining expressions

The defining expression for entropy in the theory of statistical mechanics established by Ludwig Boltzmann and J. Willard Gibbs in the 1870s, is of the form:

$$S = -k_B \sum_i p_i \ln p_i.$$

where p_i is the probability of the microstate i taken from an equilibrium ensemble.

The defining expression for entropy in the theory of information established by Claude E. Shannon in 1948 is of the form:

$$H = -\sum_i p_i \log_b p_i.$$

where p_i is the probability of the message m_i taken from the message space M and b is the base of the logarithm used. Common values of b are 2, Euler's number e, and 10, and the unit of entropy is bit for $b = 2$, nat for $b = e$, and dit (or digit) for $b = 10$.[1]

Mathematically H may also be seen as an average information, taken over the message space, because when a certain message occurs with probability pi, the information $-\log(pi)$ will be obtained.

If all the microstates are equiprobable (a microcanonical ensemble), the statistical thermodynamic entropy reduces to the form, as given by Boltzmann,

$$S = k_B \ln W$$

Boltzmann's grave in the Zentralfriedhof, Vienna, with bust and entropy formula.

where W is the number of microstates.

If all the messages are equiprobable, the information entropy reduces to the Hartley entropy

$$H = \log_b |M|$$

where $|M|$ is the cardinality of the message space M.

The logarithm in the thermodynamic definition is the natural logarithm. It can be shown that the Gibbs entropy formula, with the natural logarithm, reproduces all of the properties of the macroscopic classical thermodynamics of Clausius. (See article: Entropy (statistical views)).

The logarithm can also be taken to the natural base in the case of information entropy. This is equivalent to choosing to measure information in nats instead of the usual bits. In practice, information entropy is almost always calculated using base 2 logarithms, but this distinction amounts to nothing other than a change in units. One nat is about 1.44 bits.

The presence of Boltzmann's constant k in the thermodynamic definitions is a historical accident, reflecting the conventional units of temperature. It is there to make sure that the statistical definition of thermodynamic entropy matches the classical entropy of Clausius, thermodynamically conjugate to temperature. For a simple compressible system that can only perform volume work, the first law of thermodynamics becomes

$$dE = -pdV + TdS$$

But one can equally well write this equation in terms of what physicists and chemists sometimes call the 'reduced' or dimensionless entropy, $\sigma = S/k$, so that

$$dE = -pdV + k_B T d\sigma$$

Just as S is conjugate to T, so σ is conjugate to kBT (the energy that is characteristic of T on a molecular scale).

11.2 Theoretical relationship

Despite the foregoing, there is a difference between the two quantities. The information entropy H can be calculated for *any* probability distribution (if the "message" is taken to be that the event i which had probability pi occurred, out of the space of the events possible), while the thermodynamic entropy S refers to thermodynamic probabilities pi specifically. The difference is more theoretical than actual, however, because any probability distribution can be approximated arbitrarily closely by some thermodynamic system.

Moreover, a direct connection can be made between the two. If the probabilities in question are the thermodynamic probabilities pi: the (reduced) Gibbs entropy σ can then be seen as simply the amount of Shannon information needed to define the detailed microscopic state of the system, given its macroscopic description. Or, in the words of G. N. Lewis writing about chemical entropy in 1930, "Gain in entropy always means loss of information, and nothing more". To be more concrete, in the discrete case using base two logarithms, the reduced Gibbs entropy is equal to the minimum number of yes–no questions needed to be answered in order to fully specify the microstate, given that we know the macrostate.

Furthermore, the prescription to find the equilibrium distributions of statistical mechanics—such as the Boltzmann distribution—by maximising the Gibbs entropy subject to appropriate constraints (the Gibbs algorithm) can be seen as something not unique to thermodynamics, but as a principle of general relevance in statistical inference, if it is desired to find a maximally uninformative probability distribution, subject to certain constraints on its averages. (These perspectives are explored further in the article Maximum entropy thermodynamics.)

11.3 Information is physical

11.3.1   Szilard's engine

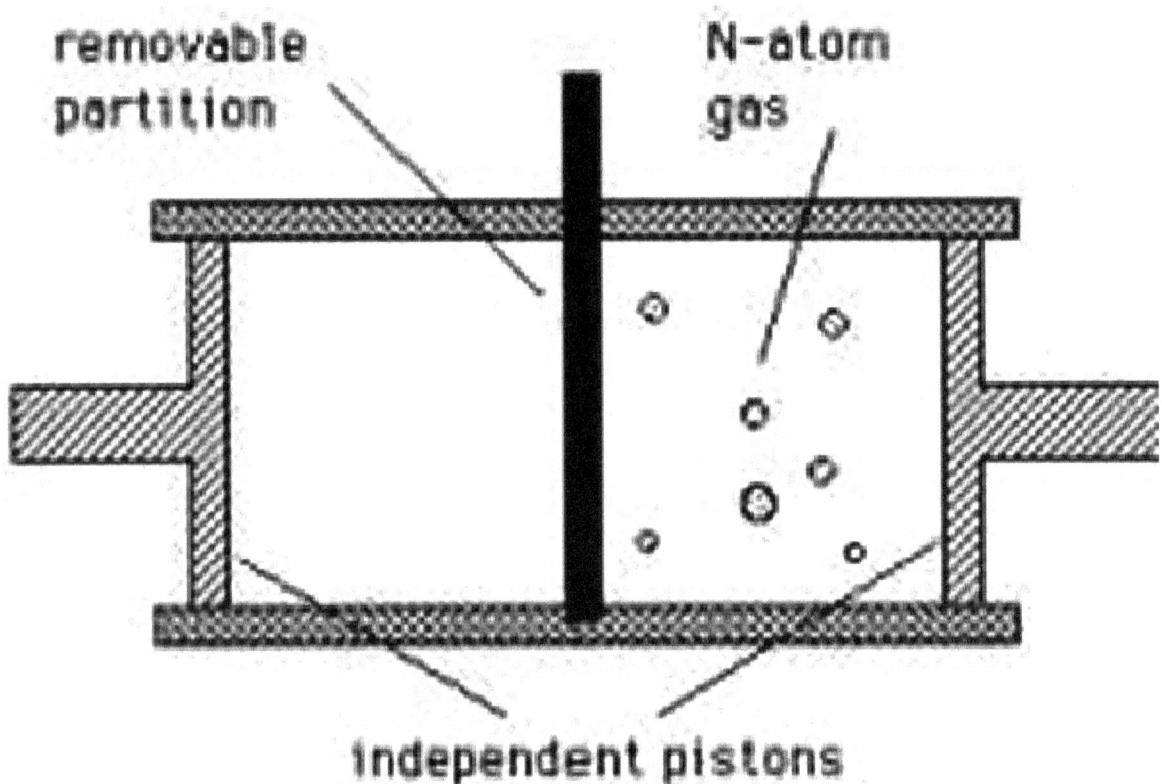

N-atom engine schematic

A physical thought experiment demonstrating how just the possession of information might in principle have thermodynamic consequences was established in 1929 by Leó Szilárd, in a refinement of the famous Maxwell's demon scenario.

Consider Maxwell's set-up, but with only a single gas particle in a box. If the supernatural demon knows which half of the box the particle is in (equivalent to a single bit of information), it can close a shutter between the two halves of the box, close a piston unopposed into the empty half of the box, and then extract $k_BT \ln 2$ joules of useful work if the shutter is opened again. The particle can then be left to isothermally expand back to its original equilibrium occupied volume. In just the right circumstances therefore, the possession of a single bit of Shannon information (a single bit of negentropy in Brillouin's term) really does correspond to a reduction in the entropy of the physical system. The global entropy is not decreased, but information to energy conversion is possible.

Using a phase-contrast microscope equipped with a high speed camera connected to a computer, as *demon*, the principle has been actually demonstrated.[2] In this experiment, information to energy conversion is performed on a Brownian particle by means of *feedback control*; that is, synchronizing the work given to the particle with the information obtained on its position. Computing energy balances for different feedback protocols, has confirmed that the Jarzynski equality requires a generalization that accounts for the amount of information involved in the feedback.

11.3.2   Landauer's principle

Main article: Landauer's principle

In fact one can generalise: any information that has a physical representation must somehow be embedded in the statistical mechanical degrees of freedom of a physical system.

Thus, Rolf Landauer argued in 1961, if one were to imagine starting with those degrees of freedom in a thermalised state, there would be a real reduction in thermodynamic entropy if they were then re-set to a known state. This can only be achieved under information-preserving microscopically deterministic dynamics if the uncertainty is somehow dumped somewhere else – i.e. if the entropy of the environment (or the non information-bearing degrees of freedom) is increased by at least an equivalent amount, as required by the Second Law, by gaining an appropriate quantity of heat: specifically $kT \ln 2$ of heat for every 1 bit of randomness erased.

On the other hand, Landauer argued, there is no thermodynamic objection to a logically reversible operation potentially being achieved in a physically reversible way in the system. It is only logically irreversible operations – for example, the erasing of a bit to a known state, or the merging of two computation paths – which must be accompanied by a corresponding entropy increase. When information is physical, all processing of its representations, i.e. generation, encoding, transmission, decoding and interpretation, are natural processes where entropy increases by consumption of free energy.[3]

Applied to the Maxwell's demon/Szilard engine scenario, this suggests that it might be possible to "read" the state of the particle into a computing apparatus with no entropy cost; but *only* if the apparatus has already been SET into a known state, rather than being in a thermalised state of uncertainty. To SET (or RESET) the apparatus into this state will cost all the entropy that can be saved by knowing the state of Szilard's particle.

11.4 Negentropy

Main article: Negentropy

Shannon entropy has been related by physicist Léon Brillouin to a concept sometimes called negentropy. In 1953, Brillouin derived a general equation[4] stating that the changing of an information bit value requires at least $kT \ln(2)$ energy. This is the same energy as the work Leo Szilard's engine produces in the idealistic case. In his book,[5] he further explored this problem concluding that any cause of a bit value change (measurement, decision about a yes/no question, erasure, display, etc.) will require the same amount, $kT \ln(2)$, of energy. Consequently, acquiring information about a system's microstates is associated with an entropy production, while erasure yields entropy production only when the bit value is changing. Setting up a bit of information in a sub-system originally in thermal equilibrium results in a local entropy reduction however there is no violation of the second law of thermodynamics, according to Brillouin, since a reduction in any local system's thermodynamic entropy results in an increase in thermodynamic entropy elsewhere. In this way, Brillouin clarified the meaning of negentropy which was considered as controversial because its earlier understanding can yield Carnot efficiency higher than one.

In 2009, Mahulikar & Herwig redefined thermodynamic negentropy as the specific entropy deficit of the dynamically ordered sub-system relative to its surroundings.[6] This definition enabled the formulation of the *Negentropy Principle*, which is mathematically shown to follow from the 2nd Law of Thermodynamics, during order existence.

11.5 Black holes

See also: Bekenstein bound

Stephen Hawking often speaks of the thermodynamic entropy of black holes in terms of their information content.[7] Do black holes destroy information? It appears that there are deep relations between the entropy of a black hole and information loss[8] See *Black hole thermodynamics* and *Black hole information paradox*.

11.6 Quantum theory

See also: Holographic principle § Energy, matter, and information equivalence and Quantum relative entropy

Hirschman showed,[9] cf. Hirschman uncertainty, that Heisenberg's uncertainty principle can be expressed as a particular lower bound on the sum of the classical distribution entropies of the *quantum observable* probability distributions of a quantum mechanical state, the square of the wave-function, in coordinate, and also momentum space, when expressed in Planck units. The resulting inequalities provide a tighter bound on the uncertainty relations of Heisenberg.

One could speak of the "joint entropy" of the position and momentum distributions in this quantity by considering them independent, but since they are not jointly observable, they cannot be considered as a joint distribution. Note that this entropy is not the accepted entropy of a quantum system, the Von Neumann entropy, $-\text{Tr}\,\varrho\,\ln\varrho = -\langle\ln\varrho\rangle$. In phase-space, the Von Neumann entropy can nevertheless be represented equivalently to Hilbert space, even though positions and momenta are quantum conjugate variables; and thus leads to a properly bounded entropy distinctly *different* (more detailed) than Hirschman's; this one accounts for the *full information content of a mixture of quantum states.*[10]

(Dissatisfaction with the Von Neumann entropy from quantum information points of view has been expressed by Stotland, Pomeransky, Bachmat and Cohen, who have introduced a yet different definition of entropy that reflects the inherent uncertainty of quantum mechanical states. This definition allows distinction between the minimum uncertainty entropy of pure states, and the excess statistical entropy of mixtures.[111])

11.7   The fluctuation theorem

See also: fluctuation theorem

The fluctuation theorem provides a mathematical justification of the second law of thermodynamics under these principles, and precisely defines the limitations of the applicability of that law for systems away from thermodynamic equilibrium.

11.8   Topics of recent research

11.8.1   Is information quantized?

In 1995, Tim Palmer signalled two unwritten assumptions about Shannon's definition of information that may make it inapplicable as such to quantum mechanics:

- The supposition that there is such a thing as an observable state (for instance the upper face of a dice or a coin) *before* the observation begins

- The fact that knowing this state does not depend on the order in which observations are made (commutativity)

Anton Zeilinger's and Caslav Brukner's article[12] synthesized and developed these remarks. The so-called Zeilinger's principle suggests that the quantization observed in QM could be bound to *information* quantization (one cannot observe less than one bit, and what is not observed is by definition "random"). Nevertheless, these claims remain quite controversial. Detailed discussions of the applicability of the Shannon information in quantum mechanics and an argument that Zeilinger's principle cannot explain quantization have been published,[13][14][15] that show that Brukner and Zeilinger change, in the middle of the calculation in their article, the numerical values of the probabilities needed to compute the Shannon entropy, so that the calculation makes little sense.

11.8.2   Extracting work from quantum information in a Szilárd engine

In 2013, a description was published[16] of a two atom version of a Szilárd engine using Quantum discord to generate work from purely quantum information.[17] Refinements in the lower temperature limit were suggested.[18]

11.9 See also

- Thermodynamic entropy

- Information entropy

- Thermodynamics

- Statistical mechanics

- Information theory

- Physical information

- Quantum entanglement

- Quantum decoherence

- Fluctuation theorem

- Black hole entropy

- Black hole information paradox

- Entropy (information theory)

- Entropy (statistical thermodynamics)

- Entropy (order and disorder)

- Orders of magnitude (entropy)

11.10 References

[1] Schneider, T.D. Information theory primer with an appendix on logarithms, National Cancer Institute, 14 April 2007.

[2] Shoichi Toyabe; Takahiro Sagawa; Masahito Ueda; Eiro Muneyuki; Masaki Sano (2010-09-29). "Information heat engine: converting information to energy by feedback control". *Nature Physics* **6** (12): 988–992. arXiv:1009.5287. Bibcode:2011NatPh...6..988T.doi:10.1038/nphys1821.We demonstrated that free energy is obtained by a feedback control using the information about thesystem;information is converted to free energy,as thefirst realization of Szilard-type Maxwell's demon.

[3] Karnani, M.; Pääkkönen, K.; Annila, A. (2009). "The physical character of information". *Proc. R. Soc. A* **465** (2107): 2155–75. Bibcode:2009RSPSA.465.2155K. doi:10.1098/rspa.2009.0063.

[4] Leon Brillouin (1953), "The negentropy principle of information". *J. Applied Physics* **24**. 1152-1163.

[5] Leon Brillouin, *Science and Information theory*, Dover, 1956

[6] Mahulikar, S.P.; Herwig, H. (August 2009). "Exact thermodynamic principles for dynamic order existence and evolution in chaos". *Chaos, Solitons & Fractals* **41** (4): 1939–48. Bibcode:2009CSF....41.1939M. doi:10.1016/j.chaos.2008.07.051.

[7] Overbye, Dennis. "Hawking's Breakthrough Is Still an Enigma". New York Times. Retrieved 19 December 2013.

[8] Schiffer M, Bekenstein JD (February 1989). "Proof of the quantum bound on specific entropy for free fields". *Physical Review D* **39** (4): 1109–15. Bibcode:1989PhRvD..39.1109S. doi:10.1103/PhysRevD.39.1109. PMID 9959747. "Black Holes and Entropy". *Physical Review D* **7** (8): 2333. Bibcode:1973PhRvD...7.2333B. doi:10.1103/PhysRevD.7.2333.Ellis, George Francis Rayner; Hawking, S. W. (1973). *The large scale structure of space-time*. Cambridge, Eng: University Press. ISBN 0-521-09906-4. von Baeyer, Christian, H. (2003). *Information — the New Language of Science*. Harvard University Press. ISBN 0-674-01387-5. Callaway DJE (April 1996). "Surface tension, hydrophobicity, and black holes: The entropic connection". *Physical Review E* **53** (4): 3738–3744. arXiv:cond-mat/9601111. Bibcode:1996PhRvE..53.3738C. doi:10.1103/PhysRevE.53.3738. PMID 9964684. Srednicki M (August 1993). "Entropy and area". *Physical Review Letters* **71** (5): 666–669. arXiv:hep-th/9303048. Bibcode:1993PhRvL..71..666S. doi:10.1103/PhysRevLett.71.666. PMID 10055336.

[9] Hirschman, Jr., I.I. (January 1957). "A note on entropy". *American Journal of Mathematics* **79** (1): 152–6. doi:10.2307/2372390. JSTOR 2372390.

[10] Zachos, C. K. (2007). "A classical bound on quantum entropy". *Journal of Physics A: Mathematical and Theoretical* **40** (21): F407. arXiv:hep-th/0609148. Bibcode:2007JPhA...40..407Z. doi:10.1088/1751-8113/40/21/F02.

[11] Alexander Stotland; Pomeransky; Eitan Bachmat; Doron Cohen (2004). "The information entropy of quantum mechanical states". *Europhysics Letters* **67** (5): 700–6. arXiv:quant-ph/0401021. Bibcode:2004EL......67..700S. doi:10.1209/epl/i2004-10110-1.

[12] "Conceptual inadequacy of the Shannon information in quantum measurement". 2001. arXiv:quant-ph/0006087. Bibcode:2001 doi:10.1103/PhysRevA.63.022113.

[13] Timpson, 2003

[14] Hall, 2000

[15] Mana, 2004

[16] Jung Jun Park, Kang-Hwan Kim, Takahiro Sagawa, Sang Wook Kim (2013). "Heat Engine Driven by Purely Quantum Information.". *Physical Review Letters* **111** (23). arXiv:1302.3011. Bibcode:2013PhRvL.111w0402P. doi:10.1103/PhysRevLett.111. 230402.Retrieved19December2013.

[17] Zyga, Lisa. "Maxwell's demon can use quantum information". Phys.org (Omicron Technology Limited). Retrieved 19 December 2013.

[18] Martin Plesch, Oscar Dahlsten, John Goold, Vlatko Vedral (September 2013). "Comment on "Quantum Szilard Engine" arXiv: 1309.4209 [quant-ph]". *arXiv:1309.4209 [quant-ph]*. arXiv:1309.4209. Bibcode:2013PhRvL.111r8901P. doi:10.1103/Phys RevLett.111.188901.Retrieved19December2013.

11.11   Additional references

- Bennett, C.H. (1973). "Logical reversibility of computation". *IBM J. Res. Develop.* **17** (6): 525–532. doi:10.1147/rd.

  176.0525. •Brillouin,Léon(2004),*Science And Information Theory*(second ed.),Dover,ISBN978-0-486-43918-1. [Republication of 1962 original.]

- Frank, Michael P. (May–June 2002). "Physical Limits of Computing". *Computing in Science and Engineering* **4** (3): 16–25. doi:10.1109/5992.998637.

- Greven, Andreas; Keller, Gerhard; Warnecke, Gerald, eds. (2003). *Entropy*. Princeton University Press. ISBN 978-0-691-11338-8. (A highly technical collection of writings giving an overview of the concept of entropy as it appears in various disciplines.)

- Kalinin, M.I.; Kononogov, S.A. (2005), "Boltzmann's constant, the energy meaning of temperature, and thermodynamic irreversibility", *Measurement Techniques* [translation of *Izmeritel'naya Tekhnika*] **48**: 632–636, doi:10.1007/s 11018-005-0195-9.

- Koutsoyiannis, D. (2011), "Hurst–Kolmogorov dynamics as a result of extremal entropy production", *Physica A* **390**: 1424–1432, Bibcode:2011PhyA..390.1424K, doi:10.1016/j.physa.2010.12.035.

- Landauer, R. (1993). "Information is Physical". *Proc. Workshop on Physics and Computation PhysComp'92*. Los Alamitos: IEEE Comp. Sci.Press. pp. 1–4. doi:10.1109/PHYCMP.1992.615478.

- Landauer, R. (1961). "Irreversibility and Heat Generation in the Computing Process". *IBM J. Res. Develop.* **5** (3): 183–191. doi:10.1147/rd.53.0183.

- Leff, H.S.; Rex, A.F., ed. (1990). *Maxwell's Demon: Entropy, Information, Computing*. Princeton NJ: Princeton University Press. ISBN 0-691-08727-X.

- Middleton, D. (1960). *An Introduction to Statistical Communication Theory*. McGraw-Hill.

- Shannon, Claude E. (July–October 1948). "A Mathematical Theory of Communication". *Bell System Technical Journal* **27** (3): 379–423. doi:10.1002/j.1538-7305.1948.tb01338.x. (as PDF)

11.12 External links

- Information Processing and Thermodynamic Entropy Stanford Encyclopedia of Philosophy.

- *An Intuitive Guide to the Concept of Entropy Arising in Various Sectors of Science* — a wikibook on the interpretation of the concept of entropy.

Chapter 12

Entropic uncertainty

In quantum mechanics, information theory, and Fourier analysis, the **entropic uncertainty** or **Hirschman uncertainty** is defined as the sum of the temporal and spectral Shannon entropies. It turns out that Heisenberg's uncertainty principle can be expressed as a lower bound on the sum of these entropies. This is *stronger* than the usual statement of the uncertainty principle in terms of the product of standard deviations.

In 1957,[1] Hirschman considered a function f and its Fourier transform g such that

$$g(y) \approx \int_{-\infty}^{\infty} \exp(-2\pi i x y) f(x)\, dx, \qquad f(x) \approx \int_{-\infty}^{\infty} \exp(2\pi i x y) g(y)\, dy ,$$

where the "$\approx$" indicates convergence in L^2, and normalized so that (by Plancherel's theorem),

$$\int_{-\infty}^{\infty} |f(x)|^2\, dx = \int_{-\infty}^{\infty} |g(y)|^2\, dy = 1 .$$

He showed that for any such functions the sum of the Shannon entropies is non-negative,

$$H(|f|^2) + H(|g|^2) \equiv -\int_{-\infty}^{\infty} |f(x)|^2 \log |f(x)|^2\, dx - \int_{-\infty}^{\infty} |g(y)|^2 \log |g(y)|^2\, dy \geq 0.$$

A tighter bound,

was conjectured by Hirschman[1] and Everett,[2] proven in 1975 by W. Beckner[3] and in the same year interpreted by as a generalized quantum mechanical uncertainty principle by Białynicki-Birula and Mycielski.[4] The equality holds in the case of Gaussian distributions.[5]

Note, however, that the above entropic uncertainty function is distinctly *different* from the quantum Von Neumann entropy represented in phase space.

12.1 Sketch of proof

The proof of this tight inequality depends on the so-called **(*q*, *p*)-norm** of the Fourier transformation. (Establishing this norm is the most difficult part of the proof.)

From this norm, one is able to establish a lower bound on the sum of the (differential) Rényi entropies, $H\alpha(|f|^2) + H\beta(|g|^2)$, where $1/\alpha + 1/\beta = 2$, which generalize the Shannon entropies. For simplicity, we consider this inequality only in one dimension; the extension to multiple dimensions is straightforward and can be found in the literature cited.

12.1.1 Babenko–Beckner inequality

The **(q, p)-norm** of the Fourier transform is defined to be[6]

$$\|\mathcal{F}\|_{q,p} = \sup_{f \in L^p(\mathbb{R})} \frac{\|\mathcal{F}f\|_q}{\|f\|_p}, \text{ where } 1 < p \le 2 \text{, and } \frac{1}{p} + \frac{1}{q} = 1.$$

In 1961, Babenko[7] found this norm for *even* integer values of q. Finally, in 1975, using Hermite functions as eigenfunctions of the Fourier transform, Beckner[3] proved that the value of this norm (in one dimension) for all $q \ge 2$ is

$$\|\mathcal{F}\|_{q,p} = \sqrt{p^{1/p}/q^{1/q}}.$$

Thus we have the **Babenko–Beckner inequality** that

$$\|\mathcal{F}f\|_q \le \left(p^{1/p}/q^{1/q} \right)^{1/2} \|f\|_p.$$

12.1.2 Rényi entropy bound

From this inequality, an expression of the uncertainty principle in terms of the Rényi entropy can be derived.[6][8]

Letting $g = \mathcal{F}f$, $2\alpha = p$, and $2\beta = q$, so that $1/\alpha + 1/\beta = 2$ and $1/2 < \alpha < 1 < \beta$, we have

$$\left(\int_{\mathbb{R}} |g(y)|^{2\beta}\, dy \right)^{1/2\beta} \le \frac{(2\alpha)^{1/4\alpha}}{(2\beta)^{1/4\beta}} \left(\int_{\mathbb{R}} |f(x)|^{2\alpha}\, dx \right)^{1/2\alpha}.$$

Squaring both sides and taking the logarithm, we get

$$\frac{1}{\beta} \log \left(\int_{\mathbb{R}} |g(y)|^{2\beta}\, dy \right) \le \frac{1}{2} \log \frac{(2\alpha)^{1/\alpha}}{(2\beta)^{1/\beta}} + \frac{1}{\alpha} \log \left(\int_{\mathbb{R}} |f(x)|^{2\alpha}\, dx \right).$$

Multiplying both sides by

$$\frac{\beta}{1-\beta} = -\frac{\alpha}{1-\alpha}$$

reverses the sense of the inequality,

$$\frac{1}{1-\beta} \log \left(\int_{\mathbb{R}} |g(y)|^{2\beta}\, dy \right) \ge \frac{\alpha}{2(\alpha-1)} \log \frac{(2\alpha)^{1/\alpha}}{(2\beta)^{1/\beta}} - \frac{1}{1-\alpha} \log \left(\int_{\mathbb{R}} |f(x)|^{2\alpha}\, dx \right).$$

Rearranging terms, finally yields an inequality in terms of the sum of the Rényi entropies,

$$\frac{1}{1-\alpha} \log \left(\int_{\mathbb{R}} |f(x)|^{2\alpha}\, dx \right) + \frac{1}{1-\beta} \log \left(\int_{\mathbb{R}} |g(y)|^{2\beta}\, dy \right) \ge \frac{\alpha}{2(\alpha-1)} \log \frac{(2\alpha)^{1/\alpha}}{(2\beta)^{1/\beta}};$$

$$H_\alpha(|f|^2) + H_\beta(|g|^2) \geq \frac{1}{2}\left(\frac{\log\alpha}{\alpha-1} + \frac{\log\beta}{\beta-1}\right) - \log 2 \ .$$

Note that this inequality is symmetric with respect to α and β: One no longer need assume that $\alpha<\beta$; only that they are positive and not both one, and that $1/\alpha + 1/\beta = 2$. To see this symmetry, simply exchange the rôles of i and $-i$ in the Fourier transform.

12.1.3   Shannon entropy bound

Taking the limit of this last inequality as $\alpha, \beta \to 1$ yields the less general Shannon entropy inequality,

$$H(|f|^2) + H(|g|^2) \geq \log\frac{e}{2}, \quad \text{where} \quad g(y) \approx \int_{\mathbb{R}} e^{-2\pi i x y} f(x)\, dx \ .$$

valid for any base of logarithm, as long as we choose an appropriate unit of information, bit, nat, etc.

The constant will be different, though, for a different normalization of the Fourier transform, (such as is usually used in physics, with normalizations chosen so that $\hbar=1$), i.e.,

$$H(|f|^2) + H(|g|^2) \geq \log(\pi e) \quad \text{for} \quad g(y) \approx \frac{1}{\sqrt{2\pi}}\int_{\mathbb{R}} e^{-ixy} f(x)\, dx \ .$$

In this case, the dilation of the Fourier transform absolute squared by a factor of 2π simply adds $\log(2\pi)$ to its entropy.

12.2   Entropy versus variance bounds

The Gaussian or normal probability distribution plays an important role in the relationship between variance and entropy: it is a problem of the calculus of variations to show that this distribution maximizes entropy for a given variance, and at the same time minimizes the variance for a given entropy. In fact, for any probability density function φ on the real line, Shannon's entropy inequality specifies:

$$H(\phi) \leq \log\sqrt{2\pi e V(\phi)},$$

where H is the Shannon entropy and V is the variance, an inequality that is saturated only in the case of a normal distribution.

Moreover, the Fourier transform of a Gaussian probability amplitude function is also Gaussian—and the absolute squares of both of these are Gaussian, too. This can then be used to derive the usual Robertson variance uncertainty inequality from the above entropic inequality, enabling *the latter to be tighter than the former*. That is (for $\hbar=1$), exponentiating the Hirschman inequality and using Shannon's expression above,

$$1/2 \leq \exp(H(|f|^2) + H(|g|^2))/(2e\pi) \leq \sqrt{V(|f|^2)V(|g|^2)} \ .$$

Hirschman[1] explained that entropy—his version of entropy was the negative of Shannon's—is a "measure of the concentration of [a probability distribution] in a set of small measure." Thus *a low or large negative Shannon entropy means that a considerable mass of the probability distribution is confined to a set of small measure*.

Note that this set of small measure need not be contiguous; a probability distribution can have several concentrations of mass in intervals of small measure, and the entropy may still be low no matter how widely scattered those intervals are. This is not the case with the variance: variance measures the concentration of mass about the mean of the distribution, and a low variance means that a considerable mass of the probability distribution is concentrated in a *contiguous interval* of small measure.

To formalize this distinction, we say that two probability density functions φ_1 and φ_2 are **equimeasurable** if

$$\forall \delta > 0, \ \mu\{x \in \mathbb{R} | \phi_1(x) \geq \delta\} = \mu\{x \in \mathbb{R} | \phi_2(x) \geq \delta\},$$

where μ is the Lebesgue measure. Any two equimeasurable probability density functions have the same Shannon entropy, and in fact the same Rényi entropy, of any order. The same is not true of variance, however. Any probability density function has a radially decreasing equimeasurable "rearrangement" whose variance is less (up to translation) than any other rearrangement of the function; and there exist rearrangements of arbitrarily high variance, (all having the same entropy.)

12.3 See also

- Inequalities in information theory
- Uncertainty principle
- Riesz–Thorin theorem
- Fourier Transform

12.4 References

[1] Hirschman, I. I., Jr. (1957), "A note on entropy", *American Journal of Mathematics* **79** (1): 152–156, doi:10.2307/2372390, JSTOR 2372390.

[2] Hugh Everett, III. The Many-Worlds Interpretation of Quantum Mechanics: the theory of the universal wave function. Everett's Dissertation

[3] Beckner, W. (1975), "Inequalities in Fourier analysis", *Annals of Mathematics* **102** (6): 159–182, doi:10.2307/1970980, JSTOR 1970980.

[4] Bialynicki-Birula, I.; Mycielski, J. (1975), "Uncertainty Relations for Information Entropy in Wave Mechanics", *Communications in Mathematical Physics* **44** (2): 129, Bibcode:1975CMaPh..44..129B, doi:10.1007/BF01608825

[5] Ozaydin, Murad; Przebinda, Tomasz (2004). "An Entropy-based Uncertainty Principle for a Locally Compact Abelian Group" (PDF). *Journal of Functional Analysis* (Elsevier Inc.) **215** (1): 241–252. doi:10.1016/j.jfa.2003.11.008. Retrieved 2011-06-23.

[6] Bialynicki-Birula, I. (2006). "Formulation of the uncertainty relations in terms of the Rényi entropies". *Physical Review A* **74** (5). doi:10.1103/PhysRevA.74.052101.

[7] K.I. Babenko. *An inequality in the theory of Fourier analysis.* Izv. Akad. Nauk SSSR, Ser. Mat. **25** (1961) pp. 531–542 English transl., Amer. Math. Soc. Transl. (2) **44**, pp. 115-128

[8] H.P. Heinig and M. Smith, *Extensions of the Heisenberg–Weil inequality.* Internat. J. Math. & Math. Sci., Vol. 9, No. 1 (1986) pp. 185–192.

12.5 Further reading

- Zozor, S.; Vignat, C. (2007). "On classes of non-Gaussian asymptotic minimizers in entropic uncertainty principles". *Physica A: Statistical Mechanics and its Applications* **375** (2): 499. doi:10.1016/j.physa.2006.09.019. arXiv:math/0605510v1

- Maassen, H.; Uffink, J. (1988). "Generalized entropic uncertainty relations". *Physical Review Letters* **60** (12): 1103–1106. Bibcode:1988PhRvL..60.1103M. doi:10.1103/PhysRevLett.60.1103. PMID 10037942.

- Ballester, M.; Wehner, S. (2007). "Entropic uncertainty relations and locking: Tight bounds for mutually unbiased bases". *Physical Review A* **75** (2). doi:10.1103/PhysRevA.75.022319.

- Ghirardi, G.; Marinatto, L.; Romano, R. (2003). "An optimal entropic uncertainty relation in a two-dimensional Hilbert space". *Physics Letters A* **317**: 32. doi:10.1016/j.physleta.2003.08.029.

Chapter 13

Entropy (arrow of time)

Entropy is the only quantity in the physical sciences (apart from certain rare interactions in particle physics: see below) that requires a particular direction for time, sometimes called an arrow of time. As one goes "forward" in time, the second law of thermodynamics says, the entropy of an isolated system can increase, but not decrease. Hence, from one perspective, entropy measurement is a way of distinguishing the past from the future. However in thermodynamic systems that are not closed, entropy can decrease with time: many systems, including living systems, reduce local entropy at the expense of an environmental increase, resulting in a net increase in entropy. Examples of such systems and phenomena include the formation of typical crystals, the workings of a refrigerator and living organisms.

Entropy, like temperature, is an abstract concept, yet, like temperature, everyone has an intuitive sense of the effects of entropy. Watching a movie, it is usually easy to determine whether it is being run forward or in reverse. When run in reverse, broken glasses spontaneously reassemble, smoke goes down a chimney, wood "unburns", cooling the environment and ice "unmelts" warming the environment. No physical laws are broken in the reverse movie except the second law of thermodynamics, which reflects the time-asymmetry of entropy. An intuitive understanding of the irreversibility of certain physical phenomena (and subsequent creation of entropy) allows one to make this determination.

By contrast, all physical processes occurring at the microscopic level, such as mechanics, do not pick out an arrow of time. Going forward in time, an atom might move to the left, whereas going backward in time the same atom might move to the right; the behavior of the atom is not *qualitatively* different in either case. It would, however, be an astronomically improbable event if a macroscopic amount of gas that originally filled a container evenly spontaneously shrunk to occupy only half the container.

Certain subatomic interactions involving the weak nuclear force violate the conservation of parity, but only very rarely. According to the CPT theorem, this means they should also be time irreversible, and so establish an arrow of time. This, however, is neither linked to the thermodynamic arrow of time, nor has anything to do with our daily experience of time irreversibility.[1]

13.1 Overview

The Second Law of Thermodynamics allows for the entropy to *remain the same* regardless of the direction of time. If the entropy is constant in either direction of time, there would be no preferred direction. However, the entropy can only be a constant if the system is in the highest possible state of disorder, such as a gas that always was, and always will be, uniformly spread out in its container. The existence of a thermodynamic arrow of time implies that the system is highly ordered in one time direction only, which would by definition be the "past". Thus this law is about the boundary conditions rather than the equations of motion of our world.

Unlike most other laws of physics, the Second Law of Thermodynamics is statistical in nature, and therefore its reliability arises from the huge number of particles present in macroscopic systems. It is not impossible, in principle, for all 6×10^{23} atoms in a mole of a gas to spontaneously migrate to one half of a container; it is only *fantastically* unlikely—so unlikely that no macroscopic violation of the Second Law has ever been observed. T Symmetry is the symmetry of physical laws under a time reversal transformation. Although in restricted contexts one may find this symmetry, the observable universe itself does not show symmetry under time reversal, primarily due to the second law of thermodynamics.

The thermodynamic arrow is often linked to the cosmological arrow of time, because it is ultimately about the boundary conditions of the early universe. According to the Big Bang theory, the Universe was initially very hot with energy distributed uniformly. For a system in which gravity is important, such as the universe, this is a low-entropy state (compared to a high-entropy state of having all matter collapsed into black holes, a state to which the system may eventually evolve). As the Universe grows, its temperature drops, which leaves less energy available to perform work in the future than was available in the past. Additionally, perturbations in the energy density grow (eventually forming galaxies and stars). Thus the Universe itself has a well-defined thermodynamic arrow of time. But this does not address the question of why the initial state of the universe was that of low entropy. If cosmic expansion were to halt and reverse due to gravity, the temperature of the Universe would once again grow hotter, but its entropy would also continue to increase due to the continued growth of perturbations and the eventual black hole formation,[2] until the latter stages of the Big Crunch when entropy would be lower than now.

13.2 An example of apparent irreversibility

Consider the situation in which a large container is filled with two separated liquids, for example a dye on one side and water on the other. With no barrier between the two liquids, the random jostling of their molecules will result in them becoming more mixed as time passes. However, if the dye and water are mixed then one does not expect them to separate out again when left to themselves. A movie of the mixing would seem realistic when played forwards, but unrealistic when played backwards.

If the large container is observed early on in the mixing process, it might be found only partially mixed. It would be reasonable to conclude that, without outside intervention, the liquid reached this state because it was more ordered in the past, when there was greater separation, and will be more disordered, or mixed, in the future.

Now imagine that the experiment is repeated, this time with only a few molecules, perhaps ten, in a very small container. One can easily imagine that by watching the random jostling of the molecules it might occur — by chance alone — that the molecules became neatly segregated, with all dye molecules on one side and all water molecules on the other. That this can be expected to occur from time to time can be concluded from the fluctuation theorem; thus it is not impossible for the molecules to segregate themselves. However, for a large numbers of molecules it is so unlikely that one would have to wait, on average, many times longer than the age of the universe for it to occur. Thus a movie that showed a large number of molecules segregating themselves as described above would appear unrealistic and one would be inclined to say that the movie was being played in reverse. See Boltzmann's Second Law as a law of disorder.

13.3 Mathematics of the arrow

The mathematics behind the *arrow of time*, entropy, and basis of the second law of thermodynamics derive from the following set-up, as detailed by Carnot (1824), Clapeyron (1832), and Clausius (1854):

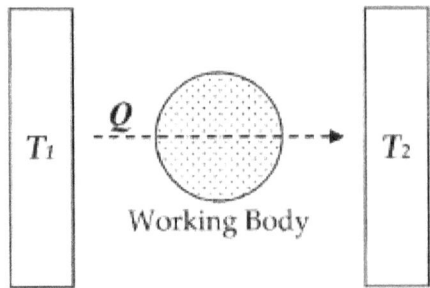

Here, as common experience demonstrates, when a hot body T_1, such as a furnace, is put into physical contact, such as being connected via a body of fluid (working body), with a cold body T_2, such as a stream of cold water, energy will invariably flow from hot to cold in the form of heat Q, and given **time** the system will reach equilibrium. Entropy, defined as Q/T, was conceived by Rudolf Clausius as a function to measure the molecular irreversibility of this process, i.e. the dissipative work the atoms and molecules do on each other during the transformation.

In this diagram, one can calculate the entropy change ΔS for the passage of the quantity of heat Q from the temperature T_1, through the "working body" of fluid (see heat engine), which was typically a body of steam, to the temperature T_2. Moreover, one could assume, for the sake of argument, that the working body contains only two molecules of water.

Next, if we make the assignment, as originally done by Clausius:

$$S = \frac{Q}{T}$$

Then the entropy change or "equivalence-value" for this transformation is:

$$\Delta S = S_{final} - S_{initial}$$

which equals:

$$\Delta S = \left(\frac{Q}{T_2} - \frac{Q}{T_1} \right)$$

and by factoring out Q, we have the following form, as was derived by Clausius:

$$\Delta S = Q \left(\frac{1}{T_2} - \frac{1}{T_1} \right)$$

Thus, for example, if Q was 50 units, T_1 was initially 100 degrees, and T_2 was initially 1 degree, then the entropy change for this process would be 49.5. Hence, entropy increased for this process, the process took a certain amount of "time", and one can correlate entropy increase with the passage of time. For this system configuration, subsequently, it is an "absolute rule". This rule is based on the fact that all natural processes are irreversible by virtue of the fact that molecules of a system, for example two molecules in a tank, not only do external work (such as to push a piston), but also do internal work on each other, in proportion to the heat used to do work (see: Mechanical equivalent of heat) during the process. Entropy accounts for the fact that internal inter-molecular friction exists.

13.4 Maxwell's demon

In 1867, James Clerk Maxwell introduced a now-famous thought experiment that highlighted the contrast between the statistical nature of entropy and the deterministic nature of the underlying physical processes. This experiment, known as Maxwell's demon, consists of a hypothetical "demon" that guards a trapdoor between two containers filled with gases at equal temperatures. By allowing fast molecules through the trapdoor in only one direction and only slow molecules in the other direction, the demon raises the temperature of one gas and lowers the temperature of the other, apparently violating the Second Law.

Maxwell's thought experiment was only resolved in the 20th century by Leó Szilárd, Charles H. Bennett, Seth Lloyd and others. The key idea is that the demon itself necessarily possesses a non-negligible amount of entropy that increases even as the gases lose entropy, so that the entropy of the system as a whole increases. This is because the demon has to contain many internal "parts" (essentially: a memory space to store information on the gas molecules) if it is to perform its job reliably, and therefore must be considered a *macroscopic* system with non-vanishing entropy. An equivalent way of saying this is that the information possessed by the demon on which atoms are considered *fast* or *slow*, can be considered a form of entropy known as information entropy.

13.5   Correlations

An important difference between the past and the future is that in any system (such as a gas of particles) its initial conditions are usually such that its different parts are uncorrelated, but as the system evolves and its different parts interact with each other, they become correlated.[3] For example, whenever dealing with a gas of particles, it is always assumed that its initial conditions are such that there is no correlation between the states of different particles (i.e. the speeds and locations of the different particles are completely random, up to the need to conform with the macrostate of the system). This is closely related to the Second Law of Thermodynamics.

Take for example (experiment A) a closed box that is, at the beginning, half-filled with ideal gas. As time passes, the gas obviously expands to fill the whole box, so that the final state is a box full of gas. This is an irreversible process, since if the box is full at the beginning (experiment B), it does not become only half-full later, except for the very unlikely situation where the gas particles have very special locations and speeds. But this is precisely because we always assume that the initial conditions are such that the particles have random locations and speeds. This is not correct for the final conditions of the system, because the particles have interacted between themselves, so that their locations and speeds have become dependent on each other, i.e. correlated. This can be understood if we look at experiment A backwards in time, which we'll call experiment C: now we begin with a box full of gas, but the particles do not have random locations and speeds; rather, their locations and speeds are so particular, that after some time they all move to one half of the box, which is the final state of the system (this is the initial state of experiment A, because now we're looking at the same experiment backwards!). The interactions between particles now do not create correlations between the particles, but in fact turn them into (at least seemingly) random, "canceling" the pre-existing correlations. The only difference between experiment C (which defies the Second Law of Thermodynamics) and experiment B (which obeys the Second Law of Thermodynamics) is that in the former the particles are uncorrelated at the end, while in the latter the particles are uncorrelated at the beginning.

In fact, if all the microscopic physical processes are reversible (see discussion below), then the Second Law of Thermodynamics can be proven for any isolated system of particles with initial conditions in which the particles states are uncorrelated. To do this, one must acknowledge the difference between the measured entropy of a system—which depends only on its macrostate (its volume, temperature etc.)—and its information entropy (also called Kolmogorov complexity),[4] which is the amount of information (number of computer bits) needed to describe the exact microstate of the system. The measured entropy is independent of correlations between particles in the system, because they do not affect its macrostate, but the information entropy **does** depend on them, because correlations lower the randomness of the system and thus lowers the amount of information needed to describe it.[5] Therefore, in the absence of such correlations the two entropies are identical, but otherwise the information entropy is smaller than the measured entropy, and the difference can be used as a measure of the amount of correlations.

Now, by Liouville's theorem, time-reversal of all microscopic processes implies that the amount of information needed to describe the exact microstate of an isolated system (its information-theoretic joint entropy) is constant in time. This joint entropy is equal to the marginal entropy (entropy assuming no correlations) plus the entropy of correlation (mutual entropy, or its negative mutual information). If we assume no correlations between the particles initially, then this joint entropy is just the marginal entropy, which is just the initial thermodynamic entropy of the system, divided by Boltzmann's constant. However, if these are indeed the initial conditions (and this is a crucial assumption), then such correlations form with time. In other words, there is a decreasing mutual entropy (or increasing mutual information), and for a time that is not too long—the correlations (mutual information) between particles only increase with time. Therefore, the thermodynamic entropy, which is proportional to the marginal entropy, must also increase with time [6] (note that "not too long" in this context is relative to the time needed, in a classical version of the system, for it to pass through all its possible microstates— a time that can be roughly estimated as τe^S , where τ is the time between particle collisions and S is the system's entropy. In any practical case this time is huge compared to everything else). Note that the correlation between particles is not a fully objective quantity. One cannot measure the mutual entropy, one can only measure its change, assuming one can measure a microstate. Thermodynamics is restricted to the case where microstates cannot be distinguished, which means that only the marginal entropy, proportional to the thermodynamic entropy, can be measured, and, in a practical sense, always increases.

13.6 The arrow of time in various phenomena

Main article: Arrow of time

All phenomena that behave differently in one time direction can ultimately be linked to the Second Law of Thermodynamics. This includes the fact that ice cubes melt in hot coffee rather than assembling themselves out of the coffee, that a block sliding on a rough surface slows down rather than speeding up, and that we can remember the past rather than the future. This last phenomenon, called the "psychological arrow of time", has deep connections with Maxwell's demon and the physics of information; In fact, it is easy to understand its link to the Second Law of Thermodynamics if one views memory as correlation between brain cells (or computer bits) and the outer world. Since the Second Law of Thermodynamics is equivalent to the growth with time of such correlations, then it states that memory is created as we move towards the future (rather than towards the past).

13.7 Current research

Current research focuses mainly on describing the thermodynamic arrow of time mathematically, either in classical or quantum systems, and on understanding its origin from the point of view of cosmological boundary conditions.

13.7.1 Dynamical systems

Some current research in dynamical systems indicates a possible "explanation" for the arrow of time. There are several ways to describe the time evolution of a dynamical system. In the classical framework, one considers a differential equation, where one of the parameters is explicitly time. By the very nature of differential equations, the solutions to such systems are inherently time-reversible. However, many of the interesting cases are either ergodic or mixing, and it is strongly suspected that mixing and ergodicity somehow underlie the fundamental mechanism of the arrow of time.

Mixing and ergodic systems do not have exact solutions, and thus proving time irreversibility in a mathematical sense is (as of 2006) impossible. Some progress can be made by studying discrete-time models or difference equations. Many discrete-time models, such as the iterated functions considered in popular fractal-drawing programs, are explicitly not time-reversible, as any given point "in the present" may have several different "pasts" associated with it: indeed, the set of all pasts is known as the Julia set. Since such systems have a built-in irreversibility, it is inappropriate to use them to explain why time is not reversible.

There are other systems that are chaotic, and are also explicitly time-reversible: among these is the baker's map, which is also exactly solvable. An interesting avenue of study is to examine solutions to such systems not by iterating the dynamical system over time, but instead, to study the corresponding Frobenius-Perron operator or transfer operator for the system. For some of these systems, it can be explicitly, mathematically shown that the transfer operators are not trace-class. This means that these operators do not have a unique eigenvalue spectrum that is independent of the choice of basis. In the case of the baker's map, it can be shown that several unique and inequivalent diagonalizations or bases exist, each with a different set of eigenvalues. It is this phenomenon that can be offered as an "explanation" for the arrow of time. That is, although the iterated, discrete-time system is explicitly time-symmetric, the transfer operator is not. Furthermore, the transfer operator can be diagonalized in one of two inequivalent ways: one that describes the forward-time evolution of the system, and one that describes the backwards-time evolution.

As of 2006, this type of time-symmetry breaking has been demonstrated for only a very small number of exactly-solvable, discrete-time systems. The transfer operator for more complex systems has not been consistently formulated, and its precise definition is mired in a variety of subtle difficulties. In particular, it has not been shown that it has a broken symmetry for the simplest exactly-solvable continuous-time ergodic systems, such as Hadamard's billiards, or the Anosov flow on the tangent space of PSL(2,R).

13.7.2   Quantum mechanics

Research on irreversibility in quantum mechanics takes several different directions. One avenue is the study of rigged Hilbert spaces, and in particular, how discrete and continuous eigenvalue spectra intermingle. For example, the rational numbers are completely intermingled with the real numbers, and yet have a unique, distinct set of properties. It is hoped that the study of Hilbert spaces with a similar inter-mingling will provide insight into the arrow of time.

Another distinct approach is through the study of quantum chaos by which attempts are made to quantize systems as classically chaotic, ergodic or mixing. The results obtained are not dissimilar from those that come from the transfer operator method. For example, the quantization of the Boltzmann gas, that is, a gas of hard (elastic) point particles in a rectangular box reveals that the eigenfunctions are space-filling fractals that occupy the entire box, and that the energy eigenvalues are very closely spaced and have an "almost continuous" spectrum (for a finite number of particles in a box, the spectrum must be, of necessity, discrete). If the initial conditions are such that all of the particles are confined to one side of the box, the system very quickly evolves into one where the particles fill the entire box. Even when all of the particles are initially on one side of the box, their wave functions do, in fact, permeate the entire box: they constructively interfere on one side, and destructively interfere on the other. Irreversibility is then argued by noting that it is "nearly impossible" for the wave functions to be "accidentally" arranged in some unlikely state: such arrangements are a set of zero measure. Because the eigenfunctions are fractals, much of the language and machinery of entropy and statistical mechanics can be imported to discuss and argue the quantum case.

13.7.3   Cosmology

Some processes that involve high energy particles and are governed by the weak force (such as K-meson decay) defy the symmetry between time directions. However, all known physical processes **do** preserve a more complicated symmetry (CPT symmetry), and are therefore unrelated to the second law of thermodynamics, or to our day-to-day experience of the arrow of time. A notable exception is the wave function collapse in quantum mechanics, which is an irreversible process. It has been conjectured that the collapse of the wave function may be the reason for the Second Law of Thermodynamics. However it is more accepted today that the opposite is correct, namely that the (possibly merely apparent) wave function collapse is a consequence of quantum decoherence, a process that is ultimately an outcome of the Second Law of Thermodynamics.

The universe was in a uniform, high density state at its very early stages, shortly after the big bang. The hot gas in the early universe was near thermodynamic equilibrium (giving rise to the horizon problem) and hence in a state of maximum entropy, given its volume. Expansion of a gas increases its entropy, however, and expansion of the universe has therefore enabled an ongoing increase in entropy. Viewed from later eras, the early universe can thus be considered to be highly ordered. The uniformity of this early near-equilibrium state has been explained by the theory of cosmic inflation.

According to this theory our universe (or, rather, its accessible part, a radius of 46 billion light years around our location) evolved from a tiny, totally uniform volume (a portion of a much bigger universe), which expanded greatly; hence it was highly ordered. Fluctuations were then created by quantum processes related to its expansion, in a manner supposed to be such that these fluctuations are uncorrelated for any practical use. This is supposed to give the desired initial conditions needed for the Second Law of Thermodynamics.

Our universe is apparently an open universe, so that its expansion will never terminate, but it is an interesting thought experiment to imagine what would have happened had our universe been closed. In such a case, its expansion would stop at a certain time in the distant future, and then begin to shrink. Moreover, a closed universe is finite. It is unclear what would happen to the Second Law of Thermodynamics in such a case. One could imagine at least three different scenarios (in fact, only the third one is plausible, since the first two require a smooth cosmic evolution, contrary to what is observed):

- A highly controversial view is that in such a case the arrow of time will reverse.[7] The quantum fluctuations—which in the meantime have evolved into galaxies and stars—will be in superposition in such a way that the whole process described above is reversed—i.e., the fluctuations are erased by destructive interference and total uniformity is achieved once again. Thus the universe ends in a big crunch, which is similar to its beginning in the big bang. Because the two are totally symmetric, and the final state is very highly ordered, entropy must decrease close to the end of the universe, so that the Second Law of Thermodynamics reverses when the universe shrinks. This can be understood as follows: in the very early universe, interactions between fluctuations created entanglement (quantum correlations) between particles spread all over the universe; during the expansion, these particles became so distant that these correlations became negligible (see quantum decoherence). At the time the expansion halts and the universe starts to shrink, such correlated particles arrive once again at contact (after circling around the universe), and the entropy starts to decrease—because highly correlated initial conditions may lead to a decrease in entropy. Another way of putting it, is that as distant particles arrive, more and more order is revealed because these particles are highly correlated with particles that arrived earlier.

- It could be that this is the crucial point where the wavefunction collapse is important: if the collapse is real, then the quantum fluctuations will not be in superposition any longer; rather they had collapsed to a particular state (a particular arrangement of galaxies and stars), thus creating a big crunch, which is very different from the big bang. Such a scenario may be viewed as adding boundary conditions (say, at the distant future) that dictate the wavefunction collapse.[8]

- The broad consensus among the scientific community today is that smooth initial conditions lead to a highly non-smooth final state, and that this is in fact the source of the thermodynamic arrow of time.[9] Highly non-smooth gravitational systems tend to collapse to black holes, so the wavefunction of the whole universe evolves from a superposition of small fluctuations to a superposition of states with many black holes in each. It may even be that it is impossible for the universe to have both a smooth beginning and a smooth ending. Note that in this scenario the energy density of the universe in the final stages of its shrinkage is much larger than in the corresponding initial stages of its expansion (there is no destructive interference, unlike in the first scenario described above), and consists of mostly black holes rather than free particles.

In the first scenario, the cosmological arrow of time is the reason for both the thermodynamic arrow of time and the quantum arrow of time. Both will slowly disappear as the universe will come to a halt, and will later be reversed.

In the second and third scenarios, it is the difference between the initial state and the final state of the universe that is responsible for the thermodynamic arrow of time. This is independent of the cosmological arrow of time. In the second scenario, the quantum arrow of time may be seen as the deep reason for this.

13.8 See also

- Entropy
- History of entropy
- Arrow of time
- Cosmic inflation
- H-theorem
- Loschmidt's paradox

13.9 References

[1] The Thermodynamic Arrow: Puzzles and Pseudo-Puzzles, Price H., Proceedings of Time and Matter, Venice, 2002

[2] Penrose, R. *The Road to Reality* pp. 686-734

[3] *Physical Origins of Time Asymmetry*, p. 109.

[4] *Physical Origins of Time Asymmetry*, p. 35.

[5] *Physical Origins of Time Asymmetry*, pp. 35-38.

[6] Some Misconceptions about Entropy

[7] Arrow of time in cosmology, Hawking S.W., Phys. Rev. D 32, 2489 - 2495 (1985)

[8] [quant-ph/0507269] Two-time interpretation of quantum mechanics

[9] scholarpedia: Time's arrow and Boltzmann's entropy

13.10   Further reading

- Halliwell, J.J. et al. (1994). *Physical Origins of Time Asymmetry*. Cambridge. ISBN 0-521-56837-4. (technical).

- Mackey, Michael C. (1992). *Time's Arrow: The Origins of Thermodynamic Behavior*. Berlin Heidelberg New York: Springer. ISBN 3-540-94093-6. OCLC 28585247. ... it is shown that for there to be a global evolution of the entropy to its maximal value ... it is *necessary and sufficient* that the system have a property known as exactness. ... these criteria suggest that all currently formulated physical laws may not be at the foundation of the thermodynamic behavior we observe every day of our lives. (page xi)
  Dover has reprinted the monograph in 2003 (ISBN 0486432432). For a short paper listing "the essential points of that argument, correcting presentation points that were confusing ... and emphasizing conclusions more forcefully than previously" see Mackey, Michael C. (2001). "Microscopic Dynamics and the Second Law of Thermodynamics" (PDF). In Mugnai, C.; Ranfagni, A.; Schulman, L.S. *Time's Arrow, Quantum Measurement and Superluminal Behavior*. Rome: Consiglio Nazionale Delle Ricerche. pp. 49–65. ISBN 88-8080-024-8.

- Sean M. Carroll, *From Eternity to Here: The Quest for the Ultimate Theory of Time*

13.11   External links

- Thermodynamic Asymmetry in Time at the online Stanford Encyclopedia of Philosophy

- Java applets simulating irreversible processes: Release of a gas from a container and Mixing of two gases

Chapter 14

Entropy rate

In the mathematical theory of probability, the **entropy rate** or **source information rate** of a stochastic process is, informally, the time density of the average information in a stochastic process. For stochastic processes with a countable index, the entropy rate $H(X)$ is the limit of the joint entropy of n members of the process Xk divided by n, as n tends to infinity:

$$H(X) = \lim_{n \to \infty} \frac{1}{n} H(X_1, X_2, \ldots X_n)$$

when the limit exists. An alternative, related quantity is:

$$H'(X) = \lim_{n \to \infty} H(X_n | X_{n-1}, X_{n-2}, \ldots X_1)$$

For strongly stationary stochastic processes, $H(X) = H'(X)$. The entropy rate can be thought of as a general property of stochastic sources; this is the asymptotic equipartition property.

14.1   Entropy rates for Markov chains

Since a stochastic process defined by a Markov chain that is irreducible and aperiodic and persistent has a stationary distribution, the entropy rate is independent of the initial distribution.

For example, for such a Markov chain Yk defined on a countable number of states, given the transition matrix Pij, $H(Y)$ is given by:

$$H(Y) = -\sum_{ij} \mu_i P_{ij} \log P_{ij}$$

where μi is the stationary distribution of the chain.

A simple consequence of this definition is that an i.i.d. stochastic process has an entropy rate that is the same as the entropy of any individual member of the process.

14.2   See also

- Information source (mathematics)

- Markov information source

- Asymptotic equipartition property

14.3   References

- Cover, T. and Thomas, J. (1991) Elements of Information Theory, John Wiley and Sons, Inc., ISBN 0-471-06259-6

14.4   External links

- Systems Analysis, Modelling and Prediction (SAMP), University of Oxford MATLAB code for estimating information-theoretic quantities for stochastic processes.

Chapter 15

Residual entropy

Residual entropy is the difference in entropy between a non-equilibrium state and crystal state of a substance close to absolute zero. This term is used in condensed matter physics to describe the entropy at zero kelvin of a glass or plastic crystal referred to the crystal state, whose entropy is zero according to Third law of thermodynamics. It occurs if a material can exist in many different states when cooled. The most common non-equilibrium state is vitreous state, glass.

A common example is the case of carbon monoxide, which has a very small dipole moment. As the carbon monoxide crystal is cooled to absolute zero, few of the carbon monoxide molecules have enough time to align themselves into a perfect crystal, (with all of the carbon monoxide molecules oriented in the same direction). Because of this, the crystal is locked into a state with 2^N different corresponding microstates, giving a residual entropy of $S = Nk \ln(2)$, rather than zero.

Another example is any amorphous solid (glass). These have residual entropy, because the atom-by-atom microscopic structure can be arranged in a huge number of different ways across a macroscopic system.

15.1 History

One of the first examples of residual entropy was pointed out by Pauling to describe water ice. In water, each oxygen atom is bonded to two hydrogen atoms. However, when water freezes it forms a tetragonal structure where each oxygen atom has four hydrogen neighbors (due to neighboring water molecules). The hydrogen atoms sitting between the oxygen atoms have some degree of freedom as long as each oxygen atom has two hydrogen atoms that are 'nearby', thus forming the traditional H_2O water molecule. However, it turns out that for a large number of water molecules in this configuration, the hydrogen atoms have a large number of possible configurations that meet the 2-in 2-out rule (each oxygen atom must have two 'near' (or 'in') hydrogen atoms, and two far (or 'out') hydrogen atoms). This freedom exists down to absolute zero, which was previously seen as an absolute one-of-a-kind configuration. The existence of these multiple configurations that meet the rules of absolute zero amounts to randomness, or in other words, entropy. Thus systems that can take multiple configurations at or near absolute zero are said to have residual entropy.

Although water ice was the first material for which residual entropy was proposed, it is generally very difficult to prepare pure defect-free crystals of water ice for studying. A great deal of research has thus been undertaken into finding other systems that exhibit residual entropy. Geometrically frustrated systems in particular often exhibit residual entropy. An important example is spin ice, which is a geometrically frustrated magnetic material where the magnetic moments of the magnetic atoms have Ising-like magnetic spins and lie on the corners of network of corner-sharing tetrahedra. This material is thus analogous to water ice, with the exception that the spins on the corners of the tetrahedra can point into or out of the tetrahedra, thereby producing the same 2-in, 2-out rule as in water ice, and therefore the same residual entropy. One of the interesting properties of geometrically frustrated magnetic materials such as spin ice is that the level of residual entropy can be controlled by the application of an external magnetic field. This property can be used to create one-shot refrigeration systems.

15.2   See also

- Proton disorder in ice

- Geometrical frustration

Chapter 16

Second law of thermodynamics

The **second law of thermodynamics** states that in every natural thermodynamic process the sum of the entropies of all participating bodies is increased. In the limiting case, for reversible processes this sum remains unchanged.

The second law is an empirical finding that has been accepted as an axiom of thermodynamic theory.

Statistical thermodynamics, classical or quantum, explains the law.

The second law has been expressed in many ways. Its first formulation is credited to the French scientist Sadi Carnot in 1824 (see Timeline of thermodynamics).

16.1   Introduction

The first law of thermodynamics provides the basic definition of thermodynamic energy, also called internal energy, associated with all thermodynamic systems, but unknown in classical mechanics, and states the rule of conservation of energy in nature.[1][2]

The concept of energy in the first law does not, however, account for the observation that natural processes have a preferred direction of progress. The first law is symmetrical with respect to the initial and final states of an evolving system. But the second law asserts that a natural process runs only in one sense, and is not reversible. For example, heat always flows spontaneously from hotter to colder bodies, and never the reverse, unless external work is performed on the system. The key concept for the explanation of this phenomenon through the second law of thermodynamics is the definition of a new physical quantity, the entropy.[3][4]

For mathematical analysis of processes, entropy is introduced as follows. In a fictive reversible process, an infinitesimal increment in the entropy (dS) of a system results from an infinitesimal transfer of heat (δQ) to a closed system divided by the common temperature (T) of the system and the surroundings which supply the heat.[5]

$$dS = \frac{\delta Q}{T}$$

The zeroth law of thermodynamics in its usual short statement allows recognition that two bodies in a relation of thermal equilibrium have the same temperature, especially that a test body has the same temperature as a reference thermometric body.[6] For a body in thermal equilibrium with another, there are indefinitely many empirical temperature scales, in general respectively depending on the properties of a particular reference thermometric body. The second law allows a distinguished temperature scale, which defines an absolute, thermodynamic temperature, independent of the properties of any particular reference thermometric body.[7][8]

16.2   Various statements of the law

The second law of thermodynamics may be expressed in many specific ways,[9] the most prominent classical statements[10] being the statement by Rudolf Clausius (1854), the statement by Lord Kelvin (1851), and the statement in axiomatic thermodynamics by Constantin Carathéodory (1909). These statements cast the law in general physical terms citing the impossibility of certain processes. The Clausius and the Kelvin statements have been shown to be equivalent.[11]

16.2.1   Carnot's principle

The historical origin of the second law of thermodynamics was in Carnot's principle. It refers to a cycle of a Carnot engine, fictively operated in the limiting mode of extreme slowness known as quasi-static, so that the heat and work transfers are between subsystems that are always in their own internal states of thermodynamic equilibrium. The Carnot engine is an idealized device of special interest to engineers who are concerned with the efficiency of heat engines. Carnot's principle was recognized by Carnot at a time when the caloric theory of heat was seriously considered, before the recognition of the first law of thermodynamics, and before the mathematical expression of the concept of entropy. Interpreted in the light of the first law, it is physically equivalent to the second law of thermodynamics, and remains valid today. It states

> The efficiency of a quasi-static or reversible Carnot cycle depends only on the temperatures of the two heat reservoirs, and is the same, whatever the working substance. A Carnot engine operated in this way is the most efficient possible heat engine using those two temperatures.[12][13][14][15][16][17][18]

16.2.2   Clausius statement

The German scientist Rudolf Clausius laid the foundation for the second law of thermodynamics in 1850 by examining the relation between heat transfer and work.[19] His formulation of the second law, which was published in German in 1854, is known as the *Clausius statement*:

> Heat can never pass from a colder to a warmer body without some other change, connected therewith, occurring at the same time.[20]

The statement by Clausius uses the concept of 'passage of heat'. As is usual in thermodynamic discussions, this means 'net transfer of energy as heat', and does not refer to contributory transfers one way and the other.

Heat cannot spontaneously flow from cold regions to hot regions without external work being performed on the system, which is evident from ordinary experience of refrigeration, for example. In a refrigerator, heat flows from cold to hot, but only when forced by an external agent, the refrigeration system.

16.2.3   Kelvin statement

Lord Kelvin expressed the second law as

> It is impossible, by means of inanimate material agency, to derive mechanical effect from any portion of matter by cooling it below the temperature of the coldest of the surrounding objects.[21]

16.2.4   Equivalence of the Clausius and the Kelvin statements

Suppose there is an engine violating the Kelvin statement: i.e., one that drains heat and converts it completely into work in a cyclic fashion without any other result. Now pair it with a reversed Carnot engine as shown by the figure. The net and sole effect of this newly created engine consisting of the two engines mentioned is transferring heat $\Delta Q = Q\left(\frac{1}{\eta} - 1\right)$ from the cooler reservoir to the hotter one, which violates the Clausius statement. Thus a violation of the Kelvin statement implies a violation of the Clausius statement, i.e. the Clausius statement implies the Kelvin statement. We can prove in a similar manner that the Kelvin statement implies the Clausius statement, and hence the two are equivalent.

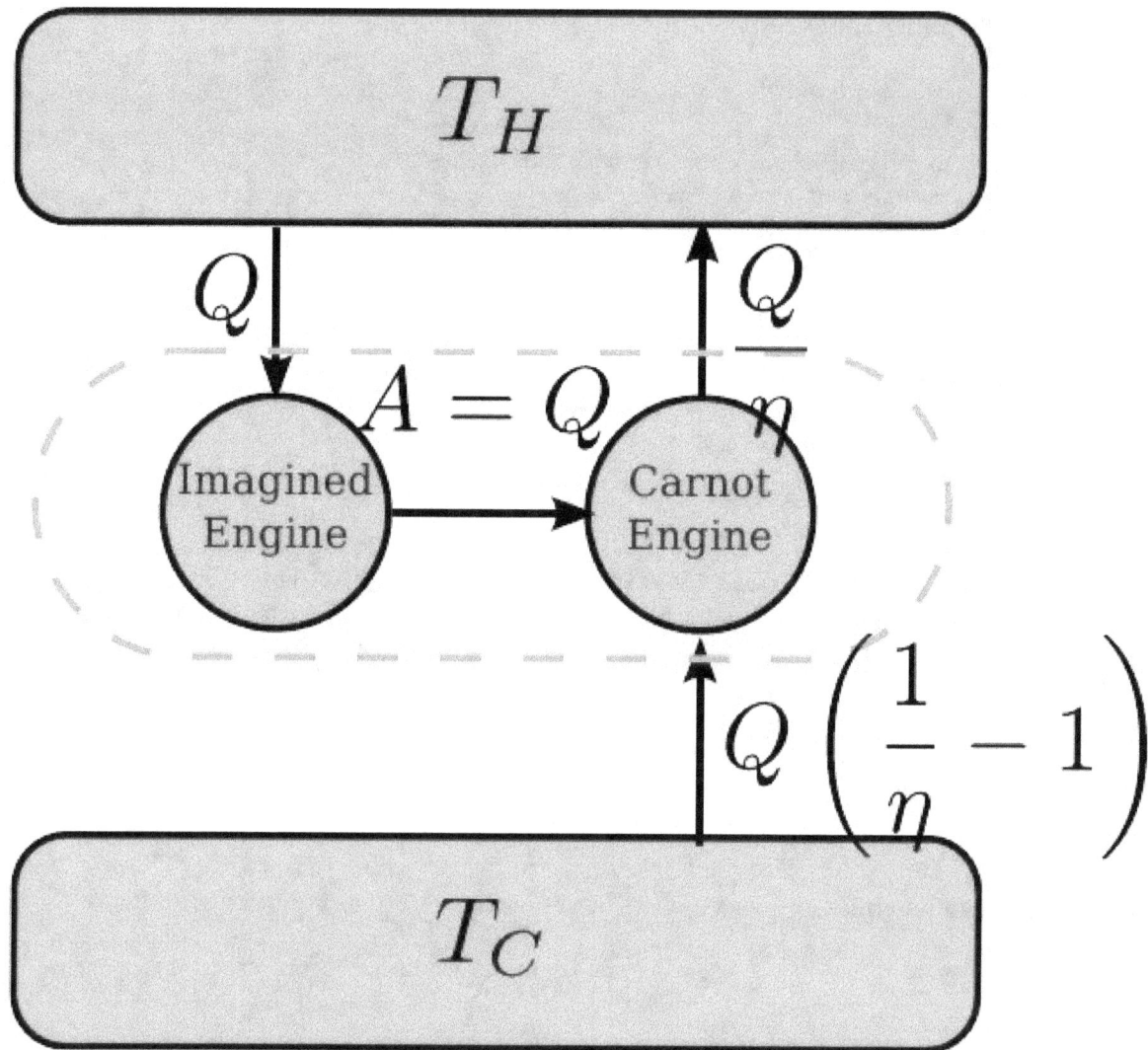

Derive Kelvin Statement from Clausius Statement

16.2.5 Planck's proposition

Planck offered the following proposition as derived directly from experience. This is sometimes regarded as his statement of the second law, but he regarded it as a starting point for the derivation of the second law.

> *It is impossible to construct an engine which will work in a complete cycle, and produce no effect except the raising of a weight and cooling of a heat reservoir.*[22][23]

16.2.6 Relation between Kelvin's statement and Planck's proposition

It is almost customary in textbooks to speak of the "Kelvin-Planck statement" of the law, as for example in the text by ter Haar and Wergeland.[24] One text gives a statement that for all the world looks like Planck's proposition, but attributes it to Kelvin without mention of Planck.[25] One monograph quotes Planck's proposition as the "Kelvin-Planck" formulation, the text naming Kelvin as its author, though it correctly cites Planck in its references.[26] The reader may compare the two statements quoted just above here.

16.2.7   Planck's statement

Planck stated the second law as follows.

> *Every process occurring in nature proceeds in the sense in which the sum of the entropies of all bodies taking part in the process is increased. In the limit, i.e. for reversible processes, the sum of the entropies remains unchanged.*[27][28][29]

16.2.8   Principle of Carathéodory

Constantin Carathéodory formulated thermodynamics on a purely mathematical axiomatic foundation. His statement of the second law is known as the Principle of Carathéodory, which may be formulated as follows:[30]

> In every neighborhood of any state S of an adiabatically enclosed system there are states inaccessible from S.[31]

With this formulation, he described the concept of adiabatic accessibility for the first time and provided the foundation for a new subfield of classical thermodynamics, often called geometrical thermodynamics. It follows from Carathéodory's principle that quantity of energy quasi-statically transferred as heat is a holonomic process function, in other words, $\delta Q = T dS$.[32]

Though it is almost customary in textbooks to say that Carathéodory's principle expresses the second law and to treat it as equivalent to the Clausius or to the Kelvin-Planck statements, such is not the case. To get all the content of the second law, Carathéodory's principle needs to be supplemented by Planck's principle, that isochoric work always increases the internal energy of a closed system that was initially in its own internal thermodynamic equilibrium.[33][34][35][36]

16.2.9   Planck's Principle

In 1926, Max Planck wrote an important paper on the basics of thermodynamics.[35][37] He indicated the principle

> The internal energy of a closed system is increased by an adiabatic process, throughout the duration of which, the volume of the system remains constant.[33][34]

This formulation does not mention heat and does not mention temperature, nor even entropy, and does not necessarily implicitly rely on those concepts, but it implies the content of the second law. A closely related statement is that "Frictional pressure never does positive work."[38] Using a now-obsolete form of words, Planck himself wrote: "The production of heat by friction is irreversible."[39][40]

Not mentioning entropy, this principle of Planck is stated in physical terms. It is very closely related to the Kelvin statement given just above.[41] Nevertheless, this principle of Planck is not actually Planck's preferred statement of the second law, which is quoted above, in a previous sub-section of the present section of this present article, and relies on the concept of entropy.

The link to Kelvin's statement is illustrated by an equivalent statement by Allahverdyan & Nieuwenhuizen, which they attribute to Kelvin: "No work can be extracted from a closed equilibrium system during a cyclic variation of a parameter by an external source."[42][43]

16.2.10   Statement for a system that has a known expression of its internal energy as a function of its extensive state variables

The second law has been shown to be equivalent to the internal energy U being a weakly convex function, when written as a function of extensive properties (mass, volume, entropy, ...).[44][45]

16.2.11 Gravitational systems

In non-gravitational systems, objects always have positive heat capacity, meaning that the temperature rises with energy. Therefore, when energy flows from a high-temperature object to a low-temperature object, the source temperature is decreased while the sink temperature is increased; hence temperature differences tend to diminish over time.

However, this is not always the case for systems in which the gravitational force is important and the general theory of relativity is required. (Apart from this, the thermodynamics of systems described by the general theory of relativity is beyond the scope of the present article.) The most striking examples are black holes, which – according to theory – have negative heat capacity. The larger the black hole, the more energy it contains, but the lower its temperature. Thus, the supermassive black hole in the center of the Milky Way is supposed to have a temperature of 10^{-14} K, much lower than the cosmic microwave background temperature of 2.7 K, but as it absorbs photons of the cosmic microwave background its mass is increasing so that its low temperature further decreases with time.

For this reason, gravitational systems tend towards non-even distribution of mass and energy. The universe in large scale is importantly a gravitational system, and the second law may therefore not apply to it.

16.3   Corollaries

16.3.1   Perpetual motion of the second kind

Main article: Perpetual motion

Before the establishment of the Second Law, many people who were interested in inventing a perpetual motion machine had tried to circumvent the restrictions of First Law of Thermodynamics by extracting the massive internal energy of the environment as the power of the machine. Such a machine is called a "perpetual motion machine of the second kind". The second law declared the impossibility of such machines.

16.3.2   Carnot theorem

Carnot's theorem (1824) is a principle that limits the maximum efficiency for any possible engine. The efficiency solely depends on the temperature difference between the hot and cold thermal reservoirs. Carnot's theorem states:

- All irreversible heat engines between two heat reservoirs are less efficient than a Carnot engine operating between the same reservoirs.

- All reversible heat engines between two heat reservoirs are equally efficient with a Carnot engine operating between the same reservoirs.

In his ideal model, the heat of caloric converted into work could be reinstated by reversing the motion of the cycle, a concept subsequently known as thermodynamic reversibility. Carnot, however, further postulated that some caloric is lost, not being converted to mechanical work. Hence, no real heat engine could realise the Carnot cycle's reversibility and was condemned to be less efficient.

Though formulated in terms of caloric (see the obsolete caloric theory), rather than entropy, this was an early insight into the second law.

16.3.3   Clausius Inequality

The Clausius Theorem (1854) states that in a cyclic process

$$\oint \frac{\delta Q}{T} \leq 0.$$

The equality holds in the reversible case[46] and the '<' is in the irreversible case. The reversible case is used to introduce the state function entropy. This is because in cyclic processes the variation of a state function is zero from state functionality.

16.3.4 Thermodynamic temperature

Main article: Thermodynamic temperature

For an arbitrary heat engine, the efficiency is:

$$\eta = \frac{A}{q_H} = \frac{q_H - q_C}{q_H} = 1 - \frac{q_C}{q_H} \qquad (1)$$

where A is the work done per cycle. Thus the efficiency depends only on qC/qH.

Carnot's theorem states that all reversible engines operating between the same heat reservoirs are equally efficient. Thus, any reversible heat engine operating between temperatures T_1 and T_2 must have the same efficiency, that is to say, the efficiency is the function of temperatures only: $\frac{q_C}{q_H} = f(T_H, T_C)$ (2).

In addition, a reversible heat engine operating between temperatures T_1 and T_3 must have the same efficiency as one consisting of two cycles, one between T_1 and another (intermediate) temperature T_2, and the second between T_2 and T_3. This can only be the case if

$$f(T_1, T_3) = \frac{q_3}{q_1} = \frac{q_2 q_3}{q_1 q_2} = f(T_1, T_2) f(T_2, T_3).$$

Now consider the case where T_1 is a fixed reference temperature: the temperature of the triple point of water. Then for any T_2 and T_3,

$$f(T_2, T_3) = \frac{f(T_1, T_3)}{f(T_1, T_2)} = \frac{273.16 \cdot f(T_1, T_3)}{273.16 \cdot f(T_1, T_2)}.$$

Therefore if thermodynamic temperature is defined by

$$T = 273.16 \cdot f(T_1, T)$$

then the function f, viewed as a function of thermodynamic temperature, is simply

$$f(T_2, T_3) = \frac{T_3}{T_2},$$

and the reference temperature T_1 will have the value 273.16. (Of course any reference temperature and any positive numerical value could be used—the choice here corresponds to the Kelvin scale.)

16.3.5 Entropy

Main article: entropy (classical thermodynamics)

According to the Clausius equality, for a reversible process

$$\oint \frac{\delta Q}{T} = 0$$

That means the line integral $\int_L \frac{\delta Q}{T}$ is path independent.

So we can define a state function S called entropy, which satisfies

$$dS = \frac{\delta Q}{T}$$

With this we can only obtain the difference of entropy by integrating the above formula. To obtain the absolute value, we need the Third Law of Thermodynamics, which states that S=0 at absolute zero for perfect crystals.

For any irreversible process, since entropy is a state function, we can always connect the initial and terminal states with an imaginary reversible process and integrating on that path to calculate the difference in entropy.

Now reverse the reversible process and combine it with the said irreversible process. Applying Clausius inequality on this loop,

$$-\Delta S + \int \frac{\delta Q}{T} = \oint \frac{\delta Q}{T} < 0$$

Thus,

$$\Delta S \geq \int \frac{\delta Q}{T}$$

where the equality holds if the transformation is reversible.

Notice that if the process is an adiabatic process, then $\delta Q = 0$, so $\Delta S \geq 0$.

16.3.6 Energy, available useful work

See also: Exergy

An important and revealing idealized special case is to consider applying the Second Law to the scenario of an isolated system (called the total system or universe), made up of two parts: a sub-system of interest, and the sub-system's surroundings. These surroundings are imagined to be so large that they can be considered as an *unlimited* heat reservoir at temperature TR and pressure PR — so that no matter how much heat is transferred to (or from) the sub-system, the temperature of the surroundings will remain TR; and no matter how much the volume of the sub-system expands (or contracts), the pressure of the surroundings will remain PR.

Whatever changes to dS and dSR occur in the entropies of the sub-system and the surroundings individually, according to the Second Law the entropy Stot of the isolated total system must not decrease:

$$dS_{tot} = dS + dS_R \geq 0$$

According to the First Law of Thermodynamics, the change dU in the internal energy of the sub-system is the sum of the heat δq added to the sub-system, *less* any work δw done *by* the sub-system, *plus* any net chemical energy entering the sub-system $d \sum \mu_i R N_i$, so that:

$$dU = \delta q - \delta w + d\left(\sum \mu_{iR} N_i\right)$$

where $\mu_i R$ are the chemical potentials of chemical species in the external surroundings.

Now the heat leaving the reservoir and entering the sub-system is

$$\delta q = T_R(-dS_R) \le T_R dS$$

where we have first used the definition of entropy in classical thermodynamics (alternatively, in statistical thermodynamics, the relation between entropy change, temperature and absorbed heat can be derived); and then the Second Law inequality from above.

It therefore follows that any net work δw done by the sub-system must obey

$$\delta w \le -dU + T_R dS + \sum \mu_{iR} dN_i$$

It is useful to separate the work δw done by the subsystem into the *useful* work δw_u that can be done *by* the sub-system, over and beyond the work $pR\, dV$ done merely by the sub-system expanding against the surrounding external pressure, giving the following relation for the useful work (exergy) that can be done:

$$\delta w_u \le -d(U - T_R S + p_R V - \sum \mu_{iR} N_i)$$

It is convenient to define the right-hand-side as the exact derivative of a thermodynamic potential, called the *availability* or *exergy* E of the subsystem.

$$E = U - T_R S + p_R V - \sum \mu_{iR} N_i$$

The Second Law therefore implies that for any process which can be considered as divided simply into a subsystem, and an unlimited temperature and pressure reservoir with which it is in contact,

$$dE + \delta w_u \le 0$$

i.e. the change in the subsystem's exergy plus the useful work done *by* the subsystem (or, the change in the subsystem's exergy less any work, additional to that done by the pressure reservoir, done *on* the system) must be less than or equal to zero.

In sum, if a proper *infinite-reservoir-like* reference state is chosen as the system surroundings in the real world, then the Second Law predicts a decrease in E for an irreversible process and no change for a reversible process.

$$dS_{tot} \ge 0 \text{ Is equivalent to } dE + \delta w_u \le 0$$

This expression together with the associated reference state permits a design engineer working at the macroscopic scale (above the thermodynamic limit) to utilize the Second Law without directly measuring or considering entropy change in a total isolated system. (*Also, see process engineer*). Those changes have already been considered by the assumption that the system under consideration can reach equilibrium with the reference state without altering the reference state. An efficiency for a process or collection of processes that compares it to the reversible ideal may also be found (*See second law efficiency*.)

This approach to the Second Law is widely utilized in engineering practice, environmental accounting, systems ecology, and other disciplines.

16.4 History

See also: History of entropy

The first theory of the conversion of heat into mechanical work is due to Nicolas Léonard Sadi Carnot in 1824. He was

Nicolas Léonard Sadi Carnot in the traditional uniform of a student of the École Polytechnique.

the first to realize correctly that the efficiency of this conversion depends on the difference of temperature between an engine and its environment.

Recognizing the significance of James Prescott Joule's work on the conservation of energy, Rudolf Clausius was the first to formulate the second law during 1850, in this form: heat does not flow *spontaneously* from cold to hot bodies. While common knowledge now, this was contrary to the caloric theory of heat popular at the time, which considered heat as a fluid. From there he was able to infer the principle of Sadi Carnot and the definition of entropy (1865).

Established during the 19th century, the Kelvin-Planck statement of the Second Law says, "It is impossible for any device that operates on a cycle to receive heat from a single reservoir and produce a net amount of work." This was shown to be equivalent to the statement of Clausius.

The ergodic hypothesis is also important for the Boltzmann approach. It says that, over long periods of time, the time spent in some region of the phase space of microstates with the same energy is proportional to the volume of this region, i.e. that all accessible microstates are equally probable over a long period of time. Equivalently, it says that time average and average over the statistical ensemble are the same.

It has been shown that not only classical systems but also quantum mechanical ones tend to maximize their entropy over time. Thus the second law follows, given initial conditions with low entropy. More precisely, it has been shown that the local von Neumann entropy is at its maximum value with a very high probability.[47] The result is valid for a large class of isolated quantum systems (e.g. a gas in a container). While the full system is pure and therefore does not have any entropy, the entanglement between gas and container gives rise to an increase of the local entropy of the gas. This result is one of the most important achievements of quantum thermodynamics.

Today, much effort in the field is attempting to understand why the initial conditions early in the universe were those of low entropy,[48][49] as this is seen as the origin of the second law (see below).

16.4.1   Informal descriptions

The second law can be stated in various succinct ways, including:

- It is impossible to produce work in the surroundings using a cyclic process connected to a single heat reservoir (Kelvin, 1851).

- It is impossible to carry out a cyclic process using an engine connected to two heat reservoirs that will have as its only effect the transfer of a quantity of heat from the low-temperature reservoir to the high-temperature reservoir (Clausius, 1854).

- If thermodynamic work is to be done at a finite rate, free energy must be expended. (Stoner, 2000)[50]

16.4.2   Mathematical descriptions

In 1856, the German physicist Rudolf Clausius stated what he called the "second fundamental theorem in the mechanical theory of heat" in the following form:[51]

$$\int \frac{\delta Q}{T} = -N$$

where Q is heat, T is temperature and N is the "equivalence-value" of all uncompensated transformations involved in a cyclical process. Later, in 1865, Clausius would come to define "equivalence-value" as entropy. On the heels of this definition, that same year, the most famous version of the second law was read in a presentation at the Philosophical Society of Zurich on April 24, in which, in the end of his presentation, Clausius concludes:

> The entropy of the universe tends to a maximum.

This statement is the best-known phrasing of the second law. Because of the looseness of its language, e.g. universe, as well as lack of specific conditions, e.g. open, closed, or isolated, many people take this simple statement to mean that the second law of thermodynamics applies virtually to every subject imaginable. This, of course, is not true; this statement is only a simplified version of a more extended and precise description.

In terms of time variation, the mathematical statement of the second law for an isolated system undergoing an arbitrary transformation is:

$$\frac{dS}{dt} \geq 0$$

where

S is the entropy of the system and

t is time.

The equality sign holds in the case that only reversible processes take place inside the system. If irreversible processes take place (which is the case in real systems in operation) the >-sign holds. An alternative way of formulating of the second law for isolated systems is:

$$\frac{dS}{dt} = \dot{S}_i \text{ with } \dot{S}_i \geq 0$$

with $\dot{S}_i$ the sum of the rate of entropy production by all processes inside the system. The advantage of this formulation is that it shows the effect of the entropy production. The rate of entropy production is a very important concept since it determines (limits) the efficiency of thermal machines. Multiplied with ambient temperature T_a it gives the so-called dissipated energy $P_{diss} = T_a \dot{S}_i$.

The expression of the second law for closed systems (so, allowing heat exchange and moving boundaries, but not exchange of matter) is:

$$\frac{dS}{dt} = \frac{\dot{Q}}{T} + \dot{S}_i \text{ with } \dot{S}_i \geq 0$$

Here

$$\dot{Q}$$

$$T$$

If heat is supplied to the system at several places we have to take the algebraic sum of the corresponding terms.

For open systems (also allowing exchange of matter):

$$\frac{dS}{dt} = \frac{\dot{Q}}{T} + \dot{S} + \dot{S}_i \text{ with } \dot{S}_i \geq 0$$

Here $\dot{S}$ is the flow of entropy into the system associated with the flow of matter entering the system. It should not be confused with the time derivative of the entropy. If matter is supplied at several places we have to take the algebraic sum of these contributions.

Statistical mechanics gives an explanation for the second law by postulating that a material is composed of atoms and molecules which are in constant motion. A particular set of positions and velocities for each particle in the system is called a microstate of the system and because of the constant motion, the system is constantly changing its microstate. Statistical mechanics postulates that, in equilibrium, each microstate that the system might be in is equally likely to occur, and when this assumption is made, it leads directly to the conclusion that the second law must hold in a statistical sense. That is, the second law will hold on average, with a statistical variation on the order of $1/\sqrt{N}$ where N is the number of particles in the system. For everyday (macroscopic) situations, the probability that the second law will be violated is practically zero. However, for systems with a small number of particles, thermodynamic parameters, including the entropy, may show significant statistical deviations from that predicted by the second law. Classical thermodynamic theory does not deal with these statistical variations.

16.5   Derivation from statistical mechanics

Further information: H-theorem

Due to Loschmidt's paradox, derivations of the Second Law have to make an assumption regarding the past, namely that the system is uncorrelated at some time in the past; this allows for simple probabilistic treatment. This assumption is usually thought as a boundary condition, and thus the second Law is ultimately a consequence of the initial conditions somewhere in the past, probably at the beginning of the universe (the Big Bang), though other scenarios have also been suggested.[52][53][54]

Given these assumptions, in statistical mechanics, the Second Law is not a postulate, rather it is a consequence of the fundamental postulate, also known as the equal prior probability postulate, so long as one is clear that simple probability arguments are applied only to the future, while for the past there are auxiliary sources of information which tell us that it was low entropy. The first part of the second law, which states that the entropy of a thermally isolated system can only increase, is a trivial consequence of the equal prior probability postulate, if we restrict the notion of the entropy to systems in thermal equilibrium. The entropy of an isolated system in thermal equilibrium containing an amount of energy of E is:

$$S = k_B \ln \left[\Omega \left(E \right) \right]$$

where $\Omega \left(E \right)$ is the number of quantum states in a small interval between E and $E + \delta E$. Here δE is a macroscopically small energy interval that is kept fixed. Strictly speaking this means that the entropy depends on the choice of δE. However, in the thermodynamic limit (i.e. in the limit of infinitely large system size), the specific entropy (entropy per unit volume or per unit mass) does not depend on δE.

Suppose we have an isolated system whose macroscopic state is specified by a number of variables. These macroscopic variables can, e.g., refer to the total volume, the positions of pistons in the system, etc. Then Ω will depend on the values of these variables. If a variable is not fixed, (e.g. we do not clamp a piston in a certain position), then because all the accessible states are equally likely in equilibrium, the free variable in equilibrium will be such that Ω is maximized as that is the most probable situation in equilibrium.

If the variable was initially fixed to some value then upon release and when the new equilibrium has been reached, the fact the variable will adjust itself so that Ω is maximized, implies that the entropy will have increased or it will have stayed the same (if the value at which the variable was fixed happened to be the equilibrium value). Suppose we start from an equilibrium situation and we suddenly remove a constraint on a variable. Then right after we do this, there are a number Ω of accessible microstates, but equilibrium has not yet been reached, so the actual probabilities of the system being in some accessible state are not yet equal to the prior probability of $1/\Omega$. We have already seen that in the final equilibrium state, the entropy will have increased or have stayed the same relative to the previous equilibrium state. Boltzmann's H-theorem, however, proves that the quantity H increases monotonically as a function of time during the intermediate out of equilibrium state.

16.5.1   Derivation of the entropy change for reversible processes

The second part of the Second Law states that the entropy change of a system undergoing a reversible process is given by:

$$dS = \frac{\delta Q}{T}$$

where the temperature is defined as:

$$\frac{1}{k_B T} \equiv \beta \equiv \frac{d \ln \left[\Omega \left(E \right) \right]}{dE}$$

See here for the justification for this definition. Suppose that the system has some external parameter, x, that can be changed. In general, the energy eigenstates of the system will depend on x. According to the adiabatic theorem of quantum mechanics, in the limit of an infinitely slow change of the system's Hamiltonian, the system will stay in the same energy eigenstate and thus change its energy according to the change in energy of the energy eigenstate it is in.

The generalized force, X, corresponding to the external variable x is defined such that $X\,dx$ is the work performed by the system if x is increased by an amount dx. E.g., if x is the volume, then X is the pressure. The generalized force for a system known to be in energy eigenstate E_r is given by:

$$X = -\frac{dE_r}{dx}$$

Since the system can be in any energy eigenstate within an interval of δE, we define the generalized force for the system as the expectation value of the above expression:

$$X = -\left\langle \frac{dE_r}{dx} \right\rangle$$

To evaluate the average, we partition the $\Omega(E)$ energy eigenstates by counting how many of them have a value for $\frac{dE_r}{dx}$ within a range between Y and $Y + \delta Y$. Calling this number $\Omega_Y(E)$, we have:

$$\Omega(E) = \sum_Y \Omega_Y(E)$$

The average defining the generalized force can now be written:

$$X = -\frac{1}{\Omega(E)} \sum_Y Y \Omega_Y(E)$$

We can relate this to the derivative of the entropy w.r.t. x at constant energy E as follows. Suppose we change x to x + dx. Then $\Omega(E)$ will change because the energy eigenstates depend on x, causing energy eigenstates to move into or out of the range between E and $E + \delta E$. Let's focus again on the energy eigenstates for which $\frac{dE_r}{dx}$ lies within the range between Y and $Y + \delta Y$. Since these energy eigenstates increase in energy by Y dx, all such energy eigenstates that are in the interval ranging from E – Y dx to E move from below E to above E. There are

$$N_Y(E) = \frac{\Omega_Y(E)}{\delta E} Y\,dx$$

such energy eigenstates. If $Y\,dx \leq \delta E$, all these energy eigenstates will move into the range between E and $E + \delta E$ and contribute to an increase in Ω. The number of energy eigenstates that move from below $E + \delta E$ to above $E + \delta E$ is, of course, given by $N_Y(E + \delta E)$. The difference

$$N_Y(E) - N_Y(E + \delta E)$$

is thus the net contribution to the increase in Ω. Note that if Y dx is larger than δE there will be the energy eigenstates that move from below E to above $E + \delta E$. They are counted in both $N_Y(E)$ and $N_Y(E + \delta E)$, therefore the above expression is also valid in that case.

Expressing the above expression as a derivative w.r.t. E and summing over Y yields the expression:

$$\left(\frac{\partial \Omega}{\partial x} \right)_E = -\sum_Y Y \left(\frac{\partial \Omega_Y}{\partial E} \right)_x = \left(\frac{\partial (\Omega X)}{\partial E} \right)_x$$

The logarithmic derivative of Ω w.r.t. x is thus given by:

$$\left(\frac{\partial \ln(\Omega)}{\partial x}\right)_E = \beta X + \left(\frac{\partial X}{\partial E}\right)_x$$

The first term is intensive, i.e. it does not scale with system size. In contrast, the last term scales as the inverse system size and will thus vanishes in the thermodynamic limit. We have thus found that:

$$\left(\frac{\partial S}{\partial x}\right)_E = \frac{X}{T}$$

Combining this with

$$\left(\frac{\partial S}{\partial E}\right)_x = \frac{1}{T}$$

Gives:

$$dS = \left(\frac{\partial S}{\partial E}\right)_x dE + \left(\frac{\partial S}{\partial x}\right)_E dx = \frac{dE}{T} + \frac{X}{T}dx = \frac{\delta Q}{T}$$

16.5.2   Derivation for systems described by the canonical ensemble

If a system is in thermal contact with a heat bath at some temperature T then, in equilibrium, the probability distribution over the energy eigenvalues are given by the canonical ensemble:

$$P_j = \frac{\exp\left(-\frac{E_j}{k_{\mathrm{B}}T}\right)}{Z}$$

Here Z is a factor that normalizes the sum of all the probabilities to 1, this function is known as the partition function. We now consider an infinitesimal reversible change in the temperature and in the external parameters on which the energy levels depend. It follows from the general formula for the entropy:

$$S = -k_{\mathrm{B}} \sum_j P_j \ln(P_j)$$

that

$$dS = -k_{\mathrm{B}} \sum_j \ln(P_j)\, dP_j$$

Inserting the formula for P_j for the canonical ensemble in here gives:

$$dS = \frac{1}{T}\sum_j E_j dP_j = \frac{1}{T}\sum_j d(E_j P_j) - \frac{1}{T}\sum_j P_j dE_j = \frac{dE + \delta W}{T} = \frac{\delta Q}{T}$$

16.5.3   General derivation from unitarity of quantum mechanics

The time development operator in quantum theory is unitary, because the Hamiltonian is hermitian. Consequently, the transition probability matrix is doubly stochastic, which implies the Second Law of Thermodynamics.[55][56] This derivation is quite general, based on the Shannon entropy, and does not require any assumptions beyond unitarity, which is universally accepted. It is a *consequence* of the irreversibility or singular nature of the general transition matrix.

16.6   Non-equilibrium states

Main article: Non-equilibrium thermodynamics

The theory of classical or equilibrium thermodynamics is idealized. A main postulate or assumption, often not even explicitly stated, is the existence of systems in their own internal states of thermodynamic equilibrium. In general, a region of space containing a physical system at a given time, that may be found in nature, is not in thermodynamic equilibrium, read in the most stringent terms. In looser terms, nothing in the entire universe is or has ever been truly in exact thermodynamic equilibrium.[57][58]

For purposes of physical analysis, it is often enough convenient to make an assumption of thermodynamic equilibrium. Such an assumption may rely on trial and error for its justification. If the assumption is justified, it can often be very valuable and useful because it makes available the theory of thermodynamics. Elements of the equilibrium assumption are that a system is observed to be unchanging over an indefinitely long time, and that there are so many particles in a system, that its particulate nature can be entirely ignored. Under such an equilibrium assumption, in general, there are no macroscopically detectable fluctuations. There is an exception, the case of critical states, which exhibit to the naked eye the phenomenon of critical opalescence. For laboratory studies of critical states, exceptionally long observation times are needed.

In all cases, the assumption of thermodynamic equilibrium, once made, implies as a consequence that no putative candidate "fluctuation" alters the entropy of the system.

It can easily happen that a physical system exhibits internal macroscopic changes that are fast enough to invalidate the assumption of the constancy of the entropy. Or that a physical system has so few particles that the particulate nature is manifest in observable fluctuations. Then the assumption of thermodynamic equilibrium is to be abandoned. There is no unqualified general definition of entropy for non-equilibrium states.[59]

Non-equilibrium thermodynamics is then appropriate. There are intermediate cases, in which the assumption of local thermodynamic equilibrium is a very good approximation,[60][61][62][63] but strictly speaking it is still an approximation, not theoretically ideal. For non-equilibrium situations in general, it may be useful to consider statistical mechanical definitions of quantities that may be conveniently called 'entropy'. These indeed belong to statistical mechanics, not to macroscopic thermodynamics.

The physics of macroscopically observable fluctuations is beyond the scope of this article.

16.7   Arrow of time

Further information: Entropy (arrow of time)
See also: Arrow of time

The second law of thermodynamics is a physical law that is not symmetric to reversal of the time direction.

The second law has been proposed to supply an explanation of the difference between moving forward and backwards in time, such as why the cause precedes the effect (the causal arrow of time).[64]

16.8   Controversies

16.8.1   Maxwell's demon

Main article: Maxwell's demon

James Clerk Maxwell imagined one container divided into two parts, A and B. Both parts are filled with the same gas at equal temperatures and placed next to each other. Observing the molecules on both sides, an imaginary demon guards a trapdoor between the two parts. When a faster-than-average molecule from A flies towards the trapdoor, the demon opens it, and the molecule will fly from A to B. The average speed of the molecules in B will have increased while in A they will have slowed down on average. Since average molecular speed corresponds to temperature, the temperature decreases in A and increases in B, contrary to the second law of thermodynamics.

One of the most famous responses to this question was suggested in 1929 by Leó Szilárd and later by Léon Brillouin. Szilárd pointed out that a real-life Maxwell's demon would need to have some means of measuring molecular speed, and that the act of acquiring information would require an expenditure of energy.

Maxwell's demon repeatedly alters the permeability of the wall between A and B. It is therefore performing thermodynamic operations, not just presiding over natural processes.

16.8.2   Loschmidt's paradox

Main article: Loschmidt's paradox

Loschmidt's paradox, also known as the reversibility paradox, is the objection that it should not be possible to deduce an irreversible process from time-symmetric dynamics. This puts the time reversal symmetry of nearly all known low-level fundamental physical processes at odds with any attempt to infer from them the second law of thermodynamics which describes the behavior of macroscopic systems. Both of these are well-accepted principles in physics, with sound observational and theoretical support, yet they seem to be in conflict; hence the paradox.

One proposed resolution of this paradox is as follows. The Loschmidt scenario refers to a strictly isolated system or to a strictly adiabatically isolated system. Heat and matter transfers are not allowed. The Loschmidt reversal times are fantastically long, far longer than any laboratory isolation of the required degree of perfection could be maintained in practice. In this sense, the Loschmidt scenario will never be subjected to empirical testing. Also in this sense, the second law, stated for an isolated system, will never be subjected to empirical testing. A system, supposedly perfectly isolated, in strictly perfect thermodynamic equilibrium, can be observed only once in its entire life, because the observation must break the isolation. Two observations would be needed to check empirically for a change of state, one initial and one final. When transfer of heat or matter are permitted, the requirements of perfection are not so tight. In practical laboratory reality, therefore, the second law can be tested only for systems with transfer of heat or matter, and not for isolated systems.

Due to this paradox, derivations of the second law have to make an assumption regarding the past, namely that the system is uncorrelated at some time in the past or, equivalently, that the entropy in the past was lower than in the future. This assumption is usually thought as a boundary condition, and thus the second Law is ultimately derived from the initial conditions of the Big Bang.[52][65]

16.8.3   Poincaré recurrence theorem

The Poincaré recurrence theorem states that certain systems will, after a sufficiently long time, return to a state very close to the initial state. The Poincaré recurrence time is the length of time elapsed until the recurrence, which is of the order of $\sim \exp(S/k)$.[66] The result applies to physical systems in which energy is conserved. The Recurrence theorem apparently contradicts the Second law of thermodynamics, which says that large dynamical systems evolve irreversibly towards the state with higher entropy, so that if one starts with a low-entropy state, the system will never return to it. There are many possible ways to resolve this paradox, but none of them is universally accepted. The most reasonable argument is that for typical thermodynamical systems the recurrence time is so large (many many times longer than the lifetime of the universe) that, for all practical purposes, one cannot observe the recurrence.

A simple example of this can be demonstrated by a thought experiment. Consider two connected spherical chambers containing four molecules of gas, initially with chamber A containing four molecules of a gas and chamber B containing nothing. The gas molecules bounce freely from one chamber to another, and according to the second law of thermodynamics, will tend to be distributed such that they are divided evenly between the chambers, with two molecules in chamber A and two molecules in chamber B, corresponding to an increase in entropy. However, assuming the conservation of energy within the chambers, it is apparent that the molecules will occasionally bounce such that all four will be in one chamber at the same time, corresponding to a spontaneous increase in order and decrease in entropy relative to the evenly distributed state.

16.9   Quotations

The law that entropy always increases holds, I think, the supreme position among the laws of Nature. If someone points out to you that your pet theory of the universe is in disagreement with Maxwell's equations — then so much the worse for Maxwell's equations. If it is found to be contradicted by observation — well, these experimentalists do bungle things sometimes. But if your theory is found to be against the second law of thermodynamics I can give you no hope; there is nothing for it but to collapse in deepest humiliation.
— Sir Arthur Stanley Eddington, *The Nature of the Physical World* (1927)

There have been nearly as many formulations of the second law as there have been discussions of it.
— Philosopher / Physicist P.W. Bridgman, (1941)

Clausius is the author of the sibyllic utterance, "The energy of the universe is constant; the entropy of the universe tends to a maximum." The objectives of continuum thermomechanics stop far short of explaining the "universe", but within that theory we may easily derive an explicit statement in some ways reminiscent of Clausius, but referring only to a modest object: an isolated body of finite size.
— Truesdell, C., Muncaster, R.G. (1980). *Fundamentals of Maxwell's Kinetic Theory of a Simple Monatomic Gas, Treated as a Branch of Rational Mechanics*, Academic Press, New York, ISBN 0-12-701350-4, p.17.

16.10   See also

- Clausius–Duhem inequality

- Fluctuation theorem

- History of thermodynamics

- Jarzynski equality

- Laws of thermodynamics

- Maximum entropy thermodynamics

- Reflections on the Motive Power of Fire

- Thermal diode

- Relativistic heat conduction

16.11   References

[1] Planck. M. (1897/1903), pp. 40–41.

[2] Munster A. (1970), pp. 8–9, 50–51.

[3] Planck. M. (1897/1903), pp. 79–107.

[4] Bailyn, M. (1994), Section 71, pp. 113–154.

[5] Bailyn, M. (1994), p. 120.

[6] J. S. Dugdale (1996). *Entropy and its Physical Meaning*. Taylor & Francis. p. 13. ISBN 0-7484-0569-0. This law is the basis of temperature.

[7] Zemansky, M.W. (1968), pp. 207–209.

[8] Quinn, T.J. (1983), p. 8.

[9] "Concept and Statements of the Second Law". web.mit.edu. Retrieved 2010-10-07.

[10] Lieb & Yngvason (1999).

[11] Rao (2004), p. 213.

[12] Carnot. S. (1824/1986).

[13] Truesdell. C. (1980), Chapter 5.

[14] Adkins, C.J. (1968/1983), pp. 56–58.

[15] Münster, A. (1970), p. 11.

[16] Kondepudi, D., Prigogine, I. (1998), pp.67–75.

[17] Lebon, G., Jou, D., Casas-Vázquez, J. (2008), p. 10.

[18] Eu, B.C. (2002), pp. 32–35.

[19] Clausius (1850).

[20] Clausius (1854), p. 86.

[21] Thomson (1851).

[22] Planck. M. (1897/1903), p. 86.

[23] Roberts, J.K., Miller, A.R. (1928/1960), p. 319.

[24] ter Haar, D., Wergeland, H. (1966), p. 17.

[25] Pippard, A.B. (1957/1966), p. 30.

[26] Čápek, V., Sheehan, D.P. (2005), p. 3

[27] Planck. M. (1897/1903), p. 100.

[28] Planck. M. (1926), p. 463, translation by Uffink, J. (2003), p. 131.

[29] Roberts, J.K., Miller, A.R. (1928/1960), p. 382. This source is partly verbatim from Planck's statement, but does not cite Planck. This source calls the statement the principle of the increase of entropy.

[30] Carathéodory, C. (1909).

[31] Buchdahl, H.A. (1966), p. 68.

[32] Sychev, V. V. (1991). *The Differential Equations of Thermodynamics*. Taylor & Francis. ISBN 978-1560321217. Retrieved 2012-11-26.

[33] Münster, A. (1970), p. 45.

[34] Lieb & Yngvason (1999), p. 49.

[35] Planck, M. (1926).

[36] Buchdahl, H.A. (1966), p. 69.

[37] Uffink, J. (2003), pp. 129–132.

[38] Truesdell, C., Muncaster, R.G. (1980). *Fundamentals of Maxwell's Kinetic Theory of a Simple Monatomic Gas, Treated as a Branch of Rational Mechanics*, Academic Press, New York, ISBN 0-12-701350-4, p. 15.

[39] Planck, M. (1897/1903), p. 81.

[40] Planck, M. (1926), p. 457, Wikipedia editor's translation.

[41] Lieb, E.H., Yngvason, J. (2003), p. 149.

[42] Allahverdyan, A.E., Nieuwenhuizen, T.H. (2001). A mathematical theorem as the basis for the second law: Thomson's formulation applied to equilibrium, http://arxiv.org/abs/cond-mat/0110422

[43] Čápek, V., Sheehan, D.P. (2005), p. 11

[44] van Gool, W.; Bruggink, J.J.C. (Eds) (1985). *Energy and time in the economic and physical sciences*. North-Holland. pp. 41–56. ISBN 0444877487.

[45] Grubbström, Robert W. (2007). "An Attempt to Introduce Dynamics Into Generalised Exergy Considerations". *Applied Energy* **84**: 701–718. doi:10.1016/j.apenergy.2007.01.003.

[46] **Clausius theorem** at Wolfram Research

[47] Gemmer, Jochen; Otte, Alexander; Mahler, Günter (2001). "Quantum Approach to a Derivation of the Second Law of Thermodynamics". *Physical Review Letters* **86** (10): 1927–1930. arXiv:quant-ph/0101140. Bibcode:2001PhRvL..86.1927G. doi:10.1103/PhysRevLett.86.1927. PMID 11289822.

[48] Sean M. Carroll; Jennifer Chen (2005). "Does Inflation Provide Natural Initial Conditions for the Universe?". *Gen.Rel.Grav.* ( ) ; *International Journal of Modern Physics D D14* ( ) – 2340 **37** (2005): 1671–1674. arXiv:gr-qc/0505037. Bibcode:2005GReGr..37.1671C. doi:10.1007/s10714-005-0148-2.

[49] Wald, R (2006). "The arrow of time and the initial conditions of the universe". *Studies in History and Philosophy of Science Part B: Studies in History and Philosophy of Modern Physics* **37** (3): 394–398. doi:10.1016/j.shpsb.2006.03.005.

[50] Stoner (2000). "Inquiries into the Nature of Free Energy and Entropy in Respect to Biochemical Thermodynamics". *Entropy* **2** (3): 106–141. arXiv:physics/0004055. Bibcode:2000Entrp...2..106S. doi:10.3390/e2030106.

[51] Clausius (1867).

[52] Hawking, SW (1985). "Arrow of time in cosmology". *Phys. Rev. D* **32** (10): 2489–2495. Bibcode:1985PhRvD..32.2489H. doi:10.1103/PhysRevD.32.2489. Retrieved 2013-02-15.

[53] Greene, Brian (2004). *The Fabric of the Cosmos*. Alfred A. Knopf. p. 171. ISBN 0-375-41288-3.

[54] Lebowitz, Joel L. (September 1993). "Boltzmann's Entropy and Time's Arrow"(PDF).*Physics Today***46**(9): 32–38.Bibcode doi:10.1063/1.881363. Retrieved 2013-02-22.

[55] Hugh Everett, "Theory of the Universal Wavefunction", Thesis, Princeton University, (1956, 1973), Appendix I, pp 121 ff, in particular equation (4.4) at the top of page 127, and the statement on page 29 that "it is known that the [Shannon] entropy [...] is a monotone increasing function of the time."

[56] Bryce Seligman DeWitt, R. Neill Graham, eds, *The Many-Worlds Interpretation of Quantum Mechanics*, Princeton Series in Physics, Princeton University Press (1973), ISBN 0-691-08131-X Contains Everett's thesis: The Theory of the Universal Wavefunction, pp 3–140.

[57] Grandy, W.T., Jr (2008), p. 151.

[58] Callen, H.B. (1960/1985), p. 15.

[59] Lieb, E.H., Yngvason, J. (2003), p. 190.

[60] Gyarmati, I. (1967/1970), pp. 4-14.

[61] Glansdorff, P., Prigogine, I. (1971).

[62] Müller, I. (1985).

[63] Müller, I. (2003).

[64] Halliwell, J.J.; et al. (1994). *Physical Origins of Time Asymmetry*. Cambridge. ISBN 0-521-56837-4. chapter 6

[65] Greene, Brian (2004). *The Fabric of the Cosmos*. Alfred A. Knopf. p. 161. ISBN 0-375-41288-3.

[66] L. Dyson, J. Lindesay and L. Susskind, *Is There Really a de Sitter/CFT Duality*, JHEP **0208**, 45 (2002)

16.11.1 Bibliography of citations

- Adkins, C.J. (1968/1983). *Equilibrium Thermodynamics*, (1st edition 1968), third edition 1983, Cambridge University Press, Cambridge UK, ISBN 0-521-25445-0.

- Attard, P. (2012). *Non-equilibrium Thermodynamics and Statistical Mechanics: Foundations and Applications*, Oxford University Press, Oxford UK, ISBN 978-0-19-966276-0.

- Bailyn, M. (1994). *A Survey of Thermodynamics*, American Institute of Physics, New York, ISBN 0-88318-797-3.

- Boltzmann, L. (1896/1964). *Lectures on Gas Theory*, translated by S.G. Brush, University of California Press, Berkeley.

- Buchdahl, H.A. (1966). *The Concepts of Classical Thermodynamics*, Cambridge University Press, Cambridge UK.

- Callen, H.B. (1960/1985). *Thermodynamics and an Introduction to Thermostatistics*, (1st edition 1960) 2nd edition 1985, Wiley, New York, ISBN 0-471-86256-8.

- Čápek, V., Sheehan, D.P. (2005). *Challenges to the Second Law of Thermodynamics: Theory and Experiment*, Springer, Dordrecht, ISBN 1-4020-3015-0.

- C. Carathéodory (1909). "Untersuchungen über die Grundlagen der Thermodynamik". *Mathematische Annalen* **67**: 355–386. doi:10.1007/bf01450409. Axiom II: In jeder beliebigen Umgebung eines willkürlich vorgeschriebenen Anfangszustandes gibt es Zustände, die durch adiabatische Zustandsänderungen nicht beliebig approximiert werden können. (p.363). A translation may be found here. Also a mostly reliable translation is to be found at Kestin, J. (1976). *The Second Law of Thermodynamics*, Dowden, Hutchinson & Ross, Stroudsburg PA.

- Carnot, S. (1824/1986). *Reflections on the motive power of fire*, Manchester University Press, Manchester UK, ISBN 0719017416. Also here.

- Chapman, S., Cowling, T.G. (1939/1970). *The Mathematical Theory of Non-uniform gases. An Account of the Kinetic Theory of Viscosity, Thermal Conduction and Diffusion in Gases*, third edition 1970, Cambridge University Press, London.

- Clausius, R. (1850). "Ueber Die Bewegende Kraft Der Wärme Und Die Gesetze, Welche Sich Daraus Für Die Wärmelehre Selbst Ableiten Lassen". *Annalen der Physik* **79**: 368–397, 500–524. Bibcode:1850AnP...155..500C. doi:10.1002/andp.18501550403. Retrieved 26 June 2012. Translated into English: Clausius, R. (July 1851). "On the Moving Force of Heat, and the Laws regarding the Nature of Heat itself which are deducible therefrom". *London, Edinburgh and Dublin Philosophical Magazine and Journal of Science*. 4th **2** (VIII): 1–21; 102–119. Retrieved 26 June 2012.

- Clausius, R. (1854). "Über eine veränderte Form des zweiten Hauptsatzes der mechanischen Wärmetheorie" (PDF).*Annalen der Physik*(Poggendoff).xciii:481–506.Bibcode:1854AnP...169..481C.doi:10.1002/andp. 18541691202.Retrieved24March2014.Translated into English:Clausius,R. (July1856)."On a Modifi ed Form of the SecondFundamental Theorem in the Mechanical Theory of Heat".*London, Edinburgh and Dublin Philosophical Magazineand Journal of Science*.4th**2**:86.Retrieved24March2014. Reprinted in:Clausius,R.(1867).*The MechanicalTheory of Heat– with its Applications to the Steam Engine and to Physical Properties of Bodies*.London:John vanVoorst.Retrieved19 June2012.

- Eu, B.C. (2002). *Generalized Thermodynamics. The Thermodynamics of Irreversible Processes and Generalized Hydrodynamics*, Kluwer Academic Publishers, Dordrecht, ISBN 1–4020–0788–4.

- Gibbs, J.W. (1876/1878). On the equilibrium of heterogeneous substances, *Trans. Conn. Acad.*, **3**: 108-248, 343-524, reprinted in *The Collected Works of J. Willard Gibbs, Ph.D. LL. D.*, edited by W.R. Longley, R.G. Van Name, Longmans, Green & Co., New York, 1928, volume 1, pp. 55–353.

- Griem, H.R. (2005). *Principles of Plasma Spectroscopy (Cambridge Monographs on Plasma Physics)*, Cambridge University Press, New York ISBN 0-521-61941-6.

- Glansdorff, P., Prigogine, I. (1971). *Thermodynamic Theory of Structure, Stability, and Fluctuations*, Wiley-Interscience, London, 1971, ISBN 0-471-30280-5.

- Grandy, W.T., Jr (2008). *Entropy and the Time Evolution of Macroscopic Systems*. Oxford University Press. ISBN 978-0-19-954617-6.

- Greven, A., Keller, G., Warnecke (editors) (2003). *Entropy*, Princeton University Press, Princeton NJ, ISBN 0-691-11338-6.

- Gyarmati, I. (1967/1970) *Non-equilibrium Thermodynamics. Field Theory and Variational Principles*, translated by E. Gyarmati and W.F. Heinz, Springer, New York.

- Kondepudi, D., Prigogine, I. (1998). *Modern Thermodynamics: From Heat Engines to Dissipative Structures*, John Wiley & Sons, Chichester, ISBN 0–471–97393–9.

- Lebon, G., Jou, D., Casas-Vázquez, J. (2008). *Understanding Non-equilibrium Thermodynamics: Foundations, Applications, Frontiers*, Springer-Verlag, Berlin, e-ISBN 978-3-540-74252-4.

- Lieb, E. H.; Yngvason, J. (1999). "The Physics and Mathematics of the Second Law of Thermodynamics" (PDF). *Physics Reports* **310**: 1–96. arXiv:cond-mat/9708200. Bibcode:1999PhR...310....1L. doi:10.1016/S0370-1573(98)00082-9. Retrieved 24 March 2014.

- Lieb, E.H., Yngvason, J. (2003). The Entropy of Classical Thermodynamics, pp. 147–195, Chapter 8 of *Entropy*, Greven, A., Keller, G., Warnecke (editors) (2003).

- Maxwell, J.C. (1867). "On the dynamical theory of gases". *Phil. Trans. Roy. Soc. London* **157**: 49–88.

- Müller, I. (1985). *Thermodynamics*, Pitman, London, ISBN 0-273-08577-8.

- Müller, I. (2003). Entropy in Nonequilibrium, pp. 79–109, Chapter 5 of *Entropy*, Greven, A., Keller, G., Warnecke (editors) (2003).

- Münster, A. (1970), *Classical Thermodynamics*, translated by E.S. Halberstadt, Wiley–Interscience, London, ISBN 0-471-62430-6.

- Pippard, A.B. (1957/1966). *Elements of Classical Thermodynamics for Advanced Students of Physics*, original publication 1957, reprint 1966, Cambridge University Press, Cambridge UK.

- Planck, M. (1897/1903). *Treatise on Thermodynamics*, translated by A. Ogg, Longmans Green, London, p. 100.

- Planck, M. (1914). *The Theory of Heat Radiation*, a translation by Masius, M. of the second German edition, P. Blakiston's Son & Co., Philadelphia.

- Planck, M. (1926). Über die Begründung des zweiten Hauptsatzes der Thermodynamik, *Sitzungsberichte der Preussischen Akademie der Wissenschaften: Physikalisch-mathematische Klasse*: 453–463.

- Quinn, T.J. (1983). *Temperature*, Academic Press, London, ISBN 0-12-569680-9.

- Rao, Y.V.C. (2004). *An Introduction to thermodynamics*. Universities Press. p. 213. ISBN 978-81-7371-461-0.

- Roberts, J.K., Miller, A.R. (1928/1960). *Heat and Thermodynamics*, (first edition 1928), fifth edition, Blackie & Son Limited, Glasgow.

- ter Haar, D., Wergeland, H. (1966). *Elements of Thermodynamics*, Addison-Wesley Publishing, Reading MA.

- Thomson, W. (1851). "On the Dynamical Theory of Heat, with numerical results deduced from Mr Joule's equivalent of a Thermal Unit, and M. Regnault's Observations on Steam". *Transactions of the Royal Society of Edinburgh* **XX** (part II): 261–268; 289–298. Also published in Thomson, W. (December 1852). "On the Dynamical Theory of Heat, with numerical results deduced from Mr Joule's equivalent of a Thermal Unit, and M. Regnault's Observations on Steam". *Philos. Mag.* 4 **IV** (22): 13. Retrieved 25 June 2012.

- Truesdell, C. (1980). *The Tragicomical History of Thermodynamics 1822–1854*, Springer, New York, ISBN 0-387-90403-4.

- Uffink, J. (2003). Irreversibility and the Second Law of Thermodynamics, Chapter 7 of *Entropy*, Greven, A., Keller, G., Warnecke (editors) (2003).

- Zemansky, M.W. (1968). *Heat and Thermodynamics. An Intermediate Textbook*, fifth edition, McGraw-Hill Book Company, New York.

16.12   Further reading

- Goldstein, Martin, and Inge F., 1993. *The Refrigerator and the Universe*. Harvard Univ. Press. Chpts. 4–9 contain an introduction to the Second Law, one a bit less technical than this entry. ISBN 978-0-674-75324-2

- Leff, Harvey S., and Rex, Andrew F. (eds.) 2003. *Maxwell's Demon 2 : Entropy, classical and quantum information, computing*. Bristol UK; Philadelphia PA: Institute of Physics. ISBN 978-0-585-49237-7

- Halliwell, J.J. (1994). *Physical Origins of Time Asymmetry*. Cambridge. ISBN 0-521-56837-4.(technical).

- Carnot, Sadi; Thurston, Robert Henry (editor and translator) (1890). *Reflections on the Motive Power of Heat and on Machines Fitted to Develop That Power*. New York: J. Wiley & Sons. (full text of 1897 ed.) (html)

- Stephen Jay Kline (1999). *The Low-Down on Entropy and Interpretive Thermodynamics*, La Cañada, CA: DCW Industries. ISBN 1928729010.

- Kostic, M (2011). "Revisiting The Second Law of Energy Degradation and Entropy Generation: From Sadi Carnot's Ingenious Reasoning to Holistic Generalization". *AIP Conf. Proc.* **1411**: 327–350. Bibcode:2011AIPC. 1411..327K.doi:10.1063/1.3665247.ISBN978-0-7354-0985-9.also at.

16.13   External links

- Stanford Encyclopedia of Philosophy: "Philosophy of Statistical Mechanics" – by Lawrence Sklar.

- *Second law of thermodynamics* in the MIT Course *Unified Thermodynamics and Propulsion* from Prof. Z. S. Spakovszky

- E.T. Jaynes, 1988, "The evolution of Carnot's principle," in G. J. Erickson and C. R. Smith (eds.)*Maximum-Entropy and Bayesian Methods in Science and Engineering, Vol* 1: p. 267.

- Caratheodory, C., "Examination of the foundations of thermodynamics," trans. by D. H. Delphenich

Rudolf Clausius

James Clerk Maxwell

Chapter 17

Irreversible process

For the concept in evolutionary theory, see Dollo's law.

In science, a process that is not reversible is called **irreversible**. This concept arises most frequently in thermodynamics.

In thermodynamics, a change in the thermodynamic state of a system and all of its surroundings cannot be precisely restored to its initial state by infinitesimal changes in some property of the system without expenditure of energy. A system that undergoes an irreversible process may still be capable of returning to its initial state; however, the impossibility occurs in restoring the environment to its own initial conditions. An irreversible process increases the entropy of the universe. However, because entropy is a state function, the change in entropy of a system is the same whether the process is reversible or irreversible. The second law of thermodynamics can be used to determine whether a process is reversible or not.

All complex natural processes are irreversible.[1][2][3][4] The phenomenon of irreversibility results from the fact that if a thermodynamic system, which is any system of sufficient complexity, of interacting molecules is brought from one thermodynamic state to another, the configuration or arrangement of the atoms and molecules in the system will change in a way that is not easily predictable.[5][6] A certain amount of "transformation energy" will be used as the molecules of the "working body" do work on each other when they change from one state to another. During this transformation, there will be a certain amount of heat energy loss or dissipation due to intermolecular friction and collisions; energy that will not be recoverable if the process is reversed.

Many biological processes that were once thought to be reversible have been found to actually be a pairing of two irreversible processes. Whereas a single enzyme was once believed to catalyze both the forward and reverse chemical changes, research has found that two separate enzymes of similar structure are typically needed to perform what results in a pair of thermodynamically irreversible processes.[7]

17.1 Absolute versus statistical reversibility

Thermodynamics defines the statistical behaviour of large numbers of entities, whose exact behavior is given by more specific laws. Since the fundamental theoretical laws of physics are all time-reversible,[8] however experimentally, probability of **real** reversibility is low, former presuppositions can be fulfilled and/or former state recovered only to higher or lower degree (see: uncertainty principle). The irreversibility of thermodynamics must be statistical in nature; that is, that it must be merely highly unlikely, but not impossible, that a system will lower in entropy.

17.2 History

The German physicist Rudolf Clausius, in the 1850s, was the first to mathematically quantify the discovery of irreversibility in nature through his introduction of the concept of entropy. In his 1854 memoir "On a Modified Form of the Second

Fundamental Theorem in the Mechanical Theory of Heat" Clausius states:

Simply, Clausius states that it is impossible for a system to transfer heat from a cooler body to a hotter body. For example, a cup of hot coffee placed in an area of room temperature (~72 °F) will transfer heat to its surroundings and thereby cool down with the temperature of the room slightly increasing (~72.3 °F). However, that same initial cup of coffee will never absorb heat from its surroundings causing it to grow even hotter with the temperature of the room decreasing (~71.7 °F). Therefore, the process of the coffee cooling down is irreversible unless extra energy is added to the system.

However, a paradox arose when attempting to reconcile microanalysis of a system with observations of its macrostate. Many processes are mathematically reversible in their microstate when analyzed using classical Newtonian mechanics. From 1872 to 1875, Ludwig Boltzmann reinforced the statistical explanation of this paradox in the form of Boltzmann's entropy formula stating that as the number of possible microstates a system might be in increases, the entropy of the system increases and it becomes less likely that the system will return to an earlier state. His formulas quantified the work done by William Thomson, 1st Baron Kelvin who had argued that:

[9][10]

Another explanation of irreversible systems was presented by French mathematician Henri Poincaré. In 1890, he published his first explanation of nonlinear dynamics, also called chaos theory. Applying the chaos theory to the second law of thermodynamics, the paradox of irreversibility can be explained in the errors associated with scaling from microstates to macrostates and the degrees of freedom used when making experimental observations. Sensitivity to initial conditions relating to the system and its environment at the microstate compounds into an exhibition of irreversible characteristics within the observable, physical realm.[11]

17.3   Examples of irreversible processes

In the physical realm, many irreversible processes are present to which the inability to achieve 100% efficiency in energy transfer can be attributed. The following is a list of spontaneous events which contribute to the irreversibility of processes.[12]

- Heat transfer through a finite temperature difference

- Friction

- Plastic deformation

- Flow of electric current through a resistance

- Magnetization or polarization with a hysteresis

- Unrestrained expansion of fluids

- Spontaneous chemical reactions

- Spontaneous mixing of matter of varying composition/states

A Joule expansion is an example of classical thermodynamics, as it is easy to work out the resulting increase in entropy. It occurs where a volume of gas is kept in one side of a thermally isolated container (via a small partition), with the other side of the container being evacuated; the partition between the two parts of the container is then opened, and the gas fills the whole container. The internal energy of the gas remains the same, while the volume increases. The original state cannot be recovered by simply compressing the gas to its original volume, since the internal energy will be increased by this compression. The original state can only be recovered by then cooling the re-compressed system, and thereby irreversibly heating the environment. The diagram to the right applies only if the first expansion is "free" (Joule expansion). i.e. there can be no atmospheric pressure outside the cylinder and no weight lifted.

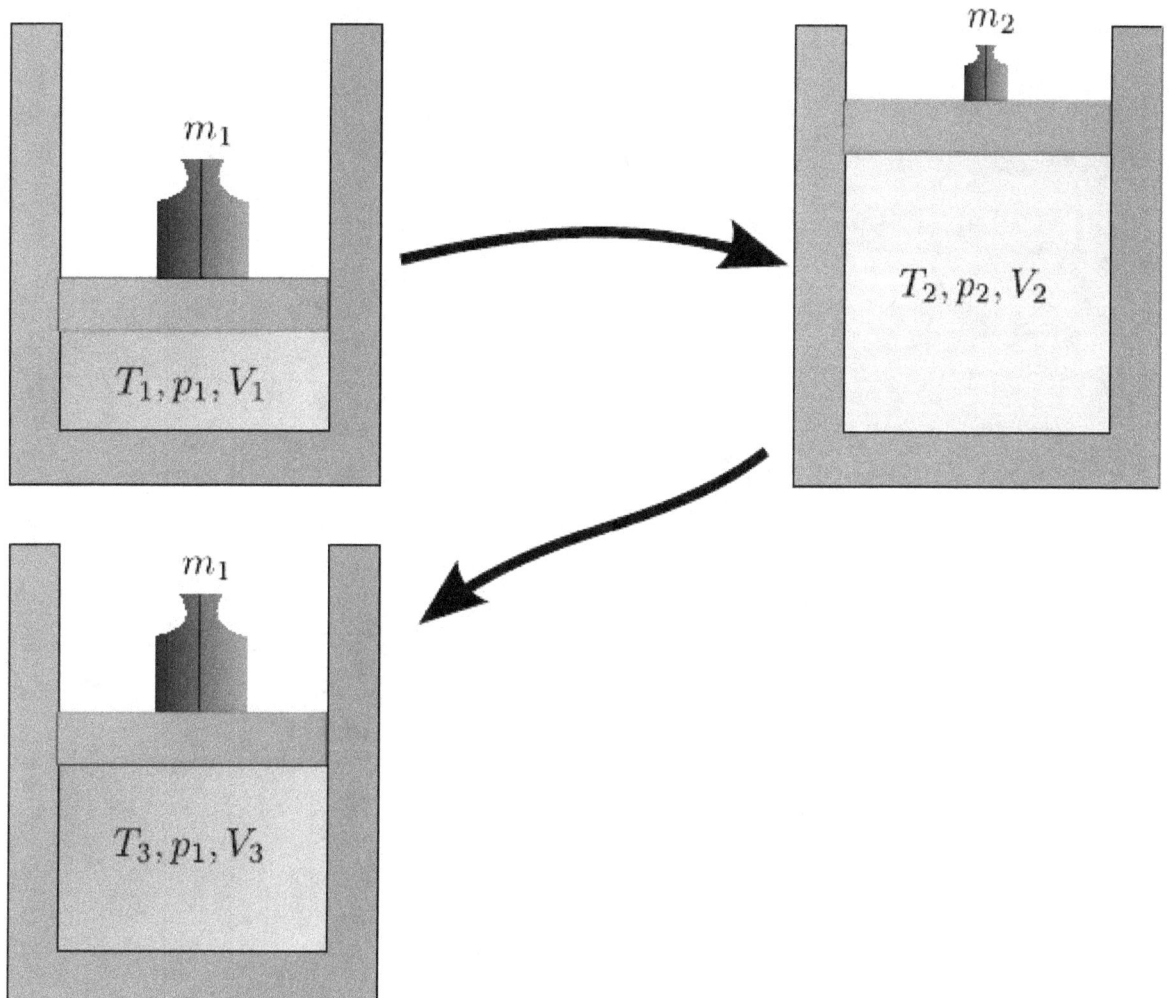

Irreversible adiabatic process: If the cylinder is a perfect insulator, the initial top-left state cannot be reached anymore after it is changed to the one on the top-right. Instead, the state on the bottom left is assumed when going back to the original pressure because energy is converted into heat.

17.4 Complex systems

The difference between reversible and irreversible events has particular explanatory value in complex systems (such as living organisms, or ecosystems). According to the biologists Humberto Maturana and Francisco Varela, living organisms are characterized by autopoiesis, which enables their continued existence. More primitive forms of self-organizing systems have been described by the physicist and chemist Ilya Prigogine. In the context of complex systems, events which lead to the end of certain self-organising processes, like death, extinction of a species or the collapse of a meteorological system can be considered as irreversible. Even if a clone with the same organizational principle (e.g. identical DNA-structure) could be developed, this would not mean that the former distinct system comes back into being. Events to which the self-organizing capacities of organisms, species or other complex systems can adapt, like minor injuries or changes in the physical environment are reversible. However, adaptation depends on import of negentropy into the organism, thereby increasing irreversible processes in its environment. Ecological principles, like those of sustainability and the precautionary principle can be defined with reference to the concept of reversibility.[13][14][15][16][17][18][19][20][21][22]

17.5   See also

- Entropy

- Entropy production

- Entropy (arrow of time)

- Exergy

- Reversible process (thermodynamics)

- One way function

- Non-equilibrium thermodynamics

17.6   References

[1] Lucia, U (1995). "Mathematical consequences and Gyarmati's principle in Rational Thermodynamics". *Il Nuovo Cimento* **B110** (10): 1227–1235.

[2] Grazzini; Lucia, U. (1997). "Global analysis of dissipations due to irreversibility". *Revue Gènèrale de Thermique* **36**: 605–609.

[3] Lucia, U. (2008). "Probability, ergodicity, irreversibility and dynamical systems". *Proceedings of the Royal Society A: Mathematical, Physical and Engineering Sciences* **464**: 1089. Bibcode:2008RSPSA.464.1089L. doi:10.1098/rspa.2007.0304.

[4] Grazzini G. e Lucia U., 2008 Evolution rate of thermodynamic systems, 1st International Workshop "Shape and Thermodynamics" – Florence 25 and 26 September 2008, pp. 1-7

[5] Lucia U., 2009, Irreversibility, entropy and incomplete information, Physica A: Statistical Mechanics and its Applications, 388, pp. 4025-4033

[6] Lucia, U (2008). "Statistical approach of the irreversible entropy variation". *Physica A: Statistical Mechanics and its Applications* **387** (14): 3454–3460. doi:10.1016/j.physa.2008.02.002.

[7] Lucia U., "Irreversible Entropy in Biological Systems", EPISTEME
Lucia, U.; Maino, G. (2003). "Thermodynamical analysis of the dynamics of tumor interaction with the host immune system". *Physica A: Statistical Mechanics and its Applications* **313** (3-4): 569–577. doi:10.1016/S0378-4371(02)00980-9.

[8] David Albert on *Time and Chance*

[9] Bishop, R.C. et al. "Irreversibility in Quantum Mechanics." Received 19 January 2004.

[10] Lebowitz, Joel. "Microscopic Reversibility and Macroscopic Behavior: Physical Explanations and Mathematical Derivations." Rutgers University. Nov 1, 1994.

[11] "The 2nd Law of Thermodynamics".Page dated 2002-2-19. Retrieved on 2010-4-01.

[12] Moran, John (2008). "Fundamentals of Engineering Thermodynamics", p. 220. John Wiley & Sons, Inc., USA. ISBN 978-0-471-78735-8.

[13] Lucia, Umberto (1998). "Maximum principle and open systems including two-phase flows". *Revue Gènèrale de Thermique* **37**: 813–817. doi:10.1016/s0035-3159(98)80007-x.

[14] Lucia U., Irreversibility and entropy in Rational Thermodynamics, Ricerche di Matematica, LI (2001) 77-87

[15] Lucia, U.; Gervino, G. (2005). "Thermoeconomic analysis of an irreversible Stirling heat pump cycle". *The European Physical Journal B*: 367–369.

[16] Lucia, Umberto; Maino, G. (2006). "The relativistic behaviour of the thermodynamic Lagrangian". *Il Nuovo Cimento B* **121** (2): 213–216. Bibcode:2006NCimB.121..213L. doi:10.1393/ncb/i2006-10035-8.

[17] Lucia, U. (2007). "Irreversible entropy variation and the problem of the trend to equilibrium". *Physica A: Statistical Mechanics and its Applications* **376**: 289–284. Bibcode:2007PhyA..376..289L. doi:10.1016/j.physa.2006.10.059.

[18] Lucia, U.; Gervino, G. (2009). "Hydrodynamics cavitation: from theory towards a new experimental approach". *Central European Journal of Physics* **7** (3): 638–644.

[19] Lucia, U (2009). "Irreversibility, entropy and incomplete information". *Physica A: Statistical Mechanics and its Applications* **388**: 4025–4033. doi:10.1016/j.physa.2009.06.027.

[20] Lucia, U. (2009). "Irreversibility, entropy and incomplete information". *Physica A: Statistical Mechanics and its Applications* **388** (19): 4025–4033. Bibcode:2009PhyA..388.4025L. doi:10.1016/j.physa.2009.06.027.

[21] Lucia U., 2009, The thermodynamic Lagrangian, in Pandalai S.G., 2009, Recent Research Developments in Physics, Vol. 8, pp. 1-5, ISBN 978-81-7895-346-5

[22] Lucia U., 2010, Maximum entropy generation and κ−exponential model, *Physica A* 389, pp. 4558-4563 Lucia, U. (2010). "Maximum entropy generation and $\kappa\kappa$-exponential model". *Physica A: Statistical Mechanics and its Applications* **389**: 4558–4563. Bibcode:2010PhyA..389.4558L. doi:10.1016/j.physa.2010.06.047.

Chapter 18

Reversible process (thermodynamics)

For other forms of reversibility, see reversibility (disambiguation).

In thermodynamics, a **reversible process** -- or *reversible cycle* if the process is cyclic -- is a process that can be "reversed" by means of infinitesimal changes in some property of the system. A reversible process does not increase entropy (of the system and surroundings). [1] During a reversible process, the system is in thermodynamic equilibrium with its surroundings throughout the entire process. Since it would take an infinite amount of time for the reversible process to finish, perfectly reversible processes are impossible. However, if the system undergoing the changes responds much faster than the applied change, the deviation from reversibility may be negligible. In a reversible cycle, the system and its surroundings will be returned to their original states if the forward cycle is followed by the reverse cycle.[2]

In thermodynamics, processes can be carried out in one of two ways: reversibly or irreversibly. Reversibility in thermodynamics refers to performing a reaction continuously at equilibrium. In an ideal thermodynamically reversible process, the energy from work performed by or on the system would be maximized, and that from heat would be minimized; heat cannot fully be converted to work and will always be lost to some degree (to the surroundings). The phenomenon of maximized work and minimized heat can be visualized on a pressure-volume curve, as the area beneath the equilibrium curve, representing work done. In order to maximize work, one must follow the equilibrium curve closely.

Irreversible processes, on the other hand, are a result of straying away from the curve, therefore decreasing the amount of overall work done; an irreversible process can be described as a thermodynamic process that leaves equilibrium. When described in terms of pressure and volume, it occurs when the pressure or the volume of a system changes so dramatically and instantaneously that the other (pressure or volume in this case) does not have time to catch up. A classic example of irreversibility is allowing a certain volume of gas to be released into a vacuum. By releasing pressure on a sample and thus allowing it to occupy a large space, the system and surroundings are not in equilibrium during the expansion process and there is little work done. However, significant work will be required, with a corresponding amount of energy dissipated as heat flow to the environment, in order to reverse the process (compressing the gas back to its original volume and temperature). [3]

An alternative definition of a *reversible process* is a process that, after it has taken place, can be reversed and, when reversed, causes no change in either the system or its surroundings. In thermodynamic terms, a process "taking place" would refer to its transition from its initial state to its final state.

18.1 Irreversibility

In an irreversible process, finite changes are made; therefore the system is not at equilibrium throughout the process. At the same point in an irreversible cycle, the system will be in the same state, but the surroundings are permanently changed after each cycle.[2]

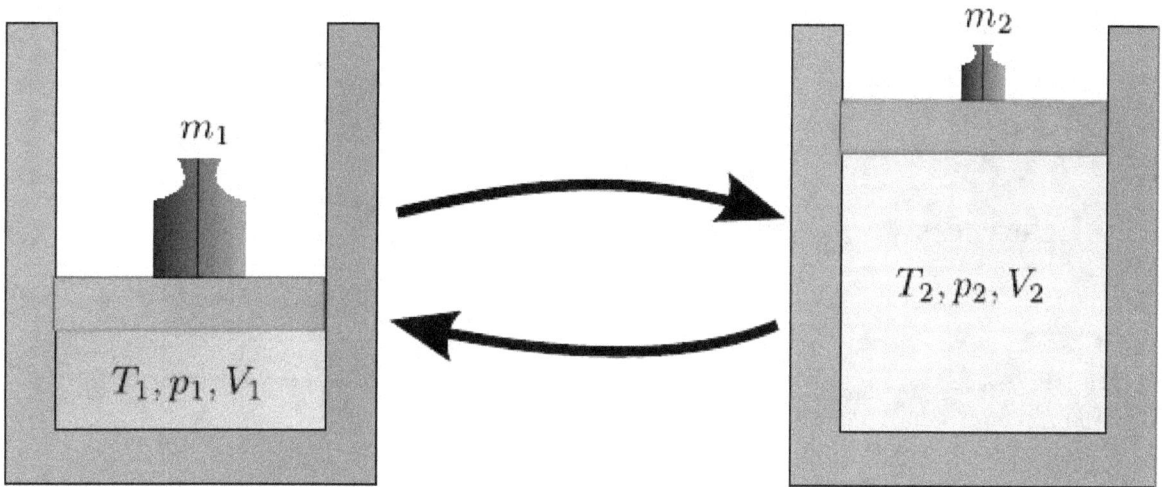

Reversible adiabatic process: The state on the left can be reached from the state on the right as well as vice versa without exchanging heat with the environment.

18.2 Boundaries and states

A reversible process changes the state of a system in such a way that the net change in the combined entropy of the system and its surroundings is zero. Reversible processes define the boundaries of how efficient heat engines can be in thermodynamics and engineering: a reversible process is one where no heat is lost from the system as "waste", and the machine is thus as efficient as it can possibly be (see Carnot cycle).

In some cases, it is important to distinguish between reversible and quasistatic processes. Reversible processes are always quasistatic, but the converse is not always true.[1] For example, an infinitesimal compression of a gas in a cylinder where there exists friction between the piston and the cylinder is a quasistatic, but not reversible process.[4] Although the system has been driven from its equilibrium state by only an infinitesimal amount, heat has been irreversibly lost due to friction, and cannot be recovered by simply moving the piston infinitesimally in the opposite direction.

18.3 Engineering archaisms

Historically, the term **Tesla principle** was used to describe (amongst other things) certain reversible processes invented by Nikola Tesla.[5] However, this phrase is no longer in conventional use. The principle stated that some systems could be reversed and operated in a complementary manner. It was developed during Tesla's research in alternating currents where the current's magnitude and direction varied cyclically. During a demonstration of the Tesla turbine, the disks revolved and machinery fastened to the shaft was operated by the engine. If the turbine's operation was reversed, the disks acted as a pump.[6]

18.4 See also

- Carnot cycle

- Entropy production

- Toffoli gate

- Time evolution

- Quantum circuit

- Reversible computing

- Maxwell's demon

- Stirling engine

18.5   References

[1] Sears, F.W. and Salinger, G.L. (1986), *Thermodynamics, Kinetic Theory, and Statistical Thermodynamics*, 3rd edition (Addison-Wesley.)

[2] Zumdahl, Steven S. (2005) "10.2 The Isothermal Expansion and Compression of an Ideal Gas." Chemical Principles. 5th Edition. (Houghton Mifflin Company)

[3] Lower, S. (2003) *Entropy Rules! What is Entropy?* Entropy

[4] Giancoli, D.C. (2000), *Physics for Scientists and Engineers (with Modern Physics)*, 3rd edition (Prentice-Hall.)

[5] *Electrical Experimenter*, January 1919. p. 615.

[6] "Tesla's New Monarch of Machines". *New York Herald Tribune*, Oct. 15, 1911. (Available online. Tesla Engine Builders Association.)

Chapter 19

Dissipation

Dissipation is the result of an irreversible process that takes place in inhomogeneous thermodynamic systems. A dissipative process is a process in which energy (internal, bulk flow kinetic, or system potential) is transformed from some initial form to some final form; the capacity of the final form to do mechanical work is less than that of the initial form. For example, heat transfer is dissipative because it is a transfer of internal energy from a hotter body to a colder one. Following the second law of thermodynamics, the entropy varies with temperature (reduces the capacity of the combination of the two bodies to do mechanical work), but never decreases in an isolated system.

These processes produce entropy (see entropy production) at a certain rate. The entropy production rate times ambient temperature gives the dissipated power. Important examples of irreversible processes are: heat flow through a thermal resistance, fluid flow through a flow resistance, diffusion (mixing), chemical reactions, and electrical current flow through an electrical resistance (Joule heating).

19.1 Definition

Thermodynamic dissipative processes are essentially irreversible. They produce entropy at a finite rate. In a process in which the temperature is locally continuously defined, the local density of rate of entropy production times local temperature gives the local density of dissipated power.[Definition needed!]

A particular occasion of occurrence of a dissipative process cannot be described by a single individual Hamiltonian formalism. A dissipative process requires a collection of admissible individual Hamiltonian descriptions, exactly which one describes the actual particular occurrence of the process of interest being unknown. This includes friction, and all similar forces that result in decoherency of energy—that is, conversion of coherent or directed energy flow into an indirected or more isotropic distribution of energy.

19.1.1 Energy

"The conversion of mechanical energy into heat is called energy dissipation." – *François Roddier*[1]

19.1.2 Physics

In computational physics, numerical dissipation (also known as "numerical diffusion") refers to certain side-effects that may occur as a result of a numerical solution to a differential equation. When the pure advection equation, which is free of dissipation, is solved by a numerical approximation method, the energy of the initial wave may be reduced in a way analogous to a diffusional process. Such a method is said to contain 'dissipation'. In some cases, "artificial dissipation" is intentionally added to improve the numerical stability characteristics of the solution.[2]

19.1.3 Mathematics

A formal, mathematical definition of dissipation, as commonly used in the mathematical study of measure-preserving dynamical systems, is given in the article *wandering set*.

19.2 Examples

19.2.1 In hydraulic engineering

Dissipation is the process of converting mechanical energy of downward-flowing water into thermal and acoustical energy. Various devices are designed in stream beds to reduce the kinetic energy of flowing waters to reduce their erosive potential on banks and river bottoms. Very often these devices look like small waterfalls or cascades, where water flows vertically or over riprap to lose some of its kinetic energy.

19.2.2 Irreversible processes

Important examples of irreversible processes are:

1. Heat flow through a thermal resistance

2. Fluid flow through a flow resistance

3. Diffusion (mixing)

4. Chemical reactions[3][4]

5. Electrical current flow through an electrical resistance (Joule heating).

19.2.3 Waves or oscillations

Waves or oscillations, lose energy over time, typically from friction or turbulence. In many cases the "lost" energy raises the temperature of the system. For example, a wave that loses amplitude is said to dissipate. The precise nature of the effects depends on the nature of the wave: an atmospheric wave, for instance, may dissipate close to the surface due to friction with the land mass, and at higher levels due to radiative cooling.

19.3 History

See also: Timeline of thermodynamics

The concept of dissipation was introduced in the field of thermodynamics by William Thomson (Lord Kelvin) in 1852.[5] Lord Kelvin deduced that a subset of the above-mentioned irreversible dissaptive processes will occur unless a process is governed by a "perfect thermodynamic engine". The processes that Lord Kelvin identified were friction, diffusion, conduction of heat and the absorption of light.

19.4 See also

-   Entropy production

-   Flood control

- Principle of maximum entropy

- Two-dimensional gas

19.5   References

[1] Degrowth and The Laws Of Thermodynamcis

[2] Thomas, J.W. Numerical Partial Differential Equation: Finite Difference Methods. Springer-Verlag. New York. (1995)

[3] Glansdorff, P., Prigogine, I. (1971). *Thermodynamic Theory of Structure, Stability, and Fluctuations*, Wiley-Interscience, London, 1971, ISBN 0-471-30280-5, p. 61.

[4] Eu, B.C. (1998). *Nonequilibrium Thermodynamics: Ensemble Method*, Kluwer Academic Publications, Dordrecht, ISBN 0-7923-4980-6, p. 49.

[5] W. Thomson *On the universal tendency in nature to the dissipation of mechanical energy* Philosophical Magazine, Ser. 4, p. 304 (1852).

Chapter 20

Internal energy

In thermodynamics, the **internal energy** is one of the two cardinal state functions of the state variables of a thermodynamic system. It refers to energy contained within the system, while excluding the kinetic energy of motion of the system as a whole and the potential energy of the system as a whole due to external force fields. It keeps account of the gains and losses of energy of the system.

The internal energy of a system can be changed by (1) heating the system, or (2) by doing work on it, or (3) by adding or taking away matter.[1] When matter transfer is prevented by impermeable walls containing the system, it is said to be closed. Then the first law of thermodynamics states that the increase in internal energy is equal to the total heat added and work done on the system by the surroundings. If the containing walls pass neither matter nor energy, the system is said to be isolated. Then its internal energy cannot change.

The internal energy of a given state of a system cannot be directly measured. It is determined through some convenient chain of thermodynamic operations and thermodynamic processes by which the given state can be prepared, starting with a reference state which is customarily assigned a reference value for its internal energy. Such a chain, or path, can be theoretically described by certain extensive state variables of the system, namely, its entropy, S, its volume, V, and its mole numbers, $\{Nj\}$. The internal energy, $U(S,V,\{Nj\})$, is a function of those. Sometimes, to that list are appended other extensive state variables, for example electric dipole moment. For practical considerations in thermodynamics and engineering it is rarely necessary or convenient to consider all energies belonging to the total intrinsic energy of a system, such as the energy given by the equivalence of mass. Typically, descriptions only include components relevant to the system and processes under study. Thermodynamics is chiefly concerned only with *changes* in the internal energy.

The internal energy is a state function of a system, because its value depends only on the current state of the system and not on the path taken or processes undergone to prepare it. It is an extensive quantity. It is the cardinal thermodynamic potential. Through it, by use of Legendre transforms, are mathematically constructed the other thermodynamic potentials. These are functions of variable lists in which some extensive variables are replaced by their conjugate intensive variables. Legendre transformation is necessary because mere substitutive replacement of extensive variables by intensive variables does not lead to thermodynamic potentials. The other cardinal function of state of a thermodynamic system is its entropy, as a function, $S(U,V,\{Nj\})$, of the same list of extensive variables of state, except that the entropy, S, is replaced in the list by the internal energy, U.[2][3][4]

Though it is a macroscopic quantity, internal energy can be explained in microscopic terms by two theoretical virtual components. One is the microscopic kinetic energy due to the microscopic motion of the system's particles (translations, rotations, vibrations). The other is the potential energy associated with the microscopic forces, including the chemical bonds, between the particles, and with the static rest mass energy of the constituents of matter. There is no simple universal relation between these quantities of microscopic energy and the quantities of energy gained or lost by the system in work, heat, or matter transfer.

The SI unit of energy is the joule (J). Sometimes it is convenient to use a corresponding density called **specific internal energy** which is internal energy per unit of mass (kilogram) of the system in question. The SI unit of specific internal energy is J/kg. If the specific internal energy is expressed relative to units of amount of substance (mol), then it is referred to as *molar internal energy* and the unit is J/mol.

From the standpoint of statistical mechanics, the internal energy is equal to the ensemble average of the total energy of the system.

20.1 Description and definition

The internal energy U of a given state of the system is determined relative to that of a standard state of the system, by adding up the macroscopic transfers of energy that accompany a change of state from the reference state to the given state:

$$\Delta U = \sum_i E_i$$

where ΔU denotes the difference between the internal energy of the given state and that of the reference state, and the Ei are the various energies transferred to the system in the steps from the reference state to the given state. It is the energy needed to create the given state of the system from the reference state.

From a non-relativistic microscopic point of view, it may be divided into microscopic potential energy, $U_{micro\ pot}$, and microscopic kinetic energy, $U_{micro\ kin}$, components:

$$U = U_{micro\ pot} + U_{micro\ kin}$$

The microscopic kinetic energy of a system arises as the sum of the motions of all the system's particles, whether it be the motion of atoms, molecules, atomic nuclei, electrons, or other particles. The microscopic potential energy algebraic summative components are those of the chemical and nuclear particle bonds, and the physical force fields within the system, such as due to internal induced electric or magnetic dipole moment, as well as the energy of deformation of solids (stress-strain). Usually, the split into microscopic kinetic and potential energies is outside the scope of macroscopic thermodynamics.

Internal energy does not include the energy due to motion or location of a system as a whole. That is to say, it excludes any kinetic or potential energy the body may have because of its motion or location in external gravitational, electrostatic, or electromagnetic fields. It does, however, include the contribution of such a field to the energy due to the coupling of the internal degrees of freedom of the object with the field. In such a case, the field is included in the thermodynamic description of the object in the form of an additional external parameter.

For practical considerations in thermodynamics or engineering, it is rarely necessary, convenient, nor even possible, to consider all energies belonging to the total intrinsic energy of a sample system, such as the energy given by the equivalence of mass. Typically, descriptions only include components relevant to the system under study. Indeed, in most systems under consideration, especially through thermodynamics, it is impossible to calculate the total internal energy.[5] Therefore, a convenient null reference point may be chosen for the internal energy.

The internal energy is an extensive property: it depends on the size of the system, or on the amount of substance it contains.

At any temperature greater than absolute zero, microscopic potential energy and kinetic energy are constantly converted into one another, but the sum remains constant in an isolated system (cf. table). In the classical picture of thermodynamics, kinetic energy vanishes at zero temperature and the internal energy is purely potential energy. However, quantum mechanics has demonstrated that even at zero temperature particles maintain a residual energy of motion, the zero point energy. A system at absolute zero is merely in its quantum-mechanical ground state, the lowest energy state available. At absolute zero a system of given composition has attained its minimum attainable entropy.

The microscopic kinetic energy portion of the internal energy gives rise to the temperature of the system. Statistical mechanics relates the pseudo-random kinetic energy of individual particles to the mean kinetic energy of the entire ensemble of particles comprising a system. Furthermore it relates the mean microscopic kinetic energy to the macroscopically observed empirical property that is expressed as temperature of the system. This energy is often referred to as the *thermal energy* of a system,[6] relating this energy, like the temperature, to the human experience of hot and cold.

Statistical mechanics considers any system to be statistically distributed across an ensemble of N microstates. Each microstate has an energy E_i and is associated with a probability p_i. The internal energy is the mean value of the system's total energy, i.e., the sum of all microstate energies, each weighted by their probability of occurrence:

$$U = \sum_{i=1}^{N} p_i \, E_i \,.$$

This is the statistical expression of the first law of thermodynamics.

20.1.1   Internal energy changes

Thermodynamics is chiefly concerned only with the changes, ΔU, in internal energy.

For a closed system, with matter transfer excluded, the changes in internal energy are due to heat transfer Q and due to work. The latter can be split into two kinds, pressure-volume work $W_{\text{pressure-volume}}$, and frictional and other kinds, such as electrical polarization, which do not alter the volume of the system, and are called isochoric, $W_{\text{isochoric}}$. Accordingly, the internal energy change ΔU for a process may be written[1]

$$\Delta U = Q + W_{\text{pressure-volume}} + W_{\text{isochoric}} \qquad \text{(closed system, no transfer of matter).} \; [\text{note 1}]$$

When a closed system receives energy as heat, this energy increases the internal energy. It is distributed between microscopic kinetic and microscopic potential energies. In general, thermodynamics does not trace this distribution. In an ideal gas all of the extra energy results in a temperature increase, as it is stored solely as microscopic kinetic energy; such heating is said to be *sensible*.

A second mechanism of change of internal energy of a closed system is the doing of work on the system, either in mechanical form by changing pressure or volume, or by other perturbations, such as directing an electric current through the system.

If the system is not closed, the third mechanism that increase the internal energy is transfer of matter into the system. This increase, ΔU_{matter} cannot be split into heat and work components. If the system is so set up physically that heat and work can be done on it by pathways separate from and independent of matter transfer, then the transfers of energy add to change the internal energy:

$$\Delta U = Q + W_{\text{pressure-volume}} + W_{\text{isochoric}} + \Delta U_{\text{matter}}$$

(separate pathway for matter transfer from heat and work transfer pathways).

If a system undergoes certain phase transformations while being heated, such as melting and vaporization, it may be observed that the temperature of the system does not change until the entire sample has completed the transformation. The energy introduced into the system while the temperature did not change is called a *latent energy*, or latent heat, in contrast to sensible heat, which is associated with temperature change.

20.2   Internal energy of the ideal gas

Thermodynamics often uses the concept of the ideal gas for teaching purposes, and as an approximation for working systems. The ideal gas is a gas of particles considered as point objects that interact only by elastic collisions and fill a volume such that their free mean path between collisions is much larger than their diameter. Such systems are approximated by the monatomic gases, helium and the other noble gases. Here the kinetic energy consists only of the translational energy of the individual atoms. Monatomic particles do not rotate or vibrate, and are not electronically excited to higher energies except at very high temperatures.

Therefore internal energy changes in an ideal gas may be described solely by changes in its kinetic energy. Kinetic energy is simply the internal energy of the perfect gas and depends entirely on its pressure, volume and thermodynamic temperature.

The internal energy of an ideal gas is proportional to its mass (number of moles) N and to its temperature T

$$U = cNT,$$

where c is the heat capacity (at constant volume) of the gas. The internal energy may be written as a function of the three extensive properties S, V, N (entropy, volume, mass) in the following way [7] [8]

$$U(S, V, N) = const \cdot e^{\frac{S}{cN}} V^{-\frac{R}{c}} N^{\frac{R+c}{c}},$$

where $const$ is an arbitrary positive constant and where R is the Universal Gas Constant. It is easily seen that U is a linearly homogeneous function of the three variables (that is, it is *extensive* in these variables), and that it is weakly convex. Knowing temperature and pressure to be the derivatives $T = \frac{\partial U}{\partial S}$, $p = -\frac{\partial U}{\partial V}$, the Ideal Gas Law $pV = RNT$ immediately follows.

20.3 Internal energy of a closed thermodynamic system

This above summation of all components of change in internal energy assume that a positive energy denotes heat added to the system or work done on the system, while a negative energy denotes work of the system on the environment.

Typically this relationship is expressed in infinitesimal terms using the differentials of each term. Only the internal energy is an exact differential. For a system undergoing only thermodynamics processes, i.e. a closed system that can exchange only heat and work, the change in the internal energy is

$$dU = \delta Q + \delta W$$

which constitutes the first law of thermodynamics.[note 1] It may be expressed in terms of other thermodynamic parameters. Each term is composed of an intensive variable (a generalized force) and its conjugate infinitesimal extensive variable (a generalized displacement).

For example, for a non-viscous fluid, the mechanical work done on the system may be related to the pressure p and volume V. The pressure is the intensive generalized force, while the volume is the extensive generalized displacement:

$$\delta W = -pdV$$

This defines the direction of work, W, to be energy flow from the working system to the surroundings, indicated by a negative term.[note 1] Taking the direction of heat transfer Q to be into the working fluid and assuming a reversible process, the heat is

$$\delta Q = TdS .$$

T is temperature
S is entropy

and the change in internal energy becomes

$$dU = TdS - pdV$$

20.3.1   Changes due to temperature and volume

The expression relating changes in internal energy to changes in temperature and volume is

$$dU = C_V\,dT + \left[T\left(\frac{\partial p}{\partial T}\right)_V - p\right]dV \ \ (1)\,.$$

This is useful if the equation of state is known.

In case of an ideal gas, we can derive that $dU = C_v\,dT$, i.e. the internal energy of an ideal gas can be written as a function that depends only on the temperature.

Proof of pressure independence for an ideal gas

The expression relating changes in internal energy to changes in temperature and volume is

$$dU = C_V\,dT + \left[T\left(\frac{\partial p}{\partial T}\right)_V - p\right]dV.$$

The equation of state is the ideal gas law

$$pV = nRT.$$

Solve for pressure:

$$p = \frac{nRT}{V}.$$

Substitute in to internal energy expression:

$$dU = C_V\,dT + \left[T\left(\frac{\partial p}{\partial T}\right)_V - \frac{nRT}{V}\right]dV.$$

Take the derivative of pressure with respect to temperature:

$$\left(\frac{\partial p}{\partial T}\right)_V = \frac{nR}{V}.$$

Replace:

$$dU = C_V\,dT + \left[\frac{nRT}{V} - \frac{nRT}{V}\right]dV.$$

And simplify:

$$dU = C_V\,dT.$$

Derivation of dU in terms of dT and dV

To express dU in terms of dT and dV, the term

$$dS = \left(\frac{\partial S}{\partial T}\right)_V dT + \left(\frac{\partial S}{\partial V}\right)_T dV$$

is substituted in the fundamental thermodynamic relation

$$dU = TdS - pdV.$$

This gives:

$$dU = T\left(\frac{\partial S}{\partial T}\right)_V dT + \left[T\left(\frac{\partial S}{\partial V}\right)_T - p\right]dV.$$

The term $T\left(\frac{\partial S}{\partial T}\right)_V$ is the heat capacity at constant volume C_V.

The partial derivative of S with respect to V can be evaluated if the equation of state is known. From the fundamental thermodynamic relation, it follows that the differential of the Helmholtz free energy A is given by:

$$dA = -SdT - pdV.$$

The symmetry of second derivatives of A with respect to T and V yields the Maxwell relation:

$$\left(\frac{\partial S}{\partial V}\right)_T = \left(\frac{\partial p}{\partial T}\right)_V.$$

This gives the expression above.

20.3.2 Changes due to temperature and pressure

When dealing with fluids or solids, an expression in terms of the temperature and pressure is usually more useful:

$$dU = (C_p - \alpha pV)\,dT + (\beta_T p - \alpha T)V\,dp$$

where it is assumed that the heat capacity at constant pressure is related to the heat capacity at constant volume according to:

$$C_p = C_V + VT\frac{\alpha^2}{\beta_T}$$

Derivation of dU in terms of dT and dP

The partial derivative of the pressure with respect to temperature at constant volume can be expressed in terms of the coefficient of thermal expansion

$$\alpha = \frac{1}{V}\left(\frac{\partial V}{\partial T}\right)_p$$

and the isothermal compressibility

$$\beta_T \equiv -\frac{1}{V} \left(\frac{\partial V}{\partial p} \right)_T$$

by writing:

$$dV = \left(\frac{\partial V}{\partial p} \right)_T dp + \left(\frac{\partial V}{\partial T} \right)_p dT = V \left(\alpha dT - \beta_T dp \right) \quad (2)$$

and equating dV to zero and solving for the ratio dp/dT. This gives:

$$\left(\frac{\partial p}{\partial T} \right)_V = -\frac{\left(\frac{\partial V}{\partial T} \right)_p}{\left(\frac{\partial V}{\partial p} \right)_T} = \frac{\alpha}{\beta_T} \quad (3)$$

Substituting (2) and (3) in (1) gives the above expression.

20.3.3   Changes due to volume at constant temperature

The internal pressure is defined as a partial derivative of the internal energy with respect to the volume at constant temperature:

$$\pi_T = \left(\frac{\partial U}{\partial V} \right)_T$$

20.4   Internal energy of multi-component systems

In addition to including the entropy S and volume V terms in the internal energy, a system is often described also in terms of the number of particles or chemical species it contains:

$$U = U(S, V, N_1, \ldots, N_n)$$

where N_j are the molar amounts of constituents of type j in the system. The internal energy is an extensive function of the extensive variables S, V, and the amounts N_j, the internal energy may be written as a linearly homogeneous function of first degree:

$$U(\alpha S, \alpha V, \alpha N_1, \alpha N_2, \ldots) = \alpha U(S, V, N_1, N_2, \ldots)$$

where α is a factor describing the growth of the system. The differential internal energy may be written as

$$dU = \frac{\partial U}{\partial S} dS + \frac{\partial U}{\partial V} dV + \sum_i \frac{\partial U}{\partial N_i} dN_i = T \, dS - p \, dV + \sum_i \mu_i dN_i$$

which shows (or defines) temperature T to be the partial derivative of U with respect to entropy S and pressure p to be the negative of the similar derivative with respect to volume V

$$T = \frac{\partial U}{\partial S},$$

$$p = -\frac{\partial U}{\partial V}.$$

and where the coefficients μ_i are the chemical potentials for the components of type i in the system. The chemical potentials are defined as the partial derivatives of the energy with respect to the variations in composition:

$$\mu_i = \left(\frac{\partial U}{\partial N_i} \right)_{S,V,N_{j \neq i}}$$

As conjugate variables to the composition $\{N_j\}$, the chemical potentials are intensive properties, intrinsically characteristic of the qualitative nature of the system, and not proportional to its extent. Because of the extensive nature of U and its independent variables, using Euler's homogeneous function theorem, the differential dU may be integrated and yields an expression for the internal energy:

$$U = TS - pV + \sum_i \mu_i N_i$$

The sum over the composition of the system is the Gibbs free energy:

$$G = \sum_i \mu_i N_i$$

that arises from changing the composition of the system at constant temperature and pressure. For a single component system, the chemical potential equals the Gibbs energy per amount of substance, i.e. particles or moles according to the original definition of the unit for $\{N_j\}$.

20.5   Internal energy in an elastic medium

For an elastic medium the mechanical energy term of the internal energy must be replaced by the more general expression involving the stress σ_{ij} and strain ε_{ij}. The infinitesimal statement is:

$$dU = TdS + V\sigma_{ij}d\varepsilon_{ij}$$

where Einstein notation has been used for the tensors, in which there is a summation over all repeated indices in the product term. The Euler theorem yields for the internal energy:[9]

$$U = TS + \frac{1}{2}\sigma_{ij}\varepsilon_{ij}$$

For a linearly elastic material, the stress is related to the strain by:

$$\sigma_{ij} = C_{ijkl}\varepsilon_{kl}$$

where the C_{ijkl} are the components of the 4th-rank elastic constant tensor of the medium.

20.6   Computational methods

The path integral Monte Carlo method is a numerical approach for determining the values of the internal energy, based on quantum dynamical principles.

20.7   History

James Joule studied the relationship between heat, work, and temperature. He observed that if he did mechanical work on a fluid, such as water, by agitating the fluid, its temperature increased. He proposed that the mechanical work he was doing on the system was converted to *thermal energy*. Specifically, he found that 4185.5 joules of energy were needed to raise the temperature of a kilogram of water by one degree Celsius.

20.8   Notes

[1] In this article we choose the sign convention of the mechanical work as typically defined in chemistry, which is different from the convention used in physics. In chemistry, work performed by the system against the environment, e.g., a system expansion, is negative, while in physics this is taken to be positive.

20.9   See also

- Calorimetry

- Enthalpy

- Exergy

- Gibbs free energy

- Helmholtz free energy

- Thermodynamic equations

- Thermodynamic potentials

20.10   References

[1] Peter Atkins, Julio de Paula (2006). *Physical Chemistry* (8 ed.). Oxford University Press. p. 9.

[2] Callen, H.B. (1960/1985). *Thermodynamics and an Introduction to Thermostatistics*, (first edition 1960), second edition 1985, John Wiley & Sons, New York, ISBN 0–471–86256–8. Chapter 5.

[3] Münster, A. (1970). *Classical Thermodynamics*, translated by E.S. Halberstadt, Wiley–Interscience, London, ISBN 0-471-62430-6, p. 6.

[4] Tschoegl, N.W. (2000). *Fundamentals of Equilibrium and Steady-State Thermodynamics*, Elsevier, Amsterdam, ISBN 0-444-50426-5, p. 17.

[5] I. Klotz, R. Rosenberg. *Chemical Thermodynamics - Basic Concepts and Methods*, 7th ed., Wiley (2008), p.39

[6] Thermal energy – Hyperphysics

[7] van Gool, W.; Bruggink, J.J.C. (Eds) (1985). *Energy and time in the economic and physical sciences*. North-Holland. pp. 41–56. ISBN 0444877487.

[8] Grubbström, Robert W. (2007). "An Attempt to Introduce Dynamics Into Generalised Exergy Considerations". *Applied Energy* **84**: 701–718. doi:10.1016/j.apenergy.2007.01.003.

[9] Landau & Lifshitz 1986

20.11 Bibliography

- Alberty, R. A. (2001). "Use of Legendre transforms in chemical thermodynamics" (PDF). *Pure Appl. Chem.* **73** (8): 1349–1380. doi:10.1351/pac200173081349.

- Lewis, Gilbert Newton; Randall, Merle: Revised by Pitzer, Kenneth S. & Brewer, Leo (1961). *Thermodynamics* (2nd ed.). New York, NY USA: McGraw-Hill Book Co. ISBN 0-07-113809-9.

- Landau, L. D.; Lifshitz, E. M. (1986). *Theory of Elasticity (Course of Theoretical Physics Volume 7).* (Translated from Russian by J.B. Sykes and W.H. Reid) (Third ed.). Boston, MA: Butterworth Heinemann. ISBN 0-7506-2633-X.

Chapter 21

Thermodynamic system

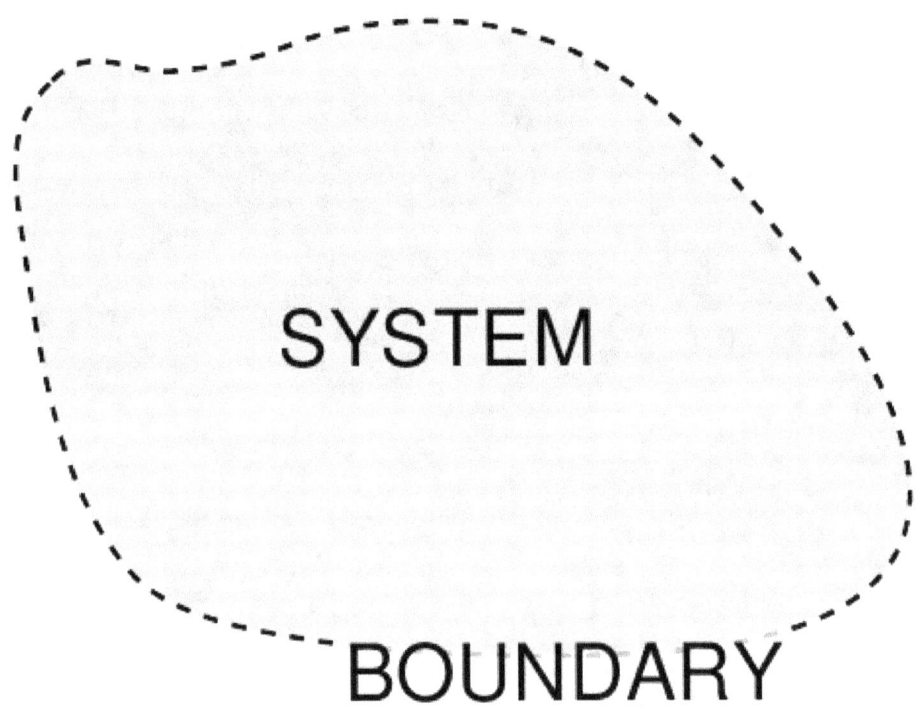

A **thermodynamic system** is the content of a macroscopic volume in space, along with its walls and surroundings; it undergoes thermodynamic processes according to the principles of thermodynamics. A physical system qualifies as a thermodynamic system only if it can be adequately described by thermodynamic variables such as temperature, entropy, internal energy and pressure.

The thermodynamic state of a thermodynamic system is its internal state as specified by its state variables. A thermodynamic account also requires a special kind of function called a state function. For example, if the state variables are internal energy, volume and mole amounts, the needed further state function is entropy. These quantities are inter-related by one or more functional relationships called equations of state. Thermodynamics defines the restrictions on the possible equations of state imposed by the laws of thermodynamics through that further function of state.

The system is delimited by walls or boundaries, either actual or notional, across which conserved (such as matter and energy) or unconserved (such as entropy) quantities can pass into and out of the system. The space outside the thermodynamic system is known as the *surroundings*, a *reservoir*, or the *environment*. The properties of the walls determine what transfers can occur. A wall that allows transfer of a quantity is said to be permeable to it, and a thermodynamic system is classified by the permeabilities of its several walls. A transfer between system and surroundings can arise by contact, such as conduction of heat, or by long-range forces such as an electric field in the surroundings.

A system with walls that prevent all transfers is said to be isolated. This is an idealized conception, because in practice some transfer is always possible, for example by gravitational forces. It is an axiom of thermodynamics that an isolated system eventually reaches internal thermodynamic equilibrium, when its state no longer changes with time.

According to the permeabilities of its walls, a system that is not isolated can be in thermodynamic equilibrium with its surroundings, or else may be in a state that is constant or precisely cyclically changing in time - a steady state that is far from equilibrium. Classical thermodynamics considers only states of thermodynamic systems in equilibrium that are either constant or precisely cycling in time.

The walls of a closed system allow transfer of energy as heat and as work, but not of matter, between it and its surroundings. The walls of an *open system* allow transfer both of matter and of energy.[1][2][3][4][5][6][7] This scheme of definition of terms is not uniformly used, though it is convenient for some purposes. In particular, some writers use 'closed system' where 'isolated system' is here used.[8][9]

In 1824 Sadi Carnot described a thermodynamic system as the **working substance** (such as the volume of steam) of any heat engine under study. The very existence of such thermodynamic systems may be considered a fundamental postulate of equilibrium thermodynamics, though it is not listed as a numbered law.[10][11] According to Bailyn, the commonly rehearsed statement of the zeroth law of thermodynamics is a consequence of this fundamental postulate.[12]

In equilibrium thermodynamics the state variables do not include fluxes because in a state of thermodynamic equilibrium all fluxes have zero values by definition. Equilibrium thermodynamic processes may of course involve fluxes but these must have ceased by the time a thermodynamic process or operation is complete bringing a system to its eventual thermodynamic state. Non-equilibrium thermodynamics allows its state variables to include non-zero fluxes, that describe transfers of matter or energy or entropy between a system and its surroundings.[13]

21.1   Overview

Thermodynamics describes the macroscopic physics of matter and energy, especially including heat transfer, by using the concept of the *thermodynamic system*, a region of the universe that is under study, specified by thermodynamic state variables, together with the kinds of transfer that may occur between it and its surroundings, as determined by the physical properties of the walls of the system.

An example system is the system of hot liquid water and solid table salt in a sealed, insulated test tube held in a vacuum (the surroundings). The test tube constantly loses heat in the form of black-body radiation, but the heat loss progresses very slowly. If there is another process going on in the test tube, for example the dissolution of the salt crystals, it probably occurs so quickly that any heat lost to the test tube during that time can be neglected. Thermodynamics in general does not measure time, but it does sometimes accept limitations on the time frame of a process.

21.2 History

The first to develop the concept of a thermodynamic system was the French physicist Sadi Carnot whose 1824 *Reflections on the Motive Power of Fire* studied what he called the *working substance*, e.g., typically a body of water vapor, in steam engines, in regards to the system's ability to do work when heat is applied to it. The working substance could be put in contact with either a heat reservoir (a boiler), a cold reservoir (a stream of cold water), or a piston (to which the working body could do work by pushing on it). In 1850, the German physicist Rudolf Clausius generalized this picture to include the concept of the surroundings, and began referring to the system as a "working body." In his 1850 manuscript *On the Motive Power of Fire*, Clausius wrote:

The article Carnot heat engine shows the original piston-and-cylinder diagram used by Carnot in discussing his ideal engine; below, we see the Carnot engine as is typically modeled in current use:

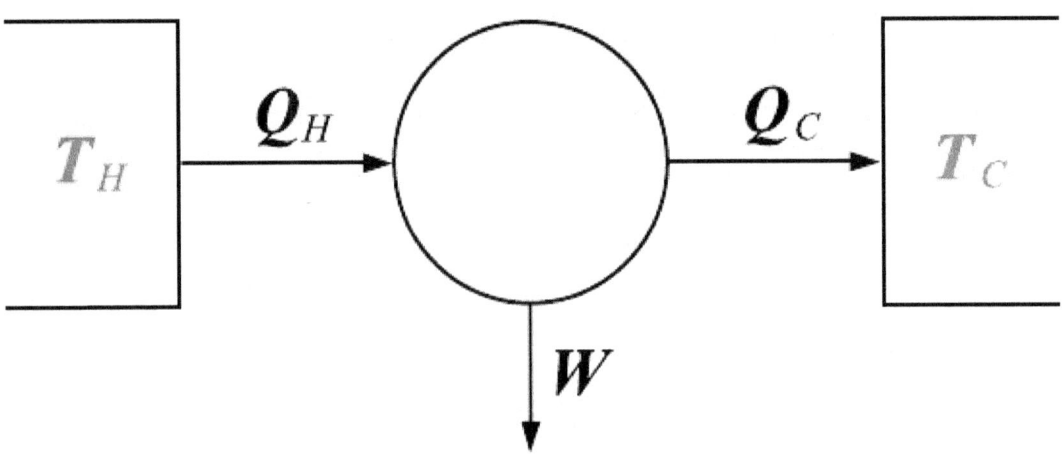

Carnot engine diagram (modern) - where heat flows from a high temperature TH furnace through the fluid of the "working body" (working substance) and into the cold sink TC, thus forcing the working substance to do mechanical work W on the surroundings, via cycles of contractions and expansions.

In the diagram shown, the "working body" (system), a term introduced by Clausius in 1850, can be any fluid or vapor body through which heat Q can be introduced or transmitted through to produce work. In 1824, Sadi Carnot, in his famous paper *Reflections on the Motive Power of Fire*, had postulated that the fluid body could be any substance capable of expansion, such as vapor of water, vapor of alcohol, vapor of mercury, a permanent gas, or air, etc. Though, in these early years, engines came in a number of configurations, typically QH was supplied by a boiler, wherein water boiled over a furnace; QC was typically a stream of cold flowing water in the form of a condenser located on a separate part of the engine. The output work W was the movement of the piston as it turned a crank-arm, which typically turned a pulley to lift water out of flooded salt mines. Carnot defined work as "weight lifted through a height."

21.3 Walls

A system is enclosed by walls that bound it and connect it to its surroundings.[14][15][16][17][18][19] Often a wall restricts passage across it by some form of matter or energy, making the connection indirect. Sometimes a wall is no more than an imaginary two-dimensional closed surface through which the connection to the surroundings is direct. Topologically, it is often considered nearly or piecewise smoothly homeomorphic with a two-sphere (ordinary sphere like a surface that forms the boundary of a ball in three dimensions), because a system is often considered simply connected.

A wall can be fixed (e.g. a constant volume reactor) or moveable (e.g. a piston). For example, in a reciprocating engine, a fixed wall means the piston is locked at its position; then, a constant volume process may occur. In that same engine, a piston may be unlocked and allowed to move in and out. Ideally, a wall may be declared adiabatic, diathermal, impermeable, permeable, or semi-permeable. Actual physical materials that provide walls with such idealized properties are not always readily available.

Anything that passes across the boundary and effects a change in the contents of the system must be accounted for in an appropriate balance equation. The volume can be the region surrounding a single atom resonating energy, such as Max Planck defined in 1900; it can be a body of steam or air in a steam engine, such as Sadi Carnot defined in 1824. It could also be just one nuclide (i.e. a system of quarks) as hypothesized in quantum thermodynamics.

21.4   Surroundings

See also: Environment (systems)

The system is the part of the universe being studied, while the *surroundings* is the remainder of the universe that lies outside the boundaries of the system. It is also known as the *environment*, and the *reservoir*. Depending on the type of system, it may interact with the system by exchanging mass, energy (including heat and work), momentum, electric charge, or other conserved properties. The environment is ignored in analysis of the system, except in regards to these interactions.

21.5   Open system

In an open system, matter may flow in and out of some segments of the system boundaries. There may be other segments of the system boundaries that pass heat or work but not matter. Respective account is kept of the transfers of energy across those and any other several boundary segments.

21.5.1   Flow process

The region of space enclosed by open system boundaries is usually called a control volume. It may or may not correspond to physical walls. It is convenient to define the shape of the control volume so that all flow of matter, in or out, occurs perpendicular to its surface. One may consider a process in which the matter flowing into and out of the system is chemically homogeneous.[20] Then the inflowing matter performs work as if it were driving a piston of fluid into the system. Also, the system performs work as if it were driving out a piston of fluid. Through the system walls that do not pass matter, heat (δQ) and work (δW) transfers may be defined, including shaft work.

Classical thermodynamics considers processes for a system that is initially and finally in its own internal state of thermodynamic equilibrium, with no flow. This is feasible also under some restrictions, if the system is a mass of fluid flowing at a uniform rate. Then for many purposes a process, called a flow process, may be considered in accord with classical thermodynamics as if the classical rule of no flow were effective.[21] For the present introductory account, it is supposed that the kinetic energy of flow, and the potential energy of elevation in the gravity field, do not change, and that the walls, other than the matter inlet and outlet, are rigid and motionless.

Under these conditions, the first law of thermodynamics for a flow process states: *the increase in the internal energy of a system is equal to the amount of energy added to the system by matter flowing in and by heating, minus the amount lost by matter flowing out and in the form of work done by the system.* Under these conditions, the first law for a flow process is written:

$$dU = dU_{in} + \delta Q - dU_{out} - \delta W$$

where U_{in} and U_{out} respectively denote the average internal energy entering and leaving the system with the flowing matter.

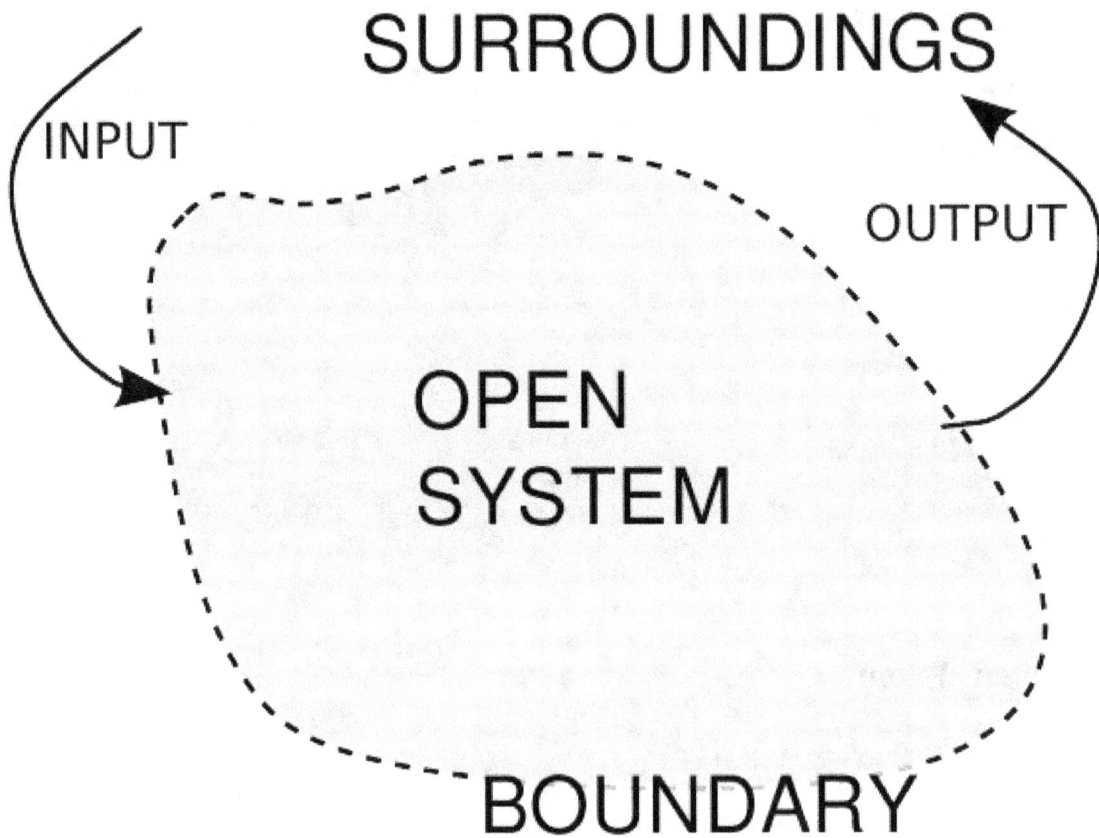

Generic open system scheme. Exchanges of matter or energy with system's surroundings are represented by input and output flows.

There are then two types of work performed: 'flow work' described above, which is performed on the fluid in the control volume (this is also often called 'PV work'), and 'shaft work', which may be performed by the fluid in the control volume on some mechanical device with a shaft. These two types of work are expressed in the equation:

$$\delta W = d(P_{out}V_{out}) - d(P_{in}V_{in}) + \delta W_{shaft}$$

Substitution into the equation above for the control volume cv yields:

$$dU_{cv} = dU_{in} + d(P_{in}V_{in}) - dU_{out} - d(P_{out}V_{out}) + \delta Q - \delta W_{shaft}$$

The definition of enthalpy, $H = U + PV$, permits us to use this thermodynamic potential to account jointly for internal energy U and PV work in fluids for a flow process:

$$dU_{cv} = dH_{in} - dH_{out} + \delta Q - \delta W_{shaft}$$

During steady-state operation of a device (*see turbine, pump, and engine*), any system property within the control volume is independent of time. Therefore, the internal energy of the system enclosed by the control volume remains constant, which implies that dU_{cv} in the expression above may be set equal to zero. This yields a useful expression for the power generation or requirement for these devices with chemical homogeneity in the absence of chemical reactions:

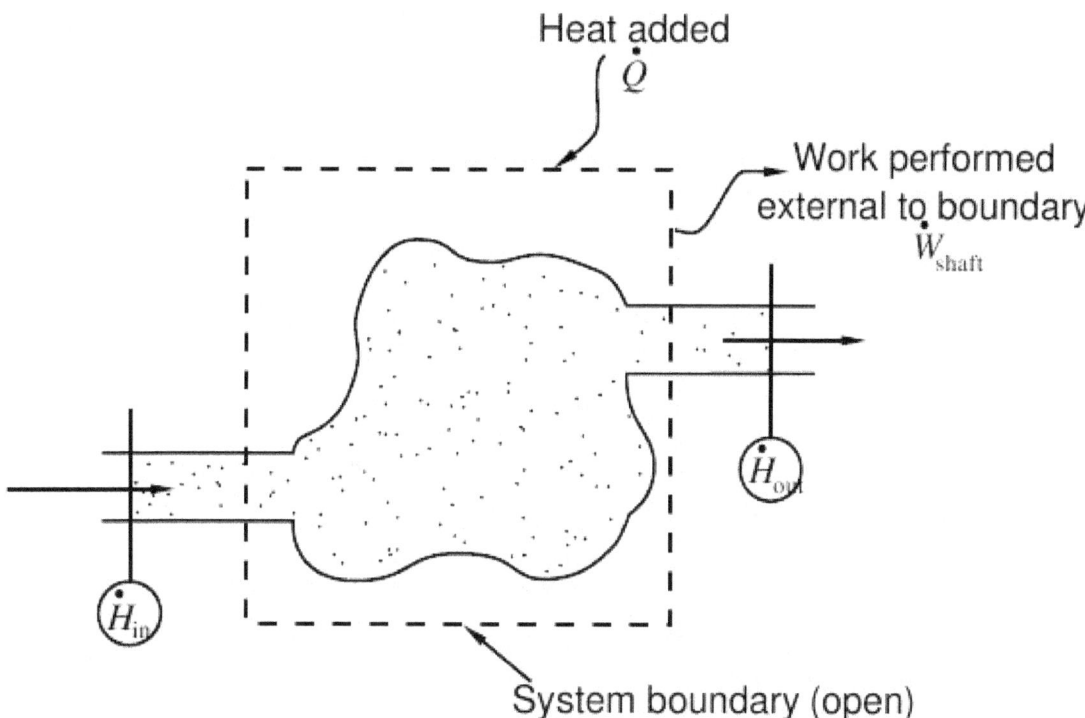

During steady, continuous operation, an energy balance applied to an open system equates shaft work performed by the system to heat added plus net enthalpy added.

$$\frac{\delta W_{shaft}}{dt} = \frac{dH_{in}}{dt} - \frac{dH_{out}}{dt} + \frac{\delta Q}{dt}$$

This expression is described by the diagram above.

21.5.2   Selective transfer of matter

For a thermodynamic process, the precise physical properties of the walls and surroundings of the system are important, because they determine the possible processes.

An open system has one or several walls that allow transfer of matter. To account for the internal energy of the open system, this requires energy transfer terms in addition to those for heat and work. It also leads to the idea of the chemical potential.

A wall selectively permeable only to a pure substance can put the system in diffusive contact with a reservoir of that pure substance in the surroundings. Then a process is possible in which that pure substance is transferred between system and surroundings. Also, across that wall a contact equilibrium with respect to that substance is possible. By suitable thermodynamic operations, the pure substance reservoir can be dealt with as a closed system. Its internal energy and its entropy can be determined as functions of its temperature, pressure, and mole number.

A thermodynamic operation can render impermeable to matter all system walls other than the contact equilibrium wall for that substance. This allows the definition of an intensive state variable, with respect to a reference state of the surroundings, for that substance. The intensive variable is called the chemical potential; for component substance i it is usually denoted μi. The corresponding extensive variable can be the number of moles Ni of the component substance in the system.

For a contact equilibrium across a wall permeable to a substance, the chemical potentials of the substance must be same on either side of the wall. This is part of the nature of thermodynamic equilibrium, and may be regarded as related to the zeroth law of thermodynamics.[22]

21.6   Closed system

Main article: Closed system § In thermodynamics

In a closed system, no mass may be transferred in or out of the system boundaries. The system always contains the same amount of matter, but heat and work can be exchanged across the boundary of the system. Whether a system can exchange heat, work, or both is dependent on the property of its boundary.

- Adiabatic boundary – not allowing any heat exchange: A thermally isolated system

- Rigid boundary – not allowing exchange of work: A mechanically isolated system

One example is fluid being compressed by a piston in a cylinder. Another example of a closed system is a bomb calorimeter, a type of constant-volume calorimeter used in measuring the heat of combustion of a particular reaction. Electrical energy travels across the boundary to produce a spark between the electrodes and initiates combustion. Heat transfer occurs across the boundary after combustion but no mass transfer takes place either way.

Beginning with the first law of thermodynamics for an open system, this is expressed as:

$$\Delta U = Q - W + m_i(h + \frac{1}{2}v^2 + gz)_i - m_e(h + \frac{1}{2}v^2 + gz)_e$$

where U is internal energy, Q is the heat added to the system, W is the work done by the system, and since no mass is transferred in or out of the system, both expressions involving mass flow are zero and the first law of thermodynamics for a closed system is derived. The first law of thermodynamics for a closed system states that the increase of internal energy of the system equals the amount of heat added to the system minus the work done by the system. For infinitesimal changes the first law for closed systems is stated by:

$$dU = \delta Q - \delta W.$$

If the work is due to a volume expansion by dV at a pressure P than:

$$\delta W = PdV.$$

For a homogeneous system, in which only reversible processes can take place, the second law of thermodynamics reads:

$$\delta Q = TdS$$

where T is the absolute temperature and S is the entropy of the system. With these relations the fundamental thermodynamic relationship, used to compute changes in internal energy, is expressed as:

$$dU = TdS - PdV.$$

For a simple system, with only one type of particle (atom or molecule), a closed system amounts to a constant number of particles. However, for systems undergoing a chemical reaction, there may be all sorts of molecules being generated and destroyed by the reaction process. In this case, the fact that the system is closed is expressed by stating that the total number of each elemental atom is conserved, no matter what kind of molecule it may be a part of. Mathematically:

$$\sum_{j=1}^{m} a_{ij} N_j = b_i^0$$

where N_j is the number of j-type molecules, a_{ij} is the number of atoms of element i in molecule j and b_i^0 is the total number of atoms of element i in the system, which remains constant, since the system is closed. There is one such equation for each element in the system.

21.7 Isolated system

Main article: Isolated system

An isolated system is more restrictive than a closed system as it does not interact with its surroundings in any way. Mass and energy remains constant within the system, and no energy or mass transfer takes place across the boundary. As time passes in an isolated system, internal differences in the system tend to even out and pressures and temperatures tend to equalize, as do density differences. A system in which all equalizing processes have gone practically to completion is in a state of thermodynamic equilibrium.

Truly isolated physical systems do not exist in reality (except perhaps for the universe as a whole), because, for example, there is always gravity between a system with mass and masses elsewhere.[23][24][25][26][27] However, real systems may behave nearly as an isolated system for finite (possibly very long) times. The concept of an isolated system can serve as a useful model approximating many real-world situations. It is an acceptable idealization used in constructing mathematical models of certain natural phenomena.

In the attempt to justify the postulate of entropy increase in the second law of thermodynamics, Boltzmann's H-theorem used equations, which assumed that a system (for example, a gas) was isolated. That is all the mechanical degrees of freedom could be specified, treating the walls simply as mirror boundary conditions. This inevitably led to Loschmidt's paradox. However, if the stochastic behavior of the molecules in actual walls is considered, along with the randomizing effect of the ambient, background thermal radiation, Boltzmann's assumption of molecular chaos can be justified.

The second law of thermodynamics for isolated systems states that the entropy of an isolated system not in equilibrium tends to increase over time, approaching maximum value at equilibrium. Overall, in an isolated system, the internal energy is constant and the entropy can never decrease. A *closed* system's entropy can decrease e.g. when heat is extracted from the system.

It is important to note that isolated systems are not equivalent to closed systems. Closed systems cannot exchange matter with the surroundings, but can exchange energy. Isolated systems can exchange neither matter nor energy with their surroundings, and as such are only theoretical and do not exist in reality (except, possibly, the entire universe).

It is worth noting that 'closed system' is often used in thermodynamics discussions when 'isolated system' would be correct - i.e. there is an assumption that energy does not enter or leave the system.

21.8 Mechanically isolated system

Main article: Mechanically isolated system

A mechanically isolated system can exchange no work energy with its environment, but may exchange heat energy and/or mass with its environment. The internal energy of a mechanically isolated system may therefore change due to the exchange of heat energy and mass. For a simple system, mechanical isolation is equivalent to constant volume and any process which occurs in such a simple system is said to be isochoric.

21.9   Systems in equilibrium

At thermodynamic equilibrium, a system's properties are, by definition, unchanging in time. Systems in equilibrium are much simpler and easier to understand than systems not in equilibrium. In some cases, when analyzing a thermodynamic process, one can assume that each intermediate state in the process is at equilibrium. This considerably simplifies the analysis.

In isolated systems it is consistently observed that as time goes on internal rearrangements diminish and stable conditions are approached. Pressures and temperatures tend to equalize, and matter arranges itself into one or a few relatively homogeneous phases. A system in which all processes of change have gone practically to completion is considered in a state of thermodynamic equilibrium. The thermodynamic properties of a system in equilibrium are unchanging in time. Equilibrium system states are much easier to describe in a deterministic manner than non-equilibrium states.

For a process to be reversible, each step in the process must be reversible. For a step in a process to be reversible, the system must be in equilibrium throughout the step. That ideal cannot be accomplished in practice because no step can be taken without perturbing the system from equilibrium, but the ideal can be approached by making changes slowly.

21.10   See also

- Physical system

21.11   References

[1] Prigogine, I., Defay, R. (1950/1954). *Chemical Thermodynamics*, Longmans, Green & Co, London, p. 66.

[2] Tisza, L. (1966). *Generalized Thermodynamics*, M.I.T Press, Cambridge MA, pp. 112–113.

[3] Guggenheim, E.A. (1949/1967). *Thermodynamics. An Advanced Treatment for Chemists and Physicists*, (1st edition 1949) 5th edition 1967, North-Holland, Amsterdam, p. 14.

[4] Münster, A. (1970). *Classical Thermodynamics*, translated by E.S. Halberstadt, Wiley–Interscience, London, pp. 6–7.

[5] Haase, R. (1971). Survey of Fundamental Laws, chapter 1 of *Thermodynamics*, pages 1–97 of volume 1, ed. W. Jost, of *Physical Chemistry. An Advanced Treatise*, ed. H. Eyring, D. Henderson, W. Jost, Academic Press, New York, lcn 73–117081, p. 3.

[6] Tschoegl, N.W. (2000). *Fundamentals of Equilibrium and Steady-State Thermodynamics*, Elsevier, Amsterdam, ISBN 0-444-50426-5, p. 5.

[7] Silbey, R.J., Alberty, R.A., Bawendi, M.G. (1955/2005). *Physical Chemistry*, fourth edition, Wiley, Hoboken NJ, p. 4.

[8] Callen, H.B. (1960/1985). *Thermodynamics and an Introduction to Thermostatistics*, (1st edition 1960) 2nd edition 1985, Wiley, New York, ISBN 0-471-86256-8, p. 17.

[9] ter Haar, D., Wergeland, H. (1966). *Elements of Thermodynamics*, Addison-Wesley Publishing, Reading MA, p. 43.

[10] Bailyn, M. (1994). *A Survey of Thermodynamics*, American Institute of Physics Press, New York, ISBN 0-88318-797-3, p. 20.

[11] Tisza, L. (1966). *Generalized Thermodynamics*, M.I.T Press, Cambridge MA, p. 119.

[12] Bailyn, M. (1994). *A Survey of Thermodynamics*, American Institute of Physics Press, New York, ISBN 0-88318-797-3, p. 22.

[13] Eu, B.C. (2002). *Generalized Thermodynamics. The Thermodynamics of Irreversible Processes and Generalized Hydrodynamics*, Kluwer Academic Publishers, Dordrecht, ISBN 1-4020-0788-4.

[14] Born, M. (1949). *Natural Philosophy of Cause and Chance*, Oxford University Press, London, p.44

[15] Tisza, L. (1966), pp. 109, 112.

[16] Haase, R. (1971), p. 7.

[17] Adkins, C.J. (1968/1975), p. 4

[18] Callen, H.B. (1960/1985), pp. 15, 17.

[19] Tschoegl, N.W. (2000), p. 5.

[20] Shavit, A., Gutfinger, C. (1995). *Thermodynamics. From Concepts to Applications*, Prentice Hall, London, ISBN 0-13-288267-1, Chapter 6.

[21] Adkins, C.J. (1968/1983). *Equilibrium Thermodynamics*, third edition, Cambridge University Press, Cambridge UK, ISBN 0-521-25445-0, pp. 46–47.

[22] Bailyn, M. (1994). *A Survey of Thermodynamics*, American Institute of Physics Press, New York, ISBN 0-88318-797-3, pp. 19–23.

[23] I.M.Kolesnikov; V.A.Vinokurov; S.I.Kolesnikov (2001). *Thermodynamics of Spontaneous and Non-Spontaneous Processes*. Nova science Publishers. p. 136. ISBN 1-56072-904-X.

[24] "A System and Its Surroundings". *ChemWiki*. University of California - Davis. Retrieved May 2012.

[25] "Hyperphysics". The Department of Physics and Astronomy of Georgia State University. Retrieved May 2012.

[26] Bryan Sanctuary. "Open, Closed and Isolated Systems in Physical Chemistry.". *Foundations of Quantum Mechanics and Physical Chemistry*. McGill University (Montreal). Retrieved May 2012.

[27] *Material and Energy Balances for Engineers and Environmentalists* (PDF). Imperial College Press. p. 7. Retrieved May 2012.

- Abbott, M.M.; van Hess, H.G. (1989). *Thermodynamics with Chemical Applications* (2nd ed.). McGraw Hill.

- Callen, H.B. (1960/1985). *Thermodynamics and an Introduction to Thermostatistics*, (1st edition 1960) 2nd edition 1985, Wiley, New York, ISBN 0-471-86256-8.

- Halliday, David; Resnick, Robert; Walker, Jearl (2008). *Fundamentals of Physics* (8th ed.). Wiley.

- Moran, Michael J.; Shapiro, Howard N. (2008). *Fundamentals of Engineering Thermodynamics* (6th ed.). Wiley.

21.12 External links

- http://www.hasdeu.bz.edu.ro/softuri/fizica/mariana/Termodinamica/1stLaw_1/close.htm

- https://www.e-education.psu.edu/png520/m14_p4.html

Chapter 22

Disgregation

In the history of thermodynamics, **disgregation** was defined in 1862 by Rudolf Clausius as the magnitude of the degree in which the molecules of a body are separated from each other.[1] This term was modeled on certain passages in French physicist Sadi Carnot's 1824 paper *On the Motive Power of Fire* that characterized the "transformations" of "working substances" (particles of a thermodynamic system) of an engine cycle, namely **"mode of aggregation"**, which was a precursor to the concept of entropy, which Clausius coined in 1865. It was also a precursor to that of Ludwig Boltzmann's 1870s theories of entropy and order and disorder.

22.1   Overview

In 1824, French physicist Sadi Carnot assumed that heat, like a substance, cannot be diminished in quantity and that it cannot increase. Specifically, he states that in a complete engine cycle 'that when a body has experienced any changes, and when after a certain number of transformations it returns to precisely its original state, that is, to that state considered in respect to density, to temperature, to mode of aggregation, let us suppose, I say that this body is found to contain the same quantity of heat that it contained at first, or else that the quantities of heat absorbed or set free in these different transformations are exactly compensated.' Furthermore, he states that 'this fact has never been called into question' and 'to deny this would overthrow the whole theory of heat to which it serves as a basis.' This famous sentence, which Clausius spent fifteen years thinking about, marks the start of thermodynamics and signals the slow transition from the older caloric theory to the newer kinetic theory, in which heat is a type of energy in transit

In 1862, Clausius defined what is now known as *entropy* or the energetic effects related to irreversibility as the "equivalence-values of transformations" in a thermodynamic cycle. Clausius then signifies the difference between "reversible" (ideal) and "irreversible" (real) processes:

22.1.1   Equivalence-values of transformations

He then states what he calls the "theorem respecting the equivalence-values of the transformations" or what is now known as the six law of thermodynamics, as such:

Quantitatively, Clausius states the mathematical expression for this theorem is as follows. Let dQ be an element of the heat given up by the body to any reservoir of heat during its own changes, heat which it may absorb from a reservoir being here reckoned as negative, and T the absolute temperature of the body at the moment of giving up this heat, then the equation:

$$\int \frac{dQ}{T} = 0$$

must be true for every reversible cyclical process, and the relation:

$$\int \frac{dQ}{T} \geq 0$$

must hold good for every cyclical process which is in any way possible.

22.1.2   Verbal justifications

Clausius then points out the inherent difficulty in the mental comprehension of this law by stating: "although the necessity of this theorem admits of strict mathematical proof if we start from the fundamental proposition above quoted, it thereby nevertheless retains an abstract form, in which it is with difficulty embraced by the mind, and we feel compelled to seek for the precise physical cause, of which this theorem is a consequence." The justification for this law, according to Clausius, is based on the following argument:

To elaborate on this, Clausius states that in all cases in which heat can perform mechanical work, these processes always admit to being reduced to the "alteration in some way or another of the arrangement of the constituent parts of the body." To exemplify this, Clausius moves into a discussion of change of state of a body, i.e. solid, liquid, gas. For instance, he states, "when bodies are expanded by heat, their molecules being thus separated from each other: in this case the mutual attractions of the molecules on the one hand, and external opposing forces on the other, insofar as any such are in operation, have to be overcome. Again, the state of aggregation of bodies is altered by heat, solid bodies rendered liquid, and both solid and liquid bodies being rendered aeriform: here likewise internal forces, and in general external forces also, have to be overcome."

22.1.3   Definition of term

Clausius then goes on to introduce the term "disgregation":

22.1.4   Ice melting

Clausius then discusses the example of the *melting of ice*, a classic example which is used in almost all chemistry books to this day, and shows how we might represent the mechanical equivalent of work related to this energetic change mathematically:

22.1.5   Measurements of disgregation

As it is difficult to obtain direct measures of the interior forces that the molecules of the body exert on each other, Clausius states that an indirect way to obtain quantitative measures of what is now called entropy is to calculate the work done in overcoming internal forces:

> In the case of the interior forces, it would accordingly be difficult—even if we did not want to measure them, but only to represent them mathematically—to find a fitting expression for them which would admit of a simple determination of the magnitude. This difficulty, however, disappears if we take into calculation, not the forces themselves, but the mechanical work which, in any change of arrangement, is required to overcome them. The expressions for the quantities of work are simpler than those for the corresponding forces; for the quantities of work can be all expressed, without further secondary statements, by the numbers which, having reference to the same unit, can be added together, or subtracted from one another, however various the forces may be to which they refer.

> It is therefore convenient to alter the form of the above law by introducing, instead of the forces themselves, the work done in overcoming them. In this form it reads as follows:

This description is an early formulation of the concept of entropy.

Ice melting on a beach in Iceland

22.2   References

[1] Clausius, Rudolf. (1862). "On the Application of the Theorem of the Equivalence of Transformations to Interior Work." Communicated to the Naturforschende Gesellschaft of Zurich, Jan. 27th, 1862; published in the Viertaljahrschrift of this Society, vol. vii. P. 48; in Poggendorff's Annalen, May 1862, vol. cxvi. p. 73; in the Philosophical Magazine, S. 4. vol. xxiv. pp. 81, 201; and in the Journal des Mathematiques of Paris, S. 2. vol. vii. P. 209.

Chapter 23

Boltzmann's entropy formula

Boltzmann's equation—carved on his gravestone.[1]

In statistical mechanics, **Boltzmann's equation** is a probability equation relating the entropy S of an ideal gas to the quantity W, which is the number of microstates corresponding to a given macrostate:

$$S = k_B \ln W \,(1)$$

where kB is the Boltzmann's constant, (also written with k) which is equal to 1.38065×10^{-23} J/K.

In short, the Boltzmann formula shows the relationship between entropy and the number of ways the atoms or molecules of a thermodynamic system can be arranged. In 1934, Swiss physical chemist Werner Kuhn successfully derived a thermal equation of state for rubber molecules using Boltzmann's formula, which has since come to be known as the entropy model of rubber.

23.1 History

The equation was originally formulated by Ludwig Boltzmann between 1872 to 1875, but later put into its current form by Max Planck in about 1900.[2][3] To quote Planck, "the logarithmic connection between entropy and probability was first stated by L. Boltzmann in his kinetic theory of gases".

The value of W was originally intended to be proportional to the *Wahrscheinlichkeit* (the German word for probability) of a macroscopic state for some probability distribution of possible microstates—the collection of (unobservable) "ways" the (observable) thermodynamic state of a system can be realized by assigning different positions(x) and momenta(p) to the various molecules. Interpreted in this way, Boltzmann's formula is the most general formula for the thermodynamic

181

Boltzmann's grave in the Zentralfriedhof, Vienna, with bust and entropy formula.

entropy. However, Boltzmann's paradigm was an ideal gas of N *identical* particles, of which N_i are in the i -th microscopic condition (range) of position and momentum. For this case, the probability of each microstate of the system is equal, so it was equivalent for Boltzmann to calculate the number of microstates associated with a macrostate. W was historically misinterpreted as literally meaning the number of microstates, and that is what it usually means today. W can be counted using the formula for permutations

$$W = N! \, / \, \prod_i N_i! \; (2)$$

where i ranges over all possible molecular conditions and ! denotes factorial. The "correction" in the denominator is due to the fact that identical particles in the same condition are indistinguishable. W is sometimes called the "thermodynamic probability" since it is an integer greater than one, while mathematical probabilities are always numbers between zero and one.

23.2 Generalization

Boltzmann's formula applies to microstates of the universe as a whole, each possible microstate of which is presumed to be equally probable.

But in thermodynamics it is important to be able to make the approximation of dividing the universe into a system of interest, plus its surroundings; and then to be able to identify the entropy of the system with the system entropy in classical thermodynamics. The microstates of such a thermodynamic system are *not* equally probable—for example, high energy microstates are less probable than low energy microstates for a thermodynamic system kept at a fixed temperature by allowing contact with a heat bath.

For thermodynamic systems where microstates of the system may not have equal probabilities, the appropriate generalization, called the Gibbs entropy, is:

$$S = -k_\mathrm{B} \sum p_i \ln p_i \; (3)$$

This reduces to equation (1) if the probabilities p_i are all equal.

Boltzmann used a $\rho \ln \rho$ formula as early as 1866.[4] He interpreted ρ as a density in phase space—without mentioning probability—but since this satisfies the axiomatic definition of a probability measure we can retrospectively interpret it as a probability anyway. Gibbs gave an explicitly probabilistic interpretation in 1878.

Boltzmann himself used an expression equivalent to (3) in his later work[5] and recognized it as more general than equation (1). That is, equation (1) is a corollary of equation (3)—and not vice versa. In every situation where equation (1) is valid, equation (3) is valid also—and not vice versa.

23.3 Boltzmann entropy excludes statistical dependencies

The term **Boltzmann entropy** is also sometimes used to indicate entropies calculated based on the approximation that the overall probability can be factored into an identical separate term for each particle—i.e., assuming each particle has an identical independent probability distribution, and ignoring interactions and correlations between the particles. This is exact for an ideal gas of identical particles, and may or may not be a good approximation for other systems.[6]

The Boltzmann entropy is obtained if one assumes one can treat all the component particles of a thermodynamic system as statistically independent. The probability distribution of the system as a whole then factorises into the product of N separate identical terms, one term for each particle; and the Gibbs entropy simplifies to the Boltzmann entropy

$$S_B = -N k_\mathrm{B} \sum_i p_i \ln p_i$$

where the summation is taken over each possible state in the 6-dimensional phase space of a *single* particle (rather than the 6N-dimensional phase space of the system as a whole).

This reflects the original statistical entropy function introduced by Ludwig Boltzmann in 1872. For the special case of an ideal gas it exactly corresponds to the proper thermodynamic entropy.

However, for anything but the most dilute of real gases, it leads to increasingly wrong predictions of entropies and physical behaviours, by ignoring the interactions and correlations between different molecules. Instead one must follow Gibbs, and consider the ensemble of states of the system as a whole, rather than single particle states.

23.4   See also

- History of entropy

- Gibbs entropy

23.5   References

[1] See: photo of Boltzmann's grave in the Zentralfriedhof, Vienna, with bust and entropy formula.

[2] Boltzmann equation. Eric Weisstein's World of Physics (states the year was 1872).

[3] Perrot, Pierre (1998). *A to Z of Thermodynamics*. Oxford University Press. ISBN 0-19-856552-6. (states the year was 1875)

[4] Ludwig Boltzmann (1866). "Über die Mechanische Bedeutung des Zweiten Hauptsatzes der Wärmetheorie". *Wiener Berichte* **53**: 195–220.

[5] Ludwig Boltzmann (1896 and 1898). *Vorlesungen über Gastheorie*. J.A. Barth, Leipzig. Check date values in: |date= (help)

[6] Jaynes, E. T. (1965). Gibbs vs Boltzmann entropies. *American Journal of Physics*, **33**, 391-8.

23.6   External links

- Introduction to Boltzmann's Equation

Chapter 24

Tsallis entropy

In physics, the **Tsallis entropy** is a generalization of the standard Boltzmann–Gibbs entropy. It was introduced in 1988 by Constantino Tsallis[1] as a basis for generalizing the standard statistical mechanics. In the scientific literature, the physical relevance of the Tsallis entropy was occasionally debated. However, from the years 2000 on, an increasingly wide spectrum of natural, artificial and social complex systems have been identified which confirm the predictions and consequences that are derived from this nonadditive entropy, such as nonextensive statistical mechanics,[2] which generalizes the Boltzmann–Gibbs theory.

Among the various experimental verifications and applications presently available in the literature, the following ones deserve a special mention:

1. The distribution characterizing the motion of cold atoms in dissipative optical lattices, predicted in 2003[3] and observed in 2006.[4]

2. The fluctuations of the magnetic field in the solar wind enabled the calculation of the q-triplet (or Tsallis triplet).[5]

3. The velocity distributions in driven dissipative dusty plasma.[6]

4. Spin glass relaxation.[7]

5. Trapped ion interacting with a classical buffer gas.[8]

6. High energy collisional experiments at LHC/CERN (CMS, ATLAS and ALICE detectors)[9][10] and RHIC/Brookhaven (STAR and PHENIX detectors).[11]

Among the various available theoretical results which clarify the physical conditions under which Tsallis entropy and associated statistics apply, the following ones can be selected:

1. Anomalous diffusion.[12][13]

2. Uniqueness theorem.[14]

3. Sensitivity to initial conditions and entropy production at the edge of chaos.[15][16]

4. Probability sets which make the nonadditive Tsallis entropy to be extensive in the thermodynamical sense.[17]

5. Strongly quantum entangled systems and thermodynamics.[18]

6. Thermostatistics of overdamped motion of interacting particles.[19][20]

7. Nonlinear generalizations of the Schroedinger, Klein-Gordon and Dirac equations.[21]

For further details a bibliography is available at http://tsallis.cat.cbpf.br/biblio.htm

Given a discrete set of probabilities $\{p_i\}$ with the condition $\sum_i p_i = 1$, and q any real number, the **Tsallis entropy** is defined as

$$S_q(p_i) = \frac{k}{q-1}\left(1 - \sum_i p_i^q\right),$$

where q is a real parameter sometimes called *entropic-index*. In the limit as $q \to 1$, the usual Boltzmann–Gibbs entropy is recovered, namely

$$S_{BG} = S_1(p) = -k\sum_i p_i \ln p_i.$$

For continuous probability distributions, we define the entropy as

$$S_q[p] = \frac{1}{q-1}\left(1 - \int (p(x))^q\, dx\right),$$

where $p(x)$ is a probability density function.

The Tsallis Entropy has been used along with the Principle of maximum entropy to derive the Tsallis distribution.

24.1   Various relationships

The discrete Tsallis entropy satisfies

$$S_q = -\lim_{x \to 1} D_q \sum_i p_i^x$$

where Dq is the q-derivative with respect to x. This may be compared to the standard entropy formula:

$$S = -\lim_{x \to 1} \frac{d}{dx} \sum_i p_i^x$$

24.2   Non-additivity

Given two independent systems A and B, for which the joint probability density satisfies

$$p(A, B) = p(A)p(B).$$

the Tsallis entropy of this system satisfies

$$S_q(A, B) = S_q(A) + S_q(B) + (1 - q)S_q(A)S_q(B).$$

From this result, it is evident that the parameter $|1 - q|$ is a measure of the departure from additivity. In the limit when $q = 1$,

$$S(A, B) = S(A) + S(B).$$

which is what is expected for an additive system. This property is sometimes referred to as "pseudo-additivity".

24.3 Exponential families

Many common distributions like the normal distribution belongs to the statistical exponential families. Tsallis entropy for an exponential family can be written [22] as

$$H_q^T(p_F(x; \theta)) = \frac{1}{1-q} \left(\left(e^{F(q\theta) - qF(\theta)} \right) E_p[e^{(q-1)k(x)}] - 1 \right)$$

where F is log-normalizer and k the term indicating the carrier measure. For multivariate normal, term k is zero, and therefore the Tsallis entropy is in closed-form.

24.4 Generalised entropies

A number of interesting physical systems[23] abide to entropic functionals that are more general than the standard Tsallis entropy. Therefore, several physically meaningful generalisations have been introduced. The two most general of those are notably: Superstatistics, introduced by C. Beck and E.G.D. Cohen in 2003[24] and Spectral Statistics, introduced by G.A. Tsekouras and Constantino Tsallis in 2005.[25] Both these entropic forms have Tsallis and Boltzmann–Gibbs statistics as special cases; Spectral Statistics has been proven to at least contain Superstatistics and it has been conjectured to also cover some additional cases.

24.5 See also

- Rényi entropy

- Tsallis distribution

24.6 References

[1] Tsallis, C. (1988). "Possible generalization of Boltzmann-Gibbs statistics". *Journal of Statistical Physics* **52**: 479–487. Bibcode: doi:10.1007/BF01016429.

[2] Tsallis, Constantino (2009). *Introduction to nonextensive statistical mechanics : approaching a complex world* (Online-Ausg. ed.). New York: Springer. ISBN 978-0-387-85358-1.

[3] Lutz, E. (2003). "Anomalous diffusion and Tsallis statistics in an optical lattice". *Physical Review A* **67** (5). arXiv:cond-mat/0210022. Bibcode:2003PhRvA..67e1402L. doi:10.1103/PhysRevA.67.051402.

[4] Douglas, P.; Bergamini, S.; Renzoni, F. (2006). "Tunable Tsallis Distributions in Dissipative Optical Lattices". *Physical Review Letters* **96** (11). doi:10.1103/PhysRevLett.96.110601.

[5] Burlaga, L. F.; - Viñas, A. F. (2005). "Triangle for the entropic index q of non-extensive statistical mechanics observed by Voyager 1 in the distant heliosphere". *Physica A: Statistical Mechanics and its Applications* **356** (2–4): 375. arXiv:physics/0507212. Bibcode:2005PhyA..356..375B. doi:10.1016/j.physa.2005.06.065.

[6] Liu, B.; Goree, J. (2008). "Superdiffusion and Non-Gaussian Statistics in a Driven-Dissipative 2D Dusty Plasma". *Physical Review Letters* **100** (5). arXiv:0801.3991. Bibcode:2008PhRvL.100e5003L. doi:10.1103/PhysRevLett.100.055003.

[7] Pickup, R.; Cywinski, R.; Pappas, C.; Farago, B.; Fouquet, P. (2009). "Generalized Spin-Glass Relaxation". *Physical Review Letters* **102** (9). Bibcode:2009PhRvL.102i7202P. doi:10.1103/PhysRevLett.102.097202.

[8] Devoe, R. (2009). "Power-Law Distributions for a Trapped Ion Interacting with a Classical Buffer Gas". *Physical Review Letters* **102** (6). arXiv:0903.0637. Bibcode:2009PhRvL.102f3001D. doi:10.1103/PhysRevLett.102.063001.

[9] Khachatryan, V.; Sirunyan, A.; Tumasyan, A.; Adam, W.; Bergauer, T.; Dragicevic, M.; Erö, J.; Fabjan, C.; Friedl, M.; Frühwirth, R.; Ghete, V. M.; Hammer, J.; Hänsel, S.; Hoch, M.; Hörmann, N.; Hrubec, J.; Jeitler, M.; Kasieczka, G.; Kiesenhofer, W.; Krammer, M.; Liko, D.; Mikulec, I.; Pernicka, M.; Rohringer, H.; Schöfbeck, R.; Strauss, J.; Taurok, A.; Teischinger, F.; Waltenberger, W.; Walzel, G. (2010). "Transverse-Momentum and Pseudorapidity Distributions of Charged Hadrons in pp Collisions at √s=7 TeV". *Physical Review Letters* **105** (2). arXiv:1005.3299. Bibcode:2010PhRvL.105b2002K. doi:10.1103/PhysRevLett.105.022002.

[10] Chatrchyan, S.; Khachatryan, V.; Sirunyan, A. M.; Tumasyan, A.; Adam, W.; Bergauer, T.; Dragicevic, M.; Erö, J.; Fabjan, C.; Friedl, M.; Frühwirth, R.; Ghete, V. M.; Hammer, J.; Hänsel, S.; Hoch, M.; Hörmann, N.; Hrubec, J.; Jeitler, M.; Kiesenhofer, W.; Krammer, M.; Liko, D.; Mikulec, I.; Pernicka, M.; Rohringer, H.; Schöfbeck, R.; Strauss, J.; Taurok, A.; Teischinger, F.; Wagner, P.; et al. (2011). "Charged particle transverse momentum spectra in pp collisions at $ √s= 0.9 and 7 TeV". *Journal of High Energy Physics* **2011** (8). arXiv:1104.3547. Bibcode:2011JHEP...08..086C. doi:10.1007/JHEP08(2011)086.

[11] Adare, A.; Afanasiev, S.; Aidala, C.; Ajitanand, N.; Akiba, Y.; Al-Bataineh, H.; Alexander, J.; Aoki, K.; Aphecetche, L.; Armendariz, R.; Aronson, S. H.; Asai, J.; Atomssa, E. T.; Averbeck, R.; Awes, T. C.; Azmoun, B.; Babintsev, V.; Bai, M.; Baksay, G.; Baksay, L.; Baldisseri, A.; Barish, K. N.; Barnes, P. D.; Bassalleck, B.; Basye, A. T.; Bathe, S.; Batsouli, S.; Baublis, V.; Baumann, C.; Bazilevsky, A. (2011). "Measurement of neutral mesons in *p+p* collisions at √s=200 GeV and scaling properties of hadron production". *Physical Review D* **83** (5). arXiv:1005.3674. Bibcode:2011PhRvD..83e2004A. doi:10.1103/PhysRevD.83.052004.

[12] Plastino, A. R.; Plastino, A. (1995). "Non-extensive statistical mechanics and generalized Fokker-Planck equation". *Physica A: Statistical Mechanics and its Applications* **222**: 347. doi:10.1016/0378-4371(95)00211-1.

[13] Tsallis, C.; Bukman, D. (1996). "Anomalous diffusion in the presence of external forces: Exact time-dependent solutions and their thermostatistical basis". *Physical Review E* **54** (3): R2197. arXiv:cond-mat/9511007. Bibcode:1996PhRvE..54.2197T. doi:10.1103/PhysRevE.54.R2197.

[14] Abe, S. (2000). "Axioms and uniqueness theorem for Tsallis entropy". *Physics Letters A* **271**: 74–79. doi:10.1016/S0375-9601(00)00337-6.

[15] Lyra, M.; Tsallis, C. (1998). "Nonextensivity and Multifractality in Low-Dimensional Dissipative Systems". *Physical Review Letters* **80**: 53. arXiv:cond-mat/9709226. Bibcode:1998PhRvL..80...53L. doi:10.1103/PhysRevLett.80.53.

[16] Baldovin, F.; Robledo, A. (2004). "Nonextensive Pesin identity: Exact renormalization group analytical results for the dynamics at the edge of chaos of the logistic map". *Physical Review E* **69** (4). arXiv:cond-mat/0304410. Bibcode:2004PhRvE..69d5202B. doi:10.1103/PhysRevE.69.045202.

[17] Tsallis, C.; Gell-Mann, M.; Sato, Y. (2005). "Asymptotically scale-invariant occupancy of phase space makes the entropy Sq extensive". *Proceedings of the National Academy of Sciences* **102** (43): 15377. doi:10.1073/pnas.0503807102.

[18] Caruso, F.; Tsallis, C. (2008). "Nonadditive entropy reconciles the area law in quantum systems with classical thermodynamics". *Physical Review E* **78** (2). Bibcode:2008PhRvE..78b1102C. doi:10.1103/PhysRevE.78.021102.

[19] Andrade, J.; Da Silva, G.; Moreira, A.; Nobre, F.; Curado, E. (2010). "Thermostatistics of Overdamped Motion of Interacting Particles".*Physical Review Letters***105**(26).arXiv:1008.1421.Bibcode:2010PhRvL.105z0601A.doi:10.1103/PhysRevLett.10

[20] Ribeiro, M.; Nobre, F.; Curado, E. M. (2012). "Time evolution of interacting vortices under overdamped motion". *Physical Review E* **85** (2). Bibcode:2012PhRvE..85b1146R. doi:10.1103/PhysRevE.85.021146.

[21] Nobre, F.; Rego-Monteiro, M.; Tsallis, C. (2011). "Nonlinear Relativistic and Quantum Equations with a Common Type of Solution".*Physical Review Letters***106**(14).arXiv:1104.5461.Bibcode:2011PhRvL.106n0601N.doi:10.1103/PhysRevLett.106

[22] Nielsen, F.; Nock, R. (2012). "A closed-form expression for the Sharma–Mittal entropy of exponential families". *Journal of Physics A: Mathematical and Theoretical* **45** (3): 032003. doi:10.1088/1751-8113/45/3/032003.

[23] Garcia-Morales, V.; Krischer, K. (2011). "Superstatistics in nanoscale electrochemical systems". *Proceedings of the National Academy of Sciences* **108** (49): 19535–19539. Bibcode:2011PNAS..10819535G. doi:10.1073/pnas.1109844108.

[24]Beck, C.; Cohen, E. G. D. (2003). "Superstatistics".*Physica A: Statistical Mechanics and its Applications***322**: 267.doi:10.101 4371(03)00019-0.

[25] Tsekouras, G.A.; Tsallis, C. (2005). "Generalized entropy arising from a distribution of q indices". *Physical Review E* **71** (4). doi:10.1103/PhysRevE.71.046144.

24.7 External links

- Tsallis Statistics, Statistical Mechanics for Non-extensive Systems and Long-Range Interactions

Chapter 25

Standard molar entropy

In chemistry, the **standard molar entropy** is the entropy content of one mole of substance, under standard conditions (not standard temperature and pressure).

The standard molar entropy is usually given the symbol $S°$, and as units of joules per mole kelvin (J mol^{-1} K^{-1}). Unlike standard enthalpies of formation, the value of $S°$ is absolute. That is, an element in its standard state has a nonzero value of $S°$ at room temperature. The entropy of a pure crystalline structure can be 0 J mol^{-1} K^{-1} only at 0 K, according to the third law of thermodynamics. However, this presupposes that the material forms a 'perfect crystal' without any frozen in entropy (defects, dislocations), which is never completely true because crystals always grow at a finite temperature. This residual entropy is often quite negligible.

25.1 Thermodynamics

If a mole of substance were at 0 K, then warmed by its surroundings to 298 K, its total molar entropy would be the addition of all N individual contributions:

$$S° = \sum_{k=1}^{N} \Delta S_k = \sum_{k=1}^{N} \int \frac{dq_k}{T} \, dT$$

Here, dq_k/T represents a very small exchange of heat energy at temperature T. The total molar entropy is the sum of many small changes in molar entropy, where each small change can be considered a reversible process.

25.2 Chemistry

The standard molar entropy of a gas at STP includes contributions from:[1]

- The heat capacity of one mole of the solid from 0 K to the melting point (including heat absorbed in any changes between different crystal structures)

- The latent heat of fusion of the solid.

- The heat capacity of the liquid from the melting point to the boiling point.

- The latent heat of vaporization of the liquid.

- The heat capacity of the gas from the boiling point to room temperature.

Changes in entropy are associated with phase transitions and chemical reactions. Chemical equations make use of the standard molar entropy of reactants and products to find the standard entropy of reaction:[2]

$$\Delta S^\circ_{rxn} = S^\circ_{products} - S^\circ_{reactants}$$

The standard entropy of reaction helps determine whether the reaction will take place spontaneously. According to the second law of thermodynamics, a spontaneous reaction always results in an increase in total entropy of the system and its surroundings:

$$\Delta S_{total} = \Delta S_{system} + \Delta S_{surroundings} > 0$$

25.3 See also

- Entropy
- Heat
- Gibbs free energy
- Helmholtz free energy
- Third law of thermodynamics

25.4 References

[1] Kosanke, K. (2004). "Chemical Thermodynamics". *Pyrotechnic chemistry*. Journal of Pyrotechnics. p. 29. ISBN 1-889526-15-0.

[2] Chang, Raymond; Brandon Cruickshank (2005). "Entropy, Free Energy and Equilibrium". *Chemistry*. McGraw-Hill Higher Education. p. 765. ISBN 0-07-251264-4.

25.5 External links

- Free Energy and Chemical Reactions - Course notes for General Chemistry (R. Paselk, Humboldt State University)

Chapter 26

Entropy of mixing

In thermodynamics the **entropy of mixing** is the increase in the total entropy when several initially separate systems of different composition, each in a thermodynamic state of internal equilibrium, are mixed without chemical reaction by the thermodynamic operation of removal of impermeable partition(s) between them, followed by a time for establishment of a new thermodynamic state of internal equilibrium in the new unpartitioned closed system.

In general, the mixing may be constrained to occur under various prescribed conditions. In the customarily prescribed conditions, the materials are each initially at a common temperature and pressure, and the new system may change its volume, while being maintained at that same constant temperature, pressure, and chemical component masses. The volume available for each material to explore is increased, from that of its initially separate compartment, to the total common final volume. The final volume need not be the sum of the initially separate volumes, so that work can be done on or by the new closed system during the process of mixing, as well as heat being transferred to or from the surroundings, because of the maintenance of constant pressure and temperature.

The internal energy of the new closed system is equal to the sum of the internal energies of the initially separate systems. The reference values for the internal energies should be specified in a way that is constrained to make this so, maintaining also that the internal energies are respectively proportional to the masses of the systems.[1]

For concision in this article, the term 'ideal material' is used to refer to an ideal gas (mixture) or an ideal solution.

In a process of mixing of ideal materials, the final common volume is the sum of the initial separate compartment volumes. There is no heat transfer and no work is done. The entropy of mixing is entirely accounted for by the diffusive expansion of each material into a final volume not initially accessible to it.

On the mixing of non-ideal materials, the total final common volume may be different from the sum of the separate initial volumes, and there may occur transfer of work or heat, to or from the surroundings; also there may be a departure of the entropy of mixing from that of the corresponding ideal case. That departure is the main reason for interest in entropy of mixing. These energy and entropy variables and their temperature dependences provide valuable information about the properties of the materials.

On a molecular level, the entropy of mixing is of interest because it is a macroscopic variable that provides information about constitutive molecular properties. In ideal materials, intermolecular forces are the same between every pair of molecular kinds, so that a molecule feels no difference between other molecules of its own kind and of those of the other kind. In non-ideal materials, there may be differences of intermolecular forces or specific molecular effects between different species, even though they are chemically non-reacting. The entropy of mixing provides information about constitutive differences of intermolecular forces or specific molecular effects in the materials.

The statistical concept of randomness is used for statistical mechanical explanation of the entropy of mixing. Mixing of ideal materials is regarded as random at a molecular level, and, correspondingly, mixing of non-ideal materials may be non-random.

26.1 Mixing of ideal materials at constant temperature and pressure

In ideal materials, intermolecular forces are the same between every pair of molecular kinds, so that a molecule "feels" no difference between itself and its molecular neighbours. This is the reference case for examining corresponding mixings of non-ideal materials.

For example, two ideal gases, at the same temperature and pressure, are initially separated by a dividing partition.

Upon removal of the dividing partition, they expand into a final common volume (the sum of the two initial volumes), and the entropy of mixing ΔS_{mix} is given by

$$\Delta S_{mix} = -nR(x_1 \ln x_1 + x_2 \ln x_2)$$

where R is the gas constant, n the total number of moles and x_i the mole fraction of component i, which initially occupies volume $V_i = x_i V$. After the removal of the partition, the $n_i = n x_i$ moles of component i may explore the combined volume V, which causes an entropy increase equal to $n x_i R \ln(V/V_i) = -nR x_i \ln x_i$ for each component gas.

In this case, the increase in entropy is due entirely to the irreversible processes of expansion of the two gases, and involves no heat or work flow between the system and its surroundings.

26.1.1 Gibbs free energy of mixing

The Gibbs free energy change $\Delta G_{mix} = \Delta H_{mix} - T\Delta S_{mix}$ determines whether mixing at constant (absolute) temperature T and pressure p is a spontaneous process. This quantity combines two physical effects—the enthalpy of mixing, which is a measure of the energy change, and the entropy of mixing considered here.

For an ideal gas mixture or an ideal solution, there is no enthalpy of mixing (ΔH_{mix}), so that the Gibbs free energy of mixing is given by the entropy term only:

$$\Delta G_{mix} = -T\Delta S_{mix}$$

For an ideal solution, the Gibbs free energy of mixing is always negative, meaning that mixing of ideal solutions is always spontaneous. The lowest value is when the mole fraction is 0.5 for a mixture of two components, or 1/n for a mixture of n components.

26.2 Solutions and temperature dependence of miscibility

26.2.1 Ideal and regular solutions

The above equation for the entropy of mixing of ideal gases is valid also for certain liquid (or solid) solutions—those formed by completely random mixing so that the components move independently in the total volume. Such random mixing of solutions occurs if the interaction energies between unlike molecules are similar to the average interaction energies between like molecules.[2][3] The value of the entropy corresponds exactly to random mixing for ideal solutions and for regular solutions, and approximately so for many real solutions.[3][4]

For binary mixtures the entropy of random mixing can be considered as a function of the mole fraction of one component.

$$\Delta S_{mix} = -nR(x_1 \ln x_1 + x_2 \ln x_2) = -nR[x \ln x + (1 - x) \ln(1 - x)]$$

For all possible mixtures, $0 < x < 1$, so that $\ln x$ and $\ln(1 - x)$ are both negative and the entropy of mixing ΔS_{mix} is positive and favors mixing of the pure components.

Also the curvature of ΔS_{mix} as a function of x is given by the second derivative $\left(\frac{\partial^2 \Delta S_{mix}}{\partial x^2}\right)_{T,P} = -nR\left(\frac{1}{x} + \frac{1}{1-x}\right)$

This curvature is negative for all possible mixtures $(0 < x < 1)$, so that mixing two solutions to form a solution of intermediate composition also increases the entropy of the system. Random mixing therefore always favors miscibility and opposes phase separation.

For ideal solutions, the enthalpy of mixing is zero so that the components are miscible in all proportions. For regular solutions a positive enthalpy of mixing may cause incomplete miscibility (phase separation for some compositions) at temperatures below the upper critical solution temperature (UCST).[5] This is the minimum temperature at which the $-T\Delta S_{mix}$ term in the Gibbs energy of mixing is sufficient to produce miscibility in all proportions.

26.2.2   Systems with a lower critical solution temperature

Nonrandom mixing with a lower entropy of mixing can occur when the attractive interactions between unlike molecules are significantly stronger (or weaker) than the mean interactions between like molecules. For some systems this can lead to a lower critical solution temperature (LCST) or lower limiting temperature for phase separation.

For example, triethylamine and water are miscible in all proportions below 19 °C, but above this critical temperature, solutions of certain compositions separate into two phases at equilibrium with each other.[6][7] This means that ΔG_{mix} is negative for mixing of the two phases below 19 °C and positive above this temperature. Therefore $\Delta S_{mix} = -\left(\frac{\partial \Delta G_{mix}}{\partial T}\right)_P$ is negative for mixing of these two equilibrium phases. This is due to the formation of attractive hydrogen bonds between the two components that prevent random mixing. Triethylamine molecules cannot form hydrogen bonds with each other but only with water molecules, so in solution they remain associated to water molecules with loss of entropy. The mixing that occurs below 19 °C is due not to entropy but to the enthalpy of formation of the hydrogen bonds.

Lower critical solution temperatures also occur in many polymer-solvent mixtures.[8] For polar systems such as polyacrylic acid in 1,4-dioxane, this is often due to the formation of hydrogen bonds between polymer and solvent. For nonpolar systems such as polystyrene in cyclohexane, phase separation has been observed in sealed tubes (at high pressure) at temperatures approaching the liquid-vapor critical point of the solvent. At such temperatures the solvent expands much more rapidly than the polymer, whose segments are covalently linked. Mixing therefore requires contraction of the solvent for compatibility of the polymer, resulting in a loss of entropy.[8]

26.3   Statistical thermodynamical explanation of the entropy of mixing of ideal gases

Since thermodynamic entropy can be related to statistical mechanics or to information theory, it is possible to calculate the entropy of mixing using these two approaches. Here we consider the simple case of mixing ideal gases.

26.3.1   Proof from statistical mechanics

Assume that the molecules of two different substances are approximately the same size, and regard space as subdivided into a square lattice whose cells are the size of the molecules. (In fact, any lattice would do, including close packing.) This is a crystal-like conceptual model to identify the molecular centers of mass. If the two phases are liquids, there is no spatial uncertainty in each one individually. (This is, of course, an approximation. Liquids have a "free volume". This is why they are (usually) less dense than solids.) Everywhere we look in component 1, there is a molecule present, and likewise for component 2. After the two different substances are intermingled (assuming they are miscible), the liquid is still dense with molecules, but now there is uncertainty about what kind of molecule is in which location. Of course, any idea of identifying molecules in given locations is a thought experiment, not something one could do, but the calculation of the uncertainty is well-defined.

We can use Boltzmann's equation for the entropy change as applied to the mixing process

$$\Delta S_{mix} = k_B \ln \Omega$$

where k_B is Boltzmann's constant. We then calculate the number of ways Ω of arranging N_1 molecules of component 1 and N_2 molecules of component 2 on a lattice, where

$$N = N_1 + N_2$$

is the total number of molecules, and therefore the number of lattice sites. Calculating the number of permutations of N objects, correcting for the fact that N_1 of them are *identical* to one another, and likewise for N_2 ,

$$\Omega = N!/N_1!N_2!$$

After applying Stirling's approximation for the factorial of a large integer m:

$$\ln m! = \sum_k \ln k \approx \int_1^m dk \ln k = m \ln m - m$$

the result is $\Delta S_{mix} = -k_B[N_1 \ln(N_1/N) + N_2 \ln(N_2/N)] = -k_B N[x_1 \ln x_1 + x_2 \ln x_2]$

where we have introduced the mole fractions, which are also the probabilities of finding any particular component in a given lattice site.

$$x_1 = N_1/N = p_1 \text{ and } x_2 = N_2/N = p_2$$

Since the Boltzmann constant $k_B = R/N_A$, where N_A is Avogadro's number, and the number of molecules $N = n N_A$, we recover the thermodynamic expression for the mixing of two ideal gases, $\Delta S_{mix} = -nR[x_1 \ln x_1 + x_2 \ln x_2]$

This expression can be generalized to a mixture of r components, N_i , with $i = 1, 2, 3, \ldots, r$

$$\Delta S_{mix} = -k_B \sum_{i=1}^r N_i \ln(N_i/N) = -N k_B \sum_{i=1}^r x_i \ln x_i = -nR \sum_{i=1}^r x_i \ln x_i$$

26.3.2   Relationship to information theory

The entropy of mixing is also proportional to the Shannon entropy or compositional uncertainty of information theory, which is defined without requiring Stirling's approximation. Claude Shannon introduced this expression for use in information theory, but similar formulas can be found as far back as the work of Ludwig Boltzmann and J. Willard Gibbs. The Shannon uncertainty is completely unrelated to the Heisenberg uncertainty principle in quantum mechanics, and is defined by

$$H = -\sum_{i=1}^r p_i \ln(p_i)$$

To relate this quantity to the entropy of mixing, we consider that the summation is over the various chemical species, so that this is the uncertainty about which kind of molecule is in any one site. It must be multiplied by the number of sites

N to get the uncertainty for the whole system. Since the probability p_i of finding species i in a given site equals the mole fraction x_i , we again obtain the entropy of mixing on multiplying by the Boltzmann constant k_B .

$$\Delta S_{mix} = -Nk_B \sum_{i=1}^{r} x_i \ln x_i$$

26.3.3 Application to gases

In gases there is a lot more spatial uncertainty because most of their volume is merely empty space. We can regard the mixing process as allowing the contents of the two originally separate contents to expand into the combined volume of the two conjoined containers. The two lattices that allow us to conceptually localize molecular centers of mass also join. The total number of empty cells is the sum of the numbers of empty cells in the two components prior to mixing. Consequently, that part of the spatial uncertainty concerning whether *any* molecule is present in a lattice cell is the sum of the initial values, and does not increase upon "mixing".

Almost everywhere we look, we find empty lattice cells. Nevertheless, we do find molecules in a few occupied cells. When there is real mixing, for each of those few occupied cells, there is a contingent uncertainty about which kind of molecule it is. When there is no real mixing because the two substances are identical, there is no uncertainty about which kind of molecule it is. Using conditional probabilities, it turns out that the analytical problem for the small subset of occupied cells is exactly the same as for mixed liquids, and the *increase* in the entropy, or spatial uncertainty, has exactly the same form as obtained previously. Obviously the subset of occupied cells is not the same at different times. But only when there is real mixing and an occupied cell is found do we ask which kind of molecule is there.

See also: Gibbs paradox, in which it would seem that "mixing" two samples of the *same* gas would produce entropy.

26.3.4 Application to solutions

If the solute is a crystalline solid, the argument is much the same. A crystal has no spatial uncertainty at all, except for crystallographic defects, and a (perfect) crystal allows us to localize the molecules using the crystal symmetry group. The fact that volumes do not add when dissolving a solid in a liquid is not important for condensed phases. If the solute is not crystalline, we can still use a spatial lattice, as good an approximation for an amorphous solid as it is for a liquid.

The Flory–Huggins solution theory provides the entropy of mixing for polymer solutions, in which the macromolecules are huge compared to the solvent molecules. In this case, the assumption is made that each monomer subunit in the polymer chain occupies a lattice site.

Note that solids in contact with each other also slowly interdiffuse, and solid mixtures of two or more components may be made at will (alloys, semiconductors, etc.). Again, the same equations for the entropy of mixing apply, but only for homogeneous, uniform phases.

26.4 Mixing under other constraints

26.4.1 Mixing with and without change of available volume

In the established customary usage, expressed in the lead section of this article, the entropy of mixing comes from two mechanisms, the intermingling and possible interactions of the distinct molecular species, and the change in the volume available for each molecular species, or the change in concentration of each molecular species. For ideal gases, the entropy of mixing at prescribed common temperature and pressure has nothing to do with mixing in the sense of intermingling and interactions of molecular species, but is only to do with expansion into the common volume.[9]

According to Fowler and Guggenheim (1939/1965),[10] the conflating of the just-mentioned two mechanisms for the entropy of mixing is well established in customary terminology, but can be confusing unless it is borne in mind that the independent variables are the common initial and final temperature and total pressure; if the respective partial pressures or the total volume are chosen as independent variables instead of the total pressure, the description is different.

Mixing with each gas kept at constant partial volume, with changing total volume

In contrast to the established customary usage, "mixing" might be conducted reversibly at constant volume for each of two fixed masses of gases of equal volume, being mixed by gradually merging their initially separate volumes by use of two ideal semipermeable membranes, each permeable only to one of the respective gases, so that the respective volumes available to each gas remain constant during the merge. Either one of the common temperature or the common pressure is chosen to be independently controlled by the experimenter, the other being allowed to vary so as to maintain constant volume for each mass of gas. In this kind of "mixing", the final common volume is equal to each of the respective separate initial volumes, and each gas finally occupies the same volume as it did initially.[11][12][13][14][15][16]

This constant volume kind of "mixing", in the special case of perfect gases, is referred to in what is sometimes called Gibbs' theorem.[11][14][16] It states that the entropy of such "mixing" of perfect gases is zero.

Mixing at constant total volume and changing partial volumes, with mechanically controlled varying pressure, and constant temperature

An experimental demonstration may be considered. The two distinct gases, in a cylinder of constant total volume, are at first separated by two contiguous pistons made respectively of two suitably specific ideal semipermeable membranes. Ideally slowly and fictively reversibly, at constant temperature, the gases are allowed to mix in the volume between the separating membranes, forcing them apart, thereby supplying work to an external system. The energy for the work comes from the heat reservoir that keeps the temperature constant. Then, by externally forcing ideally slowly the separating membranes together, back to contiguity, work is done on the mixed gases, fictively reversibly separating them again, so that heat is returned to the heat reservoir at constant temperature. Because the mixing and separation are ideally slow and fictively reversible, the work supplied by the gases as they mix is equal to the work done in separating them again. Passing from fictive reversibility to physical reality, some amount of additional work, that remains external to the gases and the heat reservoir, must be provided from an external source for this cycle, as required by the second law of thermodynamics, because this cycle has only one heat reservoir at constant temperature, and the external provision of work cannot be completely efficient.[12]

26.5 Gibbs' paradox: "mixing" of identical species

Main article: Gibbs paradox

The molecular species must be constitutively detectably distinct for the entropy of mixing to exist. Thus arises the so-called *Gibbs paradox*, as follows. If molecular species are identical, there is no entropy change, because, defined in thermodynamic terms, there is no process and no mixing. Yet the slightest detectable difference in constitutive properties between the two species yields a thermodynamically recognized process of mixing and a considerable entropy change, namely the entropy of mixing.

The "paradox" is that a slightest detectable constitutive distinction can lead to a considerably large change in amount of entropy: on the hypothesis that the distinction might tend continuously to zero, with a continuous change in the properties of the materials, the entropy would change discontinuously.[17]

This is paradoxical from a general physical viewpoint, but not from a specifically thermodynamic viewpoint. The distinguishability of two materials is constitutive difference, not thermodynamic difference, for materials universally obey the laws of thermodynamics.[18] It does not seem paradoxical from the viewpoint of thermodynamics, because in that discipline the constitutive difference is not questioned; Gibbs himself did not see it as paradoxical.

Physically, however, the constitutive difference between any two chemical substances cannot be continuously decreased till it actually vanishes.[19] For example, ortho- and para-hydrogen differ by a finite amount. A difference smaller than that is hard to think of. The hypothesis, that the distinction might tend continuously to zero, is unphysical. This is neither examined nor explained by thermodynamics. Differences of constitution are explained by quantum mechanics.[20]

For a detectable distinction, some means should be physically available. One theoretical means would be through an ideal semi-permeable membrane.[13] It should allow passage, backwards and forwards, of one species, while passage of the other is prevented entirely. The entirety of prevention should include perfect efficacy over a practically infinite time, in view of the nature of thermodynamic equilibrium. Even the slightest departure from ideality, as assessed over a finite time, would extend to utter non-ideality, as assessed over a practically infinite time. Such quantum phenomena as tunneling ensure that nature does not allow the membrane ideality that would support the theoretically demanded continuous decrease, to zero, of detectable distinction.

For ideal gases, the entropy of mixing does not depend on the degree of difference between the distinct molecular species, but only on the fact that they are distinct; for non-ideal gases, the entropy of mixing can depend on the degree of difference of the distinct molecular species. But the suggested 'mixing' of identical molecular species is not in thermodynamic terms a mixing at all, because thermodynamics refers to states specified by state variables, and does not permit an imaginary labelling of particles. Only if the molecular species are different is there mixing in the thermodynamic sense.[21][22][23][24][25][26]

26.6   References

[1] Prigogine, I. (1955/1967). *Introduction to Thermodynamics of Irreversible Processes*, third edition, Interscience Publishers, New York, p. 12.

[2] Atkins & de Paula (2006), page 149.

[3] K. Denbigh, "The Principles of Chemical Equilibrium" (3rd ed., Cambridge University Press 1971) p.432

[4] P.A. Rock "Chemical Thermodynamics. Principles and Applications.(MacMillan 1969) p.263

[5] Atkins & de Paula (2006), page 186.

[6] Atkins & de Paula (2006), page 187.

[7] M.A. White, "Properties of Materials" (Oxford University Press 1999) p.175

[8] Cowie, J.M.G. "Polymers: Chemistry and Physics of Modern Materials" (2nd edn, Blackie 1991) p.174-176

[9] Bailyn (1994), page 273.

[10] Fowler, R., Guggenheim, E.A. (1939/1965). *Statistical Thermodynamics. A version of Statistical Mechanics for Students of Physics and Chemistry*, Cambridge University Press, Cambridge UK, pages 163-164

[11] Planck, M. (1897/1903). *Treatise on Thermodynamics*, translated with the author's sanction by Alexander Ogg, Longmans, Green and Co., London, Sections 235-236.

[12] Partington, J.R. (1949), pp. 163-164.

[13] Adkins (1968/1983), page 217.

[14] Callen, H.B. (1960/1985). *Thermodynamics and an Introduction to Thermostatistics*, second edition, Wiley, New York, ISBN 981-253-185-8, pages 69-70.

[15] Buchdahl, H.A. (1966). *The Concepts of Classical Thermodynamics*, Cambridge University Press, London, pages 170-171.

[16] Iribarne, J.V., Godson, W.L. (1973/1981), *Atmospheric Thermodynamics*, second edition, D. Reidel, Kluwer Academic Publishers, Dordrecht, ISBN 90-277-1296-4, pages 48-49.

[17]   • ter Haar & Wergeland (1966), p. 87.

[18] Truesdell, C. (1969). *Rational Thermodynamics: a Course of Lectures on Selected Topics*, McGraw-Hill Book Company, New York, p. 6

[19] Partington, J.R. (1949), p. 164, who cites Larmor, J. (1929), *Mathematical and Physical Papers*, volume 2, Cambridge University Press, Cambridge UK, p. 99.

[20] Landé, A. (1955). *Foundations of Quantum Mechanics: a Study in Continuity and Symmetry*, Yale University Press, New Haven, p.10.

[21] Tolman, R.C. (1938). *The Principles of Statistical Mechanics*, Oxford University Press, Oxford, pages 626-628.

[22] Adkins (1968/1983), pages 217–218.

[23] Landsberg, P.T. (1978). *Thermodynamics and Statistical Mechanics*, Oxford University Press, Oxford, ISBN 0-19-851142-6, page 74.

[24] Bailyn (1994), pages 274, 516-517.

[25] Grandy, W.T., Jr (2008). *Entropy and the Time Evolution of Macroscopic Systems*, Oxford University Press, Oxford, ISBN 978-0-19-954617-6, pages 60-62.

[26] Kondepudi, D. (2008). *Introduction to Modern Thermodynamics*, Wiley, Chichester, ISBN 978-0-470-01598-8, pages 197-199.

26.6.1   Cited bibliography

- Adkins, C.J. (1968/1983). *Equilibrium Thermodynamics*, third edition, McGraw-Hill, London, ISBN 0-521-25445-0.

- Atkins, P.W., de Paula, J. (2006). *Atkins' Physical Chemistry*, eighth edition, W.H. Freeman, New York, ISBN 978-0-7167-8759-4.

- Bailyn, M. (1994). *A Survey of Thermodynamics*, American Institute of Physics, New York, ISBN 0-88318-797-3.

- Partington, J.R. (1949). *An Advanced Treatise on Physical Chemistry*, Volume 1, *Fundamental Principles. The Properties of Gases*, Longmans, Green, and Co., London.

- ter Haar, D., Wergeland, H. (1966). *Elements of Thermodynamics*, Addison-Wesley Publishing, Reading MA.

26.7   External links

- Online lecture

Chapter 27

Loop entropy

Loop entropy is the entropy lost upon bringing together two residues of a polymer within a prescribed distance. For a single loop, the entropy varies logarithmically with the number of residues N in the loop

$$\Delta S = \alpha k_B \ln N$$

where k_B is Boltzmann's constant and α is a coefficient that depends on the properties of the polymer. This entropy formula corresponds to a power-law distribution $P \sim N^{-\alpha}$ for the probability of the residues contacting.

The loop entropy may also vary with the position of the contacting residues. Residues near the ends of the polymer are more likely to contact (quantitatively, have a lower α) than those in the middle (i.e., far from the ends), primarily due to excluded volume effects.

27.1 Wang-Uhlenbeck entropy

The loop entropy formula becomes more complicated with multiples loops, but may be determined for a Gaussian polymer using a matrix method developed by Wang and Uhlenbeck. Let there be M contacts among the residues, which define M loops of the polymers. The Wang-Uhlenbeck matrix $\mathbf{W}$ is an $M \times M$ symmetric, real matrix whose elements W_{ij} equal the number of common residues between loops i and j . The entropy of making the specified contacts equals

$$\Delta S = \alpha k_B \ln \det \mathbf{W}$$

As an example, consider the entropy lost upon making the contacts between residues 26 and 84 and residues 58 and 110 in a polymer (cf. ribonuclease A). The first and second loops have lengths 58 (=84-26) and 52 (=110-58), respectively, and they have 26 (=84-58) residues in common. The corresponding Wang-Uhlenbeck matrix is

$$\mathbf{W} = \begin{bmatrix} 58 & 26 \\ 26 & 52 \end{bmatrix}$$

whose determinant is 2340. Taking the logarithm and multiplying by the constants αk_B gives the entropy.

27.2 References

- Wang MC and Uhlenbeck GE. (1945) *Rev. Mod. Phys.*, **17**, 323.

Chapter 28

Conformational entropy

Not to be confused with configurational entropy.

Conformational entropy is the entropy associated with the number of conformations of a molecule. The concept is most commonly applied to biological macromolecules such as proteins and RNA, but also be used for polysaccharides and other molecules. To calculate the conformational entropy, the possible conformations of the molecule may first be discretized into a finite number of states, usually characterized by unique combinations of certain structural parameters, each of which has been assigned an energy. In proteins, backbone dihedral angles and side chain rotamers are commonly used as parameters, and in RNA the base pairing pattern may be used. These characteristics are used to define the degrees of freedom (in the statistical mechanics sense of a possible "microstate"). The conformational entropy associated with a particular structure or state, such as an alpha-helix, a folded or an unfolded protein structure, is then dependent on the probability or the occupancy of that structure.

The entropy of heterogeneous random coil or denatured proteins is significantly higher than that of the folded native state tertiary structure. In particular, the conformational entropy of the amino acid side chains in a protein is thought to be a major contributor to the energetic stabilization of the denatured state and thus a barrier to protein folding.[1] However, a recent study has shown that side-chain conformational entropy can stabilize native structures among alternative compact structures.[2] The conformational entropy of RNA and proteins can be estimated; for example, empirical methods to estimate the loss of conformational entropy in a particular side chain on incorporation into a folded protein can roughly predict the effects of particular point mutations in a protein. Side-chain conformational entropies can be defined as Boltzmann sampling over all possible rotameric states:[3]

$$S = -R\Sigma_i p_i ln(p_i)$$

where R is the gas constant and p_i is the probability of a residue being in rotamer i.[3]

The limited conformational range of proline residues lowers the conformational entropy of the denatured state and thus increases the energy difference between the denatured and native states. A correlation has been observed between the thermostability of a protein and its proline residue content.[4]

28.1 References

[1] Doig AJ, Sternberg MJE. (1995). Side-chain conformational entropy in protein folding. *Protein Science* 4:2247-51.

[2] Zhang J, Liu JS (2006) On Side-Chain Conformational Entropy of Proteins. PLoS Comput Biol 2(12): e168.doi:10.1371/journa

[3] Pickett SD, Sternberg MJ. (1993). Empirical scale of side-chain conformational entropy in protein folding. *J Mol Biol* 231(3):825-39.

[4] Watanabe K., Masuda T., Ohashi H., Mihara H. & Suzuki Y. Multiple proline substitutions cumulatively thermostabilize Bacillus cereus ATCC7064 oligo-1,6-glucosidase. Irrefragable proof supporting the proline rule. *Eur J Biochem* 226,277-83 (1994).

28.2   See also

- Configuration entropy

- Folding funnel

- Molten globule

- Loop entropy

- Protein folding

Chapter 29

Entropic force

In physics, an **entropic force** acting in a system is a phenomenological force resulting from the entire system's statistical tendency to increase its entropy, rather than from a particular underlying microscopic force.[1]

29.1 Mathematical formulation

In the canonical ensemble, the entropic force $\mathbf{F}$ associated to a macrostate partition $\{\mathbf{X}\}$ is given by:[2][3]

$$\mathbf{F}(\mathbf{X_0}) = T\nabla_{\mathbf{X}}S(\mathbf{X})|_{\mathbf{X_o}}$$

where T is the temperature, $S(\mathbf{X})$ is the entropy associated to the macrostate $\mathbf{X}$ and $\mathbf{X_0}$ is the present macrostate.

29.2 Examples

29.2.1 Brownian Motion

The entropic approach to Brownian movement was initially proposed by R. M. Neumann.[2][4] Neumann derived the entropic force for a particle undergoing three-dimensional Brownian motion using the Boltzmann equation, denoting this force as a *diffusional driving force* or *radial force*. In the paper, three example systems are shown to exhibit such a force:

- electrostatic system of molten salt

- surface tension and

- elasticity of rubber.

29.2.2 Polymers

A standard example of an entropic force is the elasticity of a freely-jointed polymer molecule described by a Gaussian distribution.[4] If the molecule is pulled into an extended configuration, the system has an increased amount of predictability. But randomly coiled configurations are overwhelmingly more probable; i.e., they have greater entropy. This results in the chain eventually returning (through diffusion) to such a configuration. To the macroscopic observer, the precise origin of the microscopic forces that drive the motion is irrelevant. The observer simply sees the polymer contract into a state of higher entropy, as if driven by an elastic force.

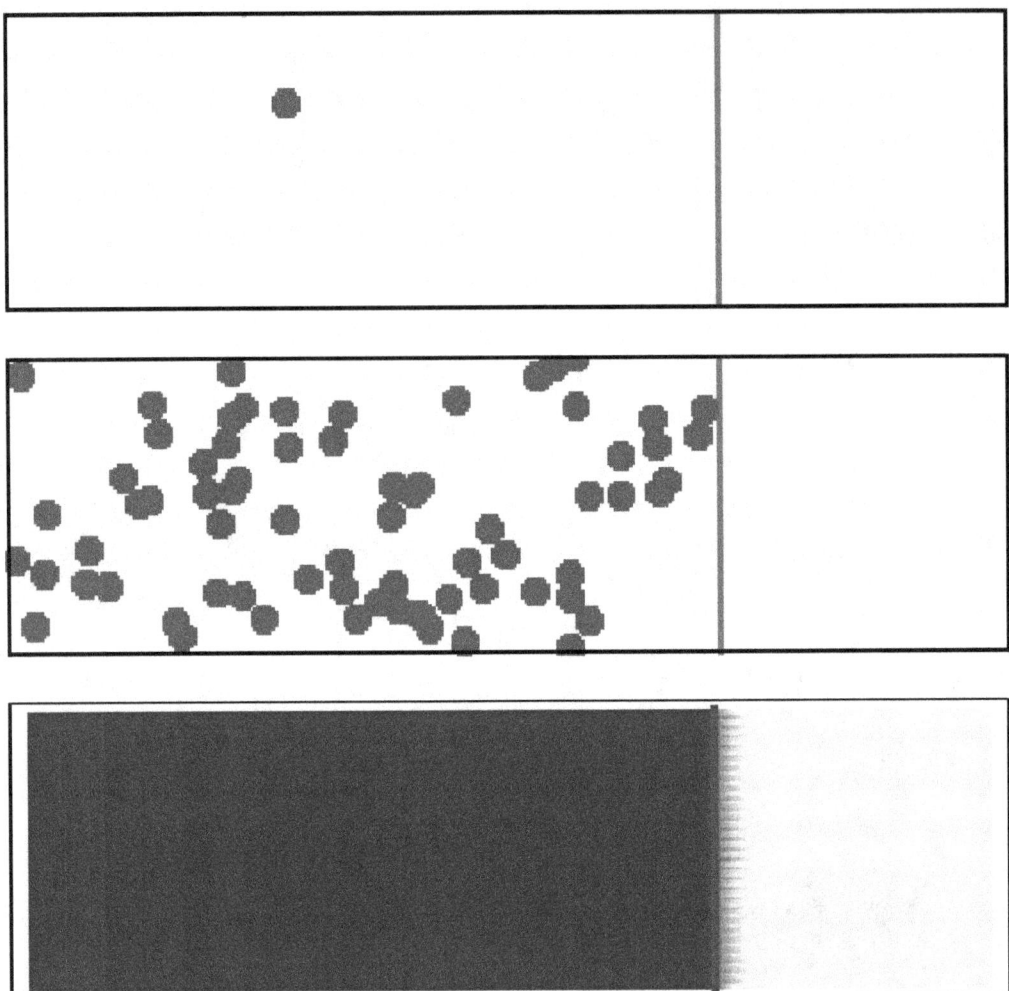

*Diffusion from a microscopic and macroscopic point of view. Initially, there are solute molecules on the left side of a barrier (purple line) and none on the right. The barrier is removed, and the solute diffuses to fill the whole container. Top: A single molecule moves around randomly. Middle: With more molecules, there is a statistical trend that the solute fills the container more and more uniformly. Bottom: With an enormous number of solute molecules, all randomness is gone: The solute appears to move smoothly and deterministically from high-concentration areas to low-concentration areas. There is no microscopic force pushing molecules rightward, but there appears to be one in the bottom panel. This fake-force is called an **entropic force**.*

29.2.3   Hydrophobic force

See also: Hydrophobic effect § The origin of hydrophobic effect

Another example of an entropic force is the hydrophobic force. At room temperature, it partly originates from the loss of entropy by the 3D network of water molecules when they interact with molecules of dissolved substance. Each water molecule is capable of

- donating two hydrogen bonds through the two protons

- accepting two more hydrogen bonds through the two sp^3-hybridized lone pairs

Therefore, water molecules can form an extended three-dimensional network. Introduction of a non-hydrogen-bonding surface disrupts this network. The water molecules rearrange themselves around the surface, so as to minimize the number of disrupted hydrogen bonds. This is in contrast to hydrogen fluoride (which can accept 3 but donate only 1) or ammonia (which can donate 3 but accept only 1), which mainly form linear chains.

Water drops on the surface of grass.

If the introduced surface had an ionic or polar nature, there would be water molecules standing upright on 1 (along the axis of an orbital for ionic bond) or 2 (along a resultant polarity axis) of the four sp^3 orbitals.[5] These orientations allow easy movement, i.e. degrees of freedom, and thus lowers entropy minimally. But a non-hydrogen-bonding surface with a moderate curvature forces the water molecule to sit tight on the surface, spreading 3 hydrogen bonds tangential to the surface, which then become locked in a clathrate-like basket shape. Water molecules involved in this clathrate-like basket around the non-hydrogen-bonding surface are constrained in their orientation. Thus, any event that would minimize such a surface is entropically favored. For example, when two such hydrophobic particles come very close, the clathrate-like baskets surrounding them merge. This releases some of the water molecules into the bulk of the water, leading to an increase in entropy. This is the basis of the so-called "attraction" between hydrophobic objects in water.

29.2.4 Directional Entropic Force

Entropic forces also occur in the physics of gases and solutions, where they generate the pressure of an ideal gas (the energy of which depends only on its temperature, not its volume), the osmotic pressure of a dilute solution, and in colloidal suspensions, where they are responsible for the crystallization of hard spheres.

In nano and colloidal science, entropic forces usually come from the effect of depletion, where small particles induce crystallization of bigger ones.

Even in the absence of depletion, however, scientist Sharon Glotzer and collaborators recently conjectured that **Directional Entropic Forces** could be responsible for the alignment of facets observed prior to the assembly and/or crystallization of systems of polyhedral nano and colloidal particles.[6] This was later proven to be correct[7][8] and allowed for the development of a roadmap for the assembly of polyhedral particles into atomic isostructures.[9]

29.3   Speculative examples

In recent years (especially since 2009) some forces that are generally regarded as conventional forces have been argued to be actually entropic in nature. These theories remain speculative and are the subject of ongoing work.

29.3.1   Gravity

Main article: Entropic gravity

It is generally believed that gravity is a microscopic force (or arguably a pseudo-force in general relativity). However, in 2009, Erik Verlinde[10] argued that gravity can be explained as an entropic force.[11]

For example, when someone throws a ball in the air, it follows a parabolic trajectory (in the absence of wind resistance). Conventionally, it is said that the ball follows a deterministic path dictated by Newton's law of gravity or general relativity. However, in the entropic theory, it is argued that the ball can follow any trajectory and picks a trajectory "at random." A calculation demonstrates that, in the collection of possible trajectories, the overwhelming majority are almost exactly the same as the parabolic trajectory; therefore, the ball is observed to follow a parabola.

29.3.2   Other forces

Other fundamental forces have been argued recently to be entropic in origin, including Coulomb's law,[12][13][14] the electroweak and strong forces,[15] and dark matter and dark energy.[16]

29.3.3   Links to Occam's Razor

A formal simultaneous connection between the mathematical structure of the discovered laws of nature, intelligence and the entropy-like measures of complexity was previously noted in 2000 by Andrei Soklakov[17] in the context of Occam's razor principle.

29.4   See also

- Colloids

- Nanomechanics

- Data clustering

- Entropic elasticity of an ideal chain

29.5   References

[1] *A history of thermodynamics: the doctrine of energy and entropy* by Ingo Müller, p115

[2] Neumann RM (1980). "Entropic approach to Brownian movement". *American Journal of Physics* **48**(5): 354. doi:10.1119/1.1

[3] On the origin of gravity and the laws of Newton. Erik Verlinde

[4] Neumann RM (1977). "The entropy of a single Gaussian macromolecule in a noninteracting solvent", *The Journal of Chemical Physics* **66** (2): 870. Bibcode:1977JChPh..66..870N. doi:10.1063/1.433923.

[5] Encyclopedia of Life Science Article on Hydrophobic Effect; See Figure 4: http://xibalba.lcg.unam.mx/~||rgalindo/bioquimica/BQPosgrado2011/I%20FQ%20repaso/HydrophobicEffect.pdf

[6] "Crystalline Assemblies and Densest Packings of a Family of Truncated Tetrahedra and the Role of Directional Entropic Forces" (PDF). ACS. Archived from the original on 2011-12-01. Retrieved 23 June 2012.

[7] "Unified Theoretical Framework for Shape Entropy in Colloids" (PDF). Archived from the original (PDF) on 2013-09-03. Retrieved 20 October 2013.

[8] "A Directional Entropic Force Approach to Assemble Anisotropic Nanoparticles into Superlattices" (PDF). Archived from the original (PDF) on 2013-09-03. Retrieved 13 Jan 2014.

[9] "Structural Diversity and the Role of Particle Shape and Dense Fluid Behavior in Assemblies of Hard Polyhedra" (PDF). Archived from the original (PDF) on 2012-02-10. Retrieved 23 June 2012.

[10] Verlinde, Eric (6 January 2010). "Title: On the Origin of Gravity and the Laws of Newton". arXiv:1001.0785 [hep-th].

[11] E.P. Verlinde. "On the Origin of Gravity and the Laws of Newton". *JHEP 04, 29 (2011)*. arXiv:1001.0785.Bibcode:2011JHE doi:10.1007/JHEP04(2011)029.

[12] http://arxiv.org//abs/1001.4965, *Coulomb Force as an Entropic Force*, T. Wang

[13] http://arxiv.org//abs/0809.4631, *Simple field theoretical approach of Coulomb systems. Entropic effects*, D. di Caprio, J.P. Badiali, M. Holovko

[14] http://arxiv.org//abs/1009.5561, *Entropic Corrections to Coulomb's Law*, A. Sheykhi, S. H. Hendi

[15] http://arxiv.org//abs/1008.4147." Emergent Gauge Fields, *Peter G.O. Freund*

[16] http://arxiv.org//abs/1009.1506 *Unification of Dark Matter and Dark Energy in a Modified Entropic Force Model*, Zhe Chang, Ming-Hua Li, Xin Li

[17] Andrei N. Soklakov, "Occam's razor as a formal basis for a physical theory" (arXiv:math-ph/0009007, September 2000; Foundations of Physics Letters, 2002), "Complexity analysis for algorithmically simple strings" (arXiv:cs/0009001, September 2000).

Chapter 30

Free entropy

A thermodynamic **free entropy** is an entropic thermodynamic potential analogous to the free energy. Also known as a Massieu, Planck, or Massieu–Planck potentials (or functions), or (rarely) free information. In statistical mechanics, free entropies frequently appear as the logarithm of a partition function. The Onsager reciprocal relations in particular, are developed in terms of entropic potentials. In mathematics, free entropy means something quite different: it is a generalization of entropy defined in the subject of free probability.

A free entropy is generated by a Legendre transform of the entropy. The different potentials correspond to different constraints to which the system may be subjected.

30.1 Examples

See also: List of thermodynamic properties

The most common examples are:

where

Note that the use of the terms "Massieu" and "Planck" for explicit Massieu-Planck potentials are somewhat obscure and ambiguous. In particular "Planck potential" has alternative meanings. The most standard notation for an entropic potential is ψ, used by both Planck and Schrödinger. (Note that Gibbs used ψ to denote the free energy.) Free entropies where invented by French engineer Francois Massieu in 1869, and actually predate Gibbs's free energy (1875).

30.2 Dependence of the potentials on the natural variables

30.2.1 Entropy

$$S = S(U, V, \{N_i\})$$

By the definition of a total differential,

$$dS = \frac{\partial S}{\partial U} dU + \frac{\partial S}{\partial V} dV + \sum_{i=1}^{s} \frac{\partial S}{\partial N_i} dN_i$$

From the equations of state,

$$dS = \frac{1}{T}dU + \frac{P}{T}dV + \sum_{i=1}^{s}(-\frac{\mu_i}{T})dN_i$$

The differentials in the above equation are all of extensive variables, so they may be integrated to yield

$$S = \frac{U}{T} + \frac{pV}{T} + \sum_{i=1}^{s}(-\frac{\mu_i N}{T})$$

30.2.2 Massieu potential / Helmholtz free entropy

$$\Phi = S - \frac{U}{T}$$

$$\Phi = \frac{U}{T} + \frac{PV}{T} + \sum_{i=1}^{s}(-\frac{\mu_i N}{T}) - \frac{U}{T}$$

$$\Phi = \frac{PV}{T} + \sum_{i=1}^{s}(-\frac{\mu_i N}{T})$$

Starting over at the definition of Φ and taking the total differential, we have via a Legendre transform (and the chain rule)

$$d\Phi = dS - \frac{1}{T}dU - Ud\frac{1}{T}$$

$$d\Phi = \frac{1}{T}dU + \frac{P}{T}dV + \sum_{i=1}^{s}(-\frac{\mu_i}{T})dN_i - \frac{1}{T}dU - Ud\frac{1}{T}$$

$$d\Phi = -Ud\frac{1}{T} + \frac{P}{T}dV + \sum_{i=1}^{s}(-\frac{\mu_i}{T})dN_i$$

The above differentials are not all of extensive variables, so the equation may not be directly integrated. From $d\Phi$ we see that

$$\Phi = \Phi(\frac{1}{T}, V, \{N_i\})$$

If reciprocal variables are not desired,[3]:222

$$d\Phi = dS - \frac{TdU - UdT}{T^2}$$

$$d\Phi = dS - \frac{1}{T}dU + \frac{U}{T^2}dT$$

$$d\Phi = \frac{1}{T}dU + \frac{P}{T}dV + \sum_{i=1}^{s}(-\frac{\mu_i}{T})dN_i - \frac{1}{T}dU + \frac{U}{T^2}dT$$

$$d\Phi = \frac{U}{T^2}dT + \frac{P}{T}dV + \sum_{i=1}^{s}(-\frac{\mu_i}{T})dN_i$$

$$\Phi = \Phi(T, V, \{N_i\})$$

30.2.3   Planck potential / Gibbs free entropy

$$\Xi = \Phi - \frac{PV}{T}$$

$$\Xi = \frac{PV}{T} + \sum_{i=1}^{s}(-\frac{\mu_i N}{T}) - \frac{PV}{T}$$

$$\Xi = \sum_{i=1}^{s}(-\frac{\mu_i N}{T})$$

Starting over at the definition of Ξ and taking the total differential, we have via a Legendre transform (and the chain rule)

$$d\Xi = d\Phi - \frac{P}{T}dV - Vd\frac{P}{T}$$

$$d\Xi = -Ud\frac{2}{T} + \frac{P}{T}dV + \sum_{i=1}^{s}(-\frac{\mu_i}{T})dN_i - \frac{P}{T}dV - Vd\frac{P}{T}$$

$$d\Xi = -Ud\frac{1}{T} - Vd\frac{P}{T} + \sum_{i=1}^{s}(-\frac{\mu_i}{T})dN_i$$

The above differentials are not all of extensive variables, so the equation may not be directly integrated. From $d\Xi$ we see that

$$\Xi = \Xi(\frac{1}{T}, \frac{P}{T}, \{N_i\})$$

If reciprocal variables are not desired,[3]:222

$$d\Xi = d\Phi - \frac{T(PdV + VdP) - PVdT}{T^2}$$

$$d\Xi = d\Phi - \frac{P}{T}dV - \frac{V}{T}dP + \frac{PV}{T^2}dT$$

$$d\Xi = \frac{U}{T^2}dT + \frac{P}{T}dV + \sum_{i=1}^{s}(-\frac{\mu_i}{T})dN_i - \frac{P}{T}dV - \frac{V}{T}dP + \frac{PV}{T^2}dT$$

$$d\Xi = \frac{U + PV}{T^2}dT - \frac{V}{T}dP + \sum_{i=1}^{s}(-\frac{\mu_i}{T})dN_i$$

$$\Xi = \Xi(T, P, \{N_i\})$$

30.3   References

[1] Antoni Planes; Eduard Vives (2000-10-24). "Entropic variables and Massieu-Planck functions". *Entropic Formulation of Statistical Mechanics*. Universitat de Barcelona. Retrieved 2007-09-18.

[2] T. Wada; A.M. Scarfone (December 2004). "Connections between Tsallis' formalisms employing the standard linear average energy and ones employing the normalized q-average energy". *Physics Letters A* **335** (5–6): 351–362. arXiv:cond-mat/0410527. Bibcode:2005PhLA..335..351W. doi:10.1016/j.physleta.2004.12.054.

[3] *The Collected Papers of Peter J. W. Debye*. New York, New York: Interscience Publishers, Inc. 1954.

30.4 Bibliography

- Massieu, M.F. (1869). "Compt. Rend" **69** (858). p. 1057.

- Callen, Herbert B. (1985). *Thermodynamics and an Introduction to Thermostatistics* (2nd ed.). New York: John Wiley & Sons. ISBN 0-471-86256-8.

Chapter 31

Entropic explosion

An **entropic explosion** is an explosion in which the reactants undergo a large change in volume without releasing a large amount of heat. The chemical decomposition of triacetone triperoxide (TATP) is an example of an entropic explosion. It is not a thermochemically highly favored event (not much energy generated in chemical bond formation in reaction products). It rather involves an entropy burst, which is the result of formation of one ozone and three acetone gas phase molecules from every molecule of TATP in the solid state.

31.1 External links and references

- Dubnikova, F.; Kosloff, R.; Almog, J.; Zeiri, Y.; Boese, R.; Itzhaky, H.; Alt, A.; Keinan, E.; J. Am. Chem. Soc.; (Article); **2005**; 127(4); 1146-1159.

- The Skaggs Institute for Chemical Biology 2004 "The calculated thermal decomposition pathway of the TATP molecule was a complicated multistep process with several highly reactive intermediates, including singlet molecular oxygen and various biradicals. Of note, the calculations predict formation of acetone and ozone as the main decomposition products and not the intuitively expected oxidation products. The key conclusion from this study is that the explosion of TATP is not a thermochemically highly favored event. Rather, the explosion involves entropy burst, which is the result of formation of 4 gas-phase molecules from every molecule of TATP in the solid state. Quite unexpectedly, the 3 isopropylidene units of the TATP molecule do not play the role of fuel that can be oxidized and release energy during the explosion. Instead, these units function as molecular scaffolds that hold the 3 peroxide units close together spatially in the appropriate orientation for the decomposition chain reaction."

- Israeli Invention Detects TATP Explosives "To our great surprise," PET's inventor, Prof. Ehud Keinan, Dean of the Technion's Faculty of Chemistry, wrote in the Journal of the American Chemical Society, "we discovered that TATP is very different from all other conventional explosives in that it does not release heat during the explosion. It explodes by rapid decomposition of every solid-state molecule to four gas-phase molecules. This rare phenomenon, scientifically known as 'Entropic Explosion', is reminiscent of the rapid reaction that produces gas in the safety airbags of cars during accidents."

- Chemical and Engineering News PDF

Chapter 32

Sackur–Tetrode equation

The **Sackur–Tetrode equation** is an expression for the entropy of a monatomic classical ideal gas which incorporates quantum considerations which give a more detailed description of its regime of validity.

The Sackur–Tetrode equation is named for Hugo Martin Tetrode[1] (1895–1931) and Otto Sackur[2] (1880–1914), who developed it independently as a solution of Boltzmann's gas statistics and entropy equations, at about the same time in 1912.

32.1 Formula

The Sackur–Tetrode equation is written:

$$S = kN \left(\ln \left[\frac{V}{N} \left(\frac{4\pi m}{3h^2} \frac{U}{N} \right)^{3/2} \right] + \frac{5}{2} \right)$$

where V is the volume of the gas, N is the number of particles in the gas, U is the internal energy of the gas, k is Boltzmann's constant, m is the mass of a gas particle, h is Planck's constant and $\ln()$ is the natural logarithm. See Gibbs paradox for a derivation of the Sackur–Tetrode equation. See also the ideal gas article for the constraints placed upon the entropy of an ideal gas by thermodynamics alone.

The Sackur–Tetrode equation can also be conveniently expressed in terms of the thermal wavelength Λ.

$$\frac{S}{kN} = \ln \left[\frac{V}{N\Lambda^3} \right] + \frac{5}{2}.$$

Note that the assumption was made that the gas is in the classical regime, and is described by Maxwell–Boltzmann statistics (with "correct Boltzmann counting"). From the definition of the thermal wavelength, this means the Sackur–Tetrode equation is only valid for

$$\frac{V}{N\Lambda^3} \gg 1$$

and in fact, the entropy predicted by the Sackur–Tetrode equation approaches negative infinity as the temperature approaches zero.

32.2   Sackur–Tetrode constant

The **Sackur–Tetrode constant**, written S_0/R, is equal to S/kN evaluated at a temperature of $T = 1$ kelvin, at standard pressure (100 kPa or 101.325 kPa, to be specified), for one mole of an ideal gas composed of particles of mass equal to one atomic mass unit ($m_u = 1.660\,538\,782(83) \times 10^{-27}$ kg). Its 2010 CODATA recommended value is:[3]

$S_0/R = -1.151\,7078(23)$ for $p^\circ = 100$ kPa

$S_0/R = -1.164\,8708(23)$ for $p^\circ = 101.325$ kPa.

32.3   **References**

[1]   H. Tetrode (1912) "Die chemische Konstante der Gase und das elementare Wirkungsquantum" (The chemical constant of gases and the elementary quantum of action), *Annalen der Physik* **38**: 434 - 442. See also: H. Tetrode (1912) "Berichtigung zu meiner Arbeit: "Die chemische Konstante der Gase und das elementare Wirkungsquantum" " (Correction to my work: "The chemical constant of gases and the elementary quantum of action"), *Annalen der Physik* **39**: 255 - 256.

[2]   Sackur published his findings in the following series of papers:

   (a)   O. Sackur (1911) "Die Anwendung der kinetischen Theorie der Gase auf chemische Probleme" (The application of the kinetic theory of gases to chemical problems), *Annalen der Physik*, **36**: 958 - 980.

   (b)   O. Sackur, "Die Bedeutung des elementaren Wirkungsquantums für die Gastheorie und die Berechnung der chemischen Konstanten" (The significance of the elementary quantum of action to gas theory and the calculation of the chemical constant), *Festschrift W. Nernst zu seinem 25jährigen Doktorjubiläum gewidmet von seinen Schülern* (Halle an der Salle, Germany: Wilhelm Knapp, 1912), pages 405 - 423.

   (c)   O. Sackur (1913) "Die universelle Bedeutung des sog. elementaren Wirkungsquantums" (The universal significance of the so-called elementary quantum of action), *Annalen der Physik* **40**: 67 - 86.

[3]   CODATA2010

Chapter 33

Entropy (computing)

In computing, **entropy** is the randomness collected by an operating system or application for use in cryptography or other uses that require random data. This randomness is often collected from hardware sources, either pre-existing ones such as mouse movements or specially provided randomness generators.

33.1 Linux kernel

The Linux kernel generates entropy from keyboard timings, mouse movements, and IDE timings and makes the random character data available to other operating system processes through the special files /dev/random and /dev/urandom. This capability was introduced in Linux version 1.3.30.[1]

There are some Linux kernel patches allowing one to use more entropy sources.[2] The audio_entropyd project, which is included in some operating systems such as Fedora, allows audio data to be used as an entropy source.[3] Also available are video_entropyd which calculates random data from a video-source and entropybroker which includes these three and can be used to distribute the entropy data to systems not capable of running any of these (e.g. virtual machines). Furthermore one can use the HAVEGE algorithm through haveged to pool entropy.[4] In some systems, network interrupts can be used as an entropy source as well.[5]

On systems using the Linux kernel, programs needing significant amounts of random data from /dev/urandom cannot co-exist with programs reading little data from /dev/random, as /dev/urandom depletes /dev/random whenever it is being read.[6]

33.2 OpenBSD kernel

Main article: OpenBSD security features

OpenBSD has integrated cryptography as one of its main goals and has always worked on increasing its entropy for encryption but also for randomising many parts of the OS, including various internal operations of its kernel. Around 2011, two of the random devices were dropped and linked into a single source as it could produce hundreds of megabytes per second of high quality random data on an average system. This made depletion of random data by userland programs impossible on OpenBSD once enough entropy has initially been gathered. This is due to OpenBSD utilising an arc4random function to maximise the efficiency or minimise the wastage of entropy that the system has gathered.

33.3 Hurd kernel

A driver ported from the Linux kernel has been made available for the Hurd kernel.[7]

33.4 Solaris

/dev/random and /dev/urandom have been available as Sun packages or patches for Solaris since Solaris 2.6,[8] and have been a standard feature since Solaris 9.[9] As of Solaris 10, administrators can remove existing entropy sources or define new ones via the kernel-level cryptographic framework.

A 3rd-party kernel module implementing /dev/random is also available for releases dating back to Solaris 2.4.[8]

33.5 OS/2

There is a software package for OS/2 that allows software processes to retrieve random data.[10]

33.6 Windows

Microsoft Windows releases newer than Windows 95 use CryptoAPI to gather entropy in a similar fashion to Linux kernel's /dev/random.[11]

Windows's CryptoAPI uses the binary registry key *HKEY_LOCAL_MACHINE\SOFTWARE\Microsoft\Cryptography\ RNG\Seed* to store a seeded value from all of its entropy sources.[12]

Because CryptoAPI is closed-source, some free and open source software applications running on the Windows platform use other measures to get randomness. For example, GnuPG, as of version 1.06, uses a variety of sources such as the number of free bytes in memory that combined with a random seed generates desired randomness it needs.[13]

Programmers using CAPI can get entropy by calling CAPI's CryptGenRandom(), after properly initializing it.[14]

33.7 Embedded Systems

Embedded Systems have real issues gathering enough entropy as they are often very simple devices with short boot times and keys are often one of the first things a system may do. A simple study demonstrated the widespread use of weak keys by finding many embedded systems such as routers using the same keys. It was thought that the number of weak keys found would have been far higher if simple and often attacker determinable one-time unique identifiers had not been incorporated into the entropy of some of these systems.

33.8 Other systems

There are some software packages that allow one to use a userspace process to gather random characters, exactly what /dev/random does, such as EGD, the Entropy Gathering Daemon.[15]

33.9 Hardware-originated entropy

Modern CPUs and hardware often feature integrated generators that can provide high-quality and high-speed entropy to operating systems. On systems based on the Linux kernel, one can read the entropy generated from such a device through

/dev/hw_random.[16] However, sometimes /dev/hw_random may be slow;[17] usually around 80 KiB/s.[18]

There are some companies manufacturing entropy generation devices, and some of them are shipped with drivers for Linux.[19]

On Debian, one can install the rng-tools package (apt-get install rng-tools) that supports the true random number generators (TRNGs) found in CPUs supporting the RdRand instruction, Trusted Platform Modules and in some Intel, AMD, or VIA chipsets,[20] effectively increasing the entropy collected into /dev/random and potentially improving the cryptographic potential. This is especially useful on headless systems that have no other sources of entropy.

33.10 Practical implications

System administrators, especially those supervising Internet servers, have to ensure that the server processes will not halt because of entropy depletion. Entropy on servers utilising the Linux kernel, or any other kernel or userspace process that generates entropy from the console and the storage subsystem, is often less than ideal because of the lack of a mouse and keyboard, thus servers have to generate their entropy from a limited set of resources such as IDE timings.

The entropy pool size in Linux is viewable through the file */proc/sys/kernel/random/entropy_avail* and should generally be at least 2000 bits (out of a maximum of 4096).[21][22] Entropy changes frequently.

Administrators responsible for systems that have low or zero entropy should not attempt to use /dev/urandom as a substitute for /dev/random as this may cause SSL/TLS connections to have lower-grade encryption.[23]

Some software systems change their Diffie-Hellman keys often, and this may in some cases help a server to continue functioning normally even with an entropy bottleneck.[24]

On servers with low entropy, a process can appear hung when it is waiting for random characters to appear in /dev/random (on Linux-based systems). For example, there was a known problem in Debian that caused exim4 to hang in some cases because of this.[25]

33.10.1 Security

Entropy sources can be used for keyboard timing attacks.[26]

Entropy can affect the cryptography (TLS/SSL) of a server: If it is too low then the regeneration of codes can take a long time to complete.

In some cases a cracker (malicious attacker) can guess some bits of entropy from the output of a pseudorandom number generator (PRNG), and this happens when not enough entropy is introduced into the PRNG.[27]

33.11 Potential sources

Commonly used entropy sources include the mouse, keyboard, and IDE timings, but there are other potential sources. For example, one could collect entropy from the computer's microphone, or by building a sensor to measure the air turbulence inside a disk drive.[28] However, microphones are usually not available in servers. Microphones are not needed, white noise can be collected from empty microphone, or line in jacks. Webcams are another source of entropy, some are noisy, others can be pointed at lava lamps. Generating entropy on servers you do not have control over is a bad idea as it can be influenced one way or another.

For Unix/BSD derivatives there exists a USB based solution that utilizes an ARM Cortex CPU for filtering / securing the bit stream generated by two entropy generator sources in the system.[29]

33.12   See also

- Entropy (information theory)

- Entropy

- Randomness

33.13   References

[1] random(4) - Linux man page (die.net)

[2] Robotic Tendencies » Missing entropy

[3] Fedora Package Database - audio-entropyd

[4] "haveged - A simple entropy daemon". Retrieved 3 April 2011.

[5] Entropy and Random Devices | LinuxLink by TimeSys - Your Embedded Linux Resource

[6] High-Entropy Randomness Generator

[7] /dev/(.u)random driver for GNU/Hurd (ibofobi.dk)

[8] Solaris /dev/random through emulation

[9] Solaris /dev/random

[10] Rexx Entropy Gathering Daemon for OS/2

[11] GPL command-line shred alternative for Windows

[12] Source for entropy on Windows platforms with CryptoAPI installed

[13] How does Windows GnuPG generate random numbers on keygen?

[14] http://www.cs.berkeley.edu/~{}daw/rnd/cryptoapi-rand http://archives.seul.org/or/cvs/Mar-2004/msg00078.html

[15] Secure Programs HOWTO - Random Numbers

[16] 'Re: SSL/TLS entropy problem,' - MARC

[17] Re: /dev/hw_random

[18] Re: /dev/hw_random

[19] http://www.std.com/~{}reinhold/truenoise.html http://random.com.hr/products/random/hg324.html

[20]

[21] Re: [exim] no reply to STARTTLS

[22] random(4) Linux man page. die.net

[23] SSL/TLS entropy problem, a.k.a. pops timeouts (was: sasl ldap problem)

[24] Josefsson. Simon: [TLS] Re: Short Ephermal Diffie-Hellman keys (ietf.org mailing list)

[25] [gnutls-dev] gnutls_rsa_params_init hangs. Is regenerating rsa-params once a day too frequent?. lists.gnupg.org

[26] Zalewski. Michal: Unix entropy source can be used for keystroke timing attacks, 2003

[27] Re: entropy depletion (was: SSL/TLS passive sniffing). 2005

[28] Build your own cryptographically safe server/client protocol - 4.8.3. Collecting entropy

[29] http://www.entropykey.co.uk

33.14 External links

- Overview of entropy and of entropy generators in Linux

Chapter 34

Heat death of the universe

For the album by Off Minor, see The Heat Death of the Universe.

The **heat death of the universe** is a historically suggested ultimate fate of the universe in which the universe has diminished to a state of no thermodynamic free energy and therefore can no longer sustain processes that consume energy (including computation and life). Heat death does not imply any particular absolute temperature; it only requires that temperature differences or other processes may no longer be exploited to perform work. In the language of physics, this is when the universe reaches thermodynamic equilibrium (maximum entropy). The hypothesis of heat death stems from the ideas of William Thomson, 1st Baron Kelvin, who in the 1850s took the theory of heat as mechanical energy loss in nature (as embodied in the first two laws of thermodynamics) and extrapolated it to larger processes on a universal scale.

In a more recent view than Kelvin's, it was asserted by a respected authority on thermodynamics,[1] Max Planck, that the phrase 'entropy of the universe' has no meaning because it admits of no accurate definition.[2]

34.1 Origins of the idea

The idea of heat death stems from the second law of thermodynamics, which states that entropy tends to increase in an isolated system. If the universe lasts for a sufficient time, it will asymptotically approach a state where all energy is evenly distributed. In other words, in nature there is a tendency to the dissipation (energy loss) of mechanical energy (motion); hence, by extrapolation, there exists the view that the mechanical movement of the universe will run down, as work is converted to heat, in time because of the second law.

The conjecture that all bodies in the universe cool off, eventually becoming too cold to support life, seems to have been first put forward by the French astronomer Jean-Sylvain Bailly in 1777 in his writings on the history of astronomy and in the ensuing correspondence with Voltaire. In Bailly's view, all planets have an internal heat and are now at some particular stage of cooling. Jupiter, for instance, is still too hot for life to arise there for thousands of years, while the Moon is already too cold. The final state, in this view, is described as one of "equilibrium" in which all motion ceases.[3]

The idea of heat death as a consequence of the laws of thermodynamics, however, was first proposed in loose terms beginning in 1851 by William Thomson, 1st Baron Kelvin, who theorized further on the mechanical energy loss views of Sadi Carnot (1824), James Joule (1843), and Rudolf Clausius (1850). Thomson's views were then elaborated on more definitively over the next decade by Hermann von Helmholtz and William Rankine.

34.1.1 History

The idea of heat death of the universe derives from discussion of the application of the first two laws of thermodynamics to universal processes. Specifically, in 1851 William Thomson (Lord Kelvin) outlined the view, as based on recent experiments on the dynamical theory of heat, that "heat is not a substance, but a dynamical form of mechanical effect, we perceive that there must be an equivalence between mechanical work and heat, as between cause and effect."[4]

Lord Kelvin originated the idea of universal heat death in 1852.

In 1852, Thomson published his "On a Universal Tendency in Nature to the Dissipation of Mechanical Energy" in which he outlined the rudiments of the second law of thermodynamics summarized by the view that mechanical motion and the energy used to create that motion will tend to dissipate or run down, naturally.[5] The ideas in this paper, in relation to their application to the age of the sun and the dynamics of the universal operation, attracted the likes of William Rankine and Hermann von Helmholtz. The three of them were said to have exchanged ideas on this subject.[6] In 1862, Thomson published "On the age of the Sun's heat", an article in which he reiterated his fundamental beliefs in the indestructibility of energy (the first law) and the universal dissipation of energy (the second law), leading to diffusion of heat, cessation of useful motion (work), and exhaustion of potential energy through the material universe while clarifying his view of the consequences for the universe as a whole. In a key paragraph, Thomson wrote:

> The result would inevitably be a state of universal rest and death, if the universe were finite and left to obey existing laws. But it is impossible to conceive a limit to the extent of matter in the universe; and therefore science points rather to an endless progress, through an endless space, of action involving the transformation of potential energy into palpable motion and hence into heat, than to a single finite mechanism, running down like a clock, and stopping for ever.[7]

In the years to follow both Thomson's 1852 and the 1865 papers, Helmholtz and Rankine both credited Thomson with the idea, but read further into his papers by publishing views stating that Thomson argued that the universe will end in a "*heat death*" (Helmholtz) which will be the "*end of all physical phenomena*" (Rankine).[6][8]

34.2 Current status

See also: Entropy § Cosmology and Entropy (arrow of time) § Cosmology

Inflationary cosmology suggests that in the early universe, before cosmic inflation, energy was uniformly distributed,[9] and the universe was thus in a state superficially similar to heat death. However, these two states are actually very different: in the early universe, gravity was a very important force, and in a gravitational system, if energy is uniformly distributed, entropy is quite low, compared to a state in which most matter has collapsed into black holes. Thus, such a state is not in thermodynamic equilibrium, as it is thermodynamically unstable.[10][11]

Proposals about the final state of the universe depend on the assumptions made about its ultimate fate, and these assumptions have varied considerably over the late 20th century and early 21st century. In a hypothesized "open" or "flat" universe that continues expanding indefinitely, a heat death is also expected to occur,[12] with the universe cooling to approach absolute zero temperature and approaching a state of maximal entropy over a very long time period. There is dispute over whether or not an expanding universe can approach maximal entropy; it has been proposed that in an expanding universe, the value of maximum entropy increases faster than the universe gains entropy, causing the universe to move progressively further away from heat death.

It is dubious whether there is a valid definition of 'the entropy of the universe'. In a view more recent than Kelvin's, Max Planck wrote that the phrase 'entropy of the universe' has no meaning because it admits of no accurate definition.[11][2] More recently, Grandy writes: "It is rather presumptuous to speak of the entropy of a universe about which we still understand so little, and we wonder how one might define thermodynamic entropy for a universe and its major constituents that have never been in equilibrium in their entire existence."[13] According to Tisza: "If an isolated system is not in equilibrium, we cannot associate an entropy with it."[14] According to Gallavotti: "... there is no universally accepted notion of entropy for systems out of equilibrium, even when in a stationary state."[15] Discussing the question of entropy for non-equilibrium states in general, Lieb and Yngvason express their opinion as follows: "Despite the fact that most physicists believe in such a nonequilibrium entropy, it has so far proved impossible to define it in a clearly satisfactory way."[16] In the opinion of Čápek and Sheehan, "*no* known formulation [of entropy] applies to *all* possible thermodynamic regimes."[17] In Landsberg's opinion, "The *third* misconception is that thermodynamics, and in particular, the concept of entropy, can without further enquiry be applied to the whole universe. ... These questions have a certain fascination, but the answers are speculations, and lie beyond the scope of this book."[18]

A recent analysis of entropy states that "The entropy of a general gravitational field is still not known," and that "gravitational entropy is difficult to quantify." The analysis considers several possible assumptions that would be needed for estimates, and suggests that the visible universe has more entropy than previously thought. This is because the analysis concludes that supermassive black holes are the largest contributor.[19] Another writer goes further; "It has long been known that gravity is important for keeping the universe out of thermal equilibrium. Gravitationally bound systems have negative specific heat—that is, the velocities of their components increase when energy is removed. ... Such a system does not evolve toward a homogeneous equilibrium state. Instead it becomes increasingly structured and heterogeneous as it fragments into subsystems."[20]

34.3 Time frame for heat death

Main article: Future of an expanding universe

From the Big Bang through the present day, matter and dark matter in the universe are thought to have been concentrated in stars, galaxies, and galaxy clusters, and are presumed to continue to be so well into the future. Therefore, the universe is not in thermodynamic equilibrium and objects can do physical work.[21], §VID. The decay time for a supermassive black hole of roughly 1 galaxy-mass (10^{11} solar masses) due to Hawking radiation is on the order of 10^{100} years,[22] so entropy can be produced until at least that time. After that time, the universe enters the so-called dark era, and is expected to consist chiefly of a dilute gas of photons and leptons.[21], §VIA. With only very diffuse matter remaining, activity in the universe will have tailed off dramatically, with extremely low energy levels and extremely long time scales. Speculatively, it is possible that the universe may enter a second inflationary epoch, or, assuming that the current vacuum state is a false vacuum, the vacuum may decay into a lower-energy state.[21], §VE. It is also possible that entropy production will cease and the universe will reach heat death.[21], §VID. Possibly another universe could be created by random quantum fluctuations or quantum tunneling in roughly $10^{10^{56}}$ years.[23] Over an infinite time, there would be a spontaneous entropy decrease via the Poincaré recurrence theorem, thermal fluctuations,[24][25] and Fluctuation theorem.[26][27]

34.4 See also

- Arrow of time

- Big Bang

- Big Bounce

- Big Crunch

- Big Rip

- Chronology of the universe

- Cyclic model

- Entropy (arrow of time)

- Graphical timeline from Big Bang to Heat Death

- Fluctuation theorem

- Heat death paradox

- Terasecond and longer

34.5 References

[1] Uffink, J. (2003). Irreversibility and the Second Law of Thermodynamics. Chapter 7 of *Entropy*, p. 129 of Greven, A., Keller, G., Warnecke (editors) (2003), *Entropy*, Princeton University Press, Princeton NJ, ISBN 0-691-11338-6. Uffink writes: "The importance of Planck's *Vorlesungen über Thermodynamik* (Planck 1897) can hardly be [over]estimated. The book has gone through 11 editions, from 1897 until 1964, and still remains the most authoritative exposition of classical thermodynamics."

[2] Planck, M. (1897/193). *Treatise on Thermodynamics*, translated by A. Ogg, p. 101.

[3] Brush, Stephen G. (1996). *A History of Modern Planetary Physics: Nebulous Earth*. Cambridge University Press. p. 77. ISBN 9780521441711.

[4] Thomson, William. (1851). "On the Dynamical Theory of Heat, with numerical results deduced from Mr Joule's equivalent of a Thermal Unit, and M. Regnault's Observations on Steam." Excerpts. [§§1–14 & §§99–100], Transactions of the Royal Society of Edinburgh, March, 1851; and Philosophical Magazine IV. 1852. [from Mathematical and Physical Papers, vol. i, art. XLVIII, pp. 174]

[5] Thomson, William (1852). "On a Universal Tendency in Nature to the Dissipation of Mechanical Energy" Proceedings of the Royal Society of Edinburgh for April 19, 1852, also Philosophical Magazine, Oct. 1852. [This version from Mathematical and Physical Papers, vol. i, art. 59, pp. 511.]

[6] Smith, Crosbie & Wise, Matthew Norton. (1989). *Energy and Empire: A Biographical Study of Lord Kelvin*. (pg. 500). Cambridge University Press.

[7] Thomson, William. (1862). "On the age of the sun's heat", Macmillan's Mag., 5, 288–93; PL, 1, 394–68.

[8] Physics Timeline (Helmholtz and Heat Death, 1854)

[9] Liddle, Andrew R. (1999). "An introduction to cosmological inflation". arXiv:astro-ph/9901124 [astro-ph].

[10] Hawking, S. (1976). "Black holes and thermodynamics". *Physical Review D* **13** (2): 191. Bibcode:1976PhRvD..13..191H. doi:10.1103/PhysRevD.13.191.

[11] Hawking, S. W.; Page, Don N. "Thermodynamics of black holes in anti-de Sitter space". *Comm. Math. Phys. 87, no. 4 (1982), 577–588.* Retrieved 2006-09-09.

[12] Plait, Philip *Death From the Skies!*. Viking Penguin, NY. ISBN 978-0-670-01997-7, p. 259

[13] Grandy, W.T. (Jr) (2008). *Entropy and the Time Evolution of Macroscopic Systems*, Oxford University Press, Oxford UK, ISBN 978-0-19-954617-6, p. 151.

[14] Tisza, L. (1966). *Generalized Thermodynamics*, M.I.T Press, Cambridge MA, p. 41.

[15] Gallavotti, G. (1999). *Short Treatise of Statistical Mechanics*, Springer, Berlin, ISBN 9783540648833, p. 290.

[16] Lieb, E.H., Yngvason, J. (2003). The entropy of classical thermodynamics, Chapter 8 of Greven, A., Keller, G., Warnecke (editors) (2003). *Entropy*. Princeton University Press, Princeton NJ, ISBN 0-691-11338-6, page 190.

[17] Čápek, V., Sheehan, D.P. (2005). *Challenges to the Second Law of Thermodynamics: Theory and Experiment*, Springer, Dordrecht, ISBN 1-4020-3015-0, p. 26.

[18] Landsberg, P.T. (1961). *Thermodynamics, with Quantum Statistical Illustrations*, Wiley, New York, p. 391.

[19] Egan, Chas A.; Lineweaver, Charles H. (2009). "A Larger Estimate of the Entropy of the Universe". arXiv:0909.3983 [astro-ph.CO].

[20] Smolin, L. (2014). "Time, laws, and future of cosmology". *Physics Today* **67**: 38–43 [42]. Bibcode:2014PhT....67c..38S. doi:10.1063/pt.3.2310.

[21] Fred C. Adams and Gregory Laughlin (1997). "A dying universe: the long-term fate and evolution of astrophysical objects". *Reviews of Modern Physics* **69** (2): 337–372. arXiv:astro-ph/9701131. Bibcode:1997RvMP...69..337A. doi:10.1103/RevModPhys.69.337..

[22] Particle emission rates from a black hole: Massless particles from an uncharged, nonrotating hole, Don N. Page, *Physical Review D* **13** (1976). pp. 198–206. doi:10.1103/PhysRevD.13.198. See in particular equation (27).

[23] Carroll, Sean M. and Chen, Jennifer (2004). "Spontaneous Inflation and Origin of the Arrow of Time". arXiv:hep-th/0410270.

[24] http://arxiv.org/pdf/astro-ph/0302131.pdf?origin=publication_detail

[25] "[1205.1046] Interplay between quantum phase transitions and the behavior of quantum correlations at finite temperatures". *arxiv.org*.

[26] Xiu-San Xing (1 November 2007). "Spontaneous entropy decrease and its statistical formula". *ResearchGate*.

[27] "Sinks in the landscape, Boltzmann brains and the cosmological constant problem - Abstract - Journal of Cosmology and Astroparticle Physics - IOPscience". *iop.org*.

34.6 Text and image sources, contributors, and licenses

34.6.1 Text

- **Introduction to entropy** *Source:* https://en.wikipedia.org/wiki/Introduction_to_entropy?oldid=672257285 *Contributors:* Edward, Kku, Tobias Bergemann, Dratman, Dave souza, Art LaPella, Army1987, Pharos, Gary, PAR, Jheald, Carcharoth, DaveApter, Vegaswikian, Fresheneesz, Loom91, Grafen, Retired username, Dhollm, Brisvegas, Light current, Serendipodous, User24, SmackBot, Bduke, T.J. Crowder, Microfrost, Xyzzyplugh, Sadi Carnot, PAS, Lazylaces, JorisvS, 16@r, Kirbytime, K. FilipeS, Cydebot, Gtxfrance, Headbomb, John254, MarshBot, Dylan Lake, Ray Eston Smith Jr, Hypergeek14, BigrTex, Davidm617617, TJKluegel, Papparolf, Adam C C, Davwillev, Zain Ebrahim111, Sesshomaru, Kbrose, ConfuciusOrnis, Dolphin51, Ac1201, Rodhullandemu, Plastikspork, Crowsnest, Yobot, Zaereth, DBBabyboydavey, Kissnmakeup, AnomieBOT, Daniele Pugliesi, Ipatrol, EryZ, Danno uk, LilHelpa, DanP4522874, FrescoBot, Vh mby, Nilock, Vrenator, Combee123, Sixtylarge2000, Drozdyuk, Wayne Slam, ClueBot NG, Michelino12, Marko Petek, Gsoverby, Prokaryotes, W. P. Uzer, Patrickrowanandrews and Anonymous: 69

- **History of entropy** *Source:* https://en.wikipedia.org/wiki/History_of_entropy?oldid=678384639 *Contributors:* Tobias Bergemann, ELApro, MBisanz, PAR, Jheald, Woohookitty, Wijnand, Ligulem, Srleffler, Blutfink, Dhollm, E Wing, SmackBot, Hmains, Colonies Chris, Tschwenn, Sadi Carnot, JzG, JHunterJ, FilipeS, Cydebot, Sam Staton, Sytelus, D.H, Tomixdf, Narssarssuaq, MER-C, Usien6, Kbrose, Hobojaks, Mari Hamilton, ClueBot, Timberframe, 718 Bot, CohesionBot, HexaChord, LemmeyBOT, OlEnglish, Yobot, Citation bot, Chjoaygame, Astrojed, Tuxedo junction, Rquine, Nerlost, W. P. Uzer and Anonymous: 19

- **Entropy** *Source:* https://en.wikipedia.org/wiki/Entropy?oldid=682883931 *Contributors:* Tobias Hoevekamp, Chenyu, CYD, Bryan Derksen, Zundark, The Anome, BlckKnght, Awaterl, XJaM, Roadrunner, Peterlin~enwiki, Jdpipe, Heron, Youandme, Olivier, Stevertigo, PhilipMW, Michael Hardy, Macvienna, Zeno Gantner, Looxix~enwiki, J'raxis, Humanoid, Darkwind, AugPi, Jiang, Kaihsu, Jani~enwiki, Mxn, Smack, Disdero, Tantalate, Timwi, Reddi, Terse, Dysprosia, Jitse Niesen, Andrewman327, Piolinfax, Tpbradbury, Saltine, J D, Atuin, Raul654, Wet-man, Lumos3, Jni, Phil Boswell, Ruudje, Robbot, Fredrik, Alrasheedan, Naddy, Sverdrup, Texture, Hadal, David Edgar, Ianml, Aetheling, Tobias Bergemann, Connelly, Paisley, Giftlite, Graeme Bartlett, DavidCary, Haeleth, BenFrantzDale, Lee J Haywood, Herbee, Xerxes314, Everyking, Anville, Dratman, Henry Flower, NotableException, Gracefool, Macrakis, Christofurio, Zeimusu, Yath, Gunnar Larsson, Karol Langner, JimWae, Mjs, H Padleckas, Pmanderson, Icairns, Arcturus, Tsemii, Edsanville, E David Moyer, Mschlindwein, Freakofnurture, Lonelsle, Rich Farmbrough, KillerChihuahua, Pjacobi, Vsmith, Dave souza, Gianluigi, Mani1, Paul August, Bender235, Kbh3rd, Kjoonlee, Geok-ing66, RJHall, Pt, El C, Laurascudder, Aaronbrick, Chuayw2000, Bobo192, Marathoner, Wisdom89, Giraffedata, VBGFscJUn3, Physicistjedi, 99of9, Obradovic Goran, Haham hanuka, Mdd, Geschichte, Gary, Mennato, Arthena, Keenan Pepper, Benjah-bmm27, Riana, Iris lorain, PAR, Melaen, Velella, Knowledge Seeker, Jheald, Count Iblis, Drat, Egg, Artur adib, Lerdsuwa, Gene Nygaard, Oleg Alexandrov, Omnist, Sand-wiches, Joriki, Velho, Simetrical, MartinSpacek, Woohookitty, Linas, TigerShark, StradivariusTV, Jacobolus, Wijnand, EnSamulili, Pkeck, Mouvement, Jwanders, Eleassar777, Tygar, SeventyThree, Jonathan48, DL5MDA, Aarghdvaark, Graham87, Marskell, V8rik, Rjwilmsi, Thechamelon, HappyCamper, Ligulem, TheIncredibleEdibleOompaLoompa, Douglace, MarnetteD, GregAsche, FlaBot, RobertG, Mathbot, Nihiltres, Gurch, Frelke, Intgr, Fresheneesz, Srleffler, Physchim62, WhyBeNormal, Chobot, DVdm, VolatileChemical, YurikBot, Wave-length, Jimp, Alpt, Kafziel, Wolfmankurd, Bobby1011, Loom91, Bhny, JabberWok, Stephenb, Gaius Cornelius, Wimt, Ugur Basak, Odysses, Shanel, NawlinWiki, SAE1962, Sitearm, Retired username, Dhollm, Ellwyz, Crasshopper, Shotgunlee, Dr. Ebola, DeadEyeArrow, Bota47, Rayc, Brisvegas, Doetoe, Ms2ger, WAS 4.250, Vadept, Light current, Enormousdude, Theodolite, Ballchef, The Fish, ChrisGriswold, Theda, Chaiken, Paganpan, Bo Jacoby, Pentasyllabic, Pipifax, DVD R W, ChemGardener, Itub, Attilios, Otheus, SmackBot, ElectricRay, Reedy, In-verseHypercube, KnowledgeOfSelf, Jim62sch, David Shear, Ixtli, Jab843, Pedrose, Edgar181, Xaosflux, Hmains, Betacommand, Skizzik, ThorinMuglindir, Kmarinas86, Oneismany, Master Jay, Kurykh, QTCaptain, Bduke, Dreg743, Complexica, Imaginaryoctopus, Basalisk, Nbarth, Sciyoshi~enwiki, Dlenmn, Colonies Chris, Darth Panda, Chrislewis.au, BW95, Zachorious, Can't sleep, clown will eat me, Ajaxkroon, ZezzaMTE, Apostolos Margaritis, Shunpiker, Homestarmy, AltheaJ, Ddon, Memming, Engwar, Nakon, J.Wolfe@unsw.edu.au, G716, Love-Monkey, Metamagician3000, Sadi Carnot, Yevgeny Kats, SashatoBot, Tsiehta, Lambiam, AThing, Oenus, Eric Hawthorne, MagnaMopus, Lakinekaki, Mbeychok, JorisvS, Mgiganteus1, Nonsuch, Dftb, Physis, Slakr, Tiogalinha~enwiki, Abjad, Dr.K., Cbuckley, HappyVR, Adodge, BranStark, HisSpaceResearch, K, Astrobayes, Paul venter, Gmaster108, RekishiEJ, Jive Dadson, JRSpriggs, Emote, Patrickwooldridge, VaughanPratt, CmdrObot, Hanspi, Jsd, Dgw, BassBone, Omnichic82, Electricmic, NE Ent, Adhanali, FilipeS, Jac16888, Cydebot, Natasha2006, Kanags, WillowW, Gtxfrance, Mike Christie, Rifleman 82, Gogo Dodo, Sam Staton, Hkyriazi, Rracecarr, Miguel de Servet, Michael C Price, Rize Above, Soumya.92, Aintsemic, Hugozam, Gurudev23, Csdidier, Abtract, Yian, Thijs!bot, Epbr123, Lg king, Opabinia regalis, Move-away00, LeBofSportif, Teh tennisman, Kahriman~enwiki, Fred t hamster, Headbomb, Neligterink, Esowteric, Electron9, EdJohnston, D.H, Dartbanks, DJ Creature, Stannered, Seaphoto, FrankLambert, Ray Eston Smith Jr, Tim Shuba, MECU, Astavats, Serpent's Choice, JAnD-bot, MER-C, Reallybored999, Physical Chemist, XerebZ, RebelRobot, Magioladitis, Bongwarrior, VoABot II, Avjoska, Bargebum, Tonyfaull, HGHSTROJAN, Dirac66, User A1, Jacobko, Glen, Steevven1, DGG, Hdt83, GuidoGer, Keith D, Ronburk, Pbroks13, Leyo, Mbweissman, Mausy5043, HEL, J. delanoy, Captain panda, Jorgenumata, Numbo3, Peter Chastain, Josterhage, Maurice Carbonaro, Thermbal, Shawn inMontreal, Camarks, Cmbreuel, Nwbeeson, Touch Of Light, Constatin666999, Pundit, Edzevallos, Juliancolton, Linshukun, DorganBot, Ris-ing*From*Ashes, Inwind, Lseixas, Izno, Idioma-bot, Fimbulfamb, Cuzkatzimhut, Ballhausflip, Larryisgood, Macedonian, Pasquale.Carelli, LokiClock, Philip Trueman, Nikhil Sanjay Bapat, TXiKiBoT, BJNartowt, Antoni Barau, Rei-bot, Anonymous Dissident, Drestros power, Hai2410, Vendrov, Leafyplant, Raymondwinn, Billgdiaz, Mwilso24, Kpedersen1, Mouse is back, Koen Van de moortel~enwiki, UffeHThyge-sen, Synthebot, Sesshomaru, Locke9k, Arcfrk, Nagy, Tennismaniac2112, Bojack727, Katzmik, EmxBot, Vbrayne, Kbrose, SieBot, Wolf.312, Moonriddengirl, Paradoctor, Gerakibot, Vanished user 82345ijgeke4tg, Arjun r acharya, Happysailor, Radon210, AngelOfSadness, Lidi-aFourdraine, Georgette2, Hamiltondaniel, WikiLaurent, Geoff Plourde, Mad540trix, Dolphin51, Emansf, ClueBot, Compdude47, Foxj, Yurko~enwiki, The Thing That Should Not Be, Ciacco, Plastikspork, Dtguelph, Riskdoc, Drmies, Bbanerje, ILikeMIDI, Josemald, Lbertolotti, DragonBot, Djr32, Awickert, PhySusie, Tnxman307, M.O.X, Wingwongdong, Revotfel, SchreiberBike, Galor612, Versus22, Edkarpov, Passwordwas1234, DumZiBoT, Tuuky, XLinkBot, Gnowor, Superkan619, BodhisattvaBot, Boob12, Ost316, Quidproquo2004, Gonfer, MilesTerrex, Subver-sive.sound, Private Pilot, WikiDao, Aunt Entropy, NCDane, JohnBonham69, Debzer, Phidus, Addbot, Eric Drexler, Tanhabot, Favonian, Ruddy9hell, Causticorulos, Mean Free Path, Dougbateman, Tide rolls, Suz115, Gatewayofintrigue, Gail, Legobot, Luckas-bot, Yobot, Za-ereth, Ht686rg90, Legobot II, Kissnmakeup, JHoffmueller~enwiki, AnomieBOT, Cantanchorus, IRP, Galoubet, Piano non troppo, Materi-alscientist, Citation bot, ArthurBot, DirlBot, Branxton, FreeRangeFrog, Xqbot, Engineering Guy, Addihockey10, Jeffrey Mall, DSisyphBot,

Necron909, Raffamaiden, Srich32977, Almabot, Munozdj, Schwijker, GrouchoBot, Tnf37, Ute in DC, Philip2357, Omnipaedista, RibotBOT, Waleswatcher, Smallman12q, Garethb1961, Mishka.medvezhonok, Chjoaygame, GT5162, Maghemite, C1t1v151on, Theowoo, Craig Pemberton, BenzolBot, Kwiki, Vh mby, MorphismOfDoom, DrilBot, Pinethicket, I dream of horses, HRoestBot, Marsiancba, Martinvl, Calmer Waters, Jschnur, RedBot, Tcnuk, Nora lives, SkyMachine, IVAN3MAN, Nobleness of Mind, Quantumechanic, TobeBot, Jschissel, Lotje, DLMcN, Dinamik-bot, Vrenator, Lordloihi, Bookbuddi, Stroppolo, Gegege13, DARTH SIDIOUS 2, Woogee, Dick Chu, Regancy42, Drpriver, Massieu, Prasadmalladi, EmausBot, John of Reading, Lea phys, 12seda78, 478jjjz, Heoigi, Netheril96, K6ka, Serketan, Capcom1116, Oceans and oceans, Akhil 0950, JSquish, Fæ, Mkratz, Lateg, Ὁ οἶστρος, Cobaltcigs, Quondum, Glockenklang1, Parodi, Music Sorter, Pachyphytum, Schurasbrat, Zueignung, Carmichael, RockMagnetist, Tritchls, GP modernus, DASHBotAV, ResearchRave, Mikhail Ryazanov, Debu334, ClueBot NG, Tschijnmotschau, Intoronto1125, Chester Markel, Marechal Ney, Widr, Natron25, Amircrypto, Helpful Pixie Bot, Art and Muscle, Jack sherrod, Ramaksoud2000, Bibcode Bot, BZTMPS, Jeffscott007, Bths83Cu87Aiu06, Juro2351, Paolo Lipparini, DIA-888, Future-Trillionaire, Zedshort, Cky2250, Uopchem2510, Uopchem2517, Eduardofeld, Justincheng12345-bot, Bobcorn123321, LEBOLTZMANN2, Toni 001, ChrisGualtieri, Adwaele, JYBot, APerson, AlecTaylor, Thinkadoodle, Webclient101, Mogism, Makecat-bot, CuriousMind01, Sfzh, Ssteve90266, KingQueenPrince, Blue3snail, Thearchontect, Spetalnick, Rjg83, Curatrice, Random Dude Who Is Cool, Sajjadha, Mattia Guerri, Probeb217, Loverthehater, TheNyleve, Rkswb, Prokaryotes, DavRosen, Damián A. Fernández Beanato, Bruce Chen 0010334, Jianhui67, W. P. Uzer, PhoenixPub, Technoalpha, ProKro, Anrnusna, Saad bin zubair, QuantumMatt101, Dragonlord Jack, Elenceq, Monkbot, Eczanne, Lamera1234, TaeYunPark, ClockWork96, Georgeciobanu, Gbkrishnappa2015, Eliodorochia, KasparBot, Ericliu shu and Anonymous: 791

- **Entropy and life** *Source:* https://en.wikipedia.org/wiki/Entropy_and_life?oldid=664083055 *Contributors:* Gabbe, Reddi, Alan Liefting, Alison, Edsanville, D6, Rich Farmbrough, Mdd, PAR, Jheald, Bobrayner, Theodork, Jrtayloriv, Abarry, Dhollm, Abb3w, CLW, Star trooper man, SmackBot, Prebys, InverseHypercube, Schnitzi, Smee, Shalom Yechiel, John C PI, EPM, Sadi Carnot, CmdrObot, Cydebot, Miguel de Servet, Eleuther, Shlomi Hillel, Pasc06, MER-C, Tonyfaull, Cardamon, TREX6662k5, Ekotkie, Mange01, Khullah-enwiki, TomS TDotO, PC78, Cmbreuel, DadaNeem, Inwind, Aymatth2, Shanata, Hamiltondaniel, Gerst20051, Sfan00 IMG, F-j123, Niceguyedc, Mattmcgyver, CohesionBot, Frongle, Fascani, Editor2020, Airplaneman, Download, NittyG, AnomieBOT, Jim1138, Citation bot, Ywaz, Skyerise, Bibliorrhea, Jandalhandler, Netheril96, Solomonfromfinland, Helpful Pixie Bot, Bibcode Bot, PhnomPencil, Panszpik, Mariraja2007, ChrisGualtieri, GenBiorics, Garamond Lethe, Mbreht, Borromi, WikiHero3, Hexidon, Monkbot, Velvel2, Neurodino and Anonymous: 26

- **Entropy (classical thermodynamics)** *Source:* https://en.wikipedia.org/wiki/Entropy_(classical_thermodynamics)?oldid=663119954 *Contributors:* Patrick, Dave souza, PAR, Jheald, Stephenb, Lokesh 2000, SmackBot, Shalom Yechiel, Sadi Carnot, JoseREMY, Jim.belk, JHunterJ, FilipeS, Cydebot, Gnfnrf, Stannered, TheSeven, Venny85, Kbrose, Plastikspork, VQuakr, CohesionBot, Sun Creator, Obnoxin, Gonfer, THEN WHO WAS PHONE?, AnomieBOT, Chjoaygame, FrescoBot, SpaceFlight89, Llewkcalbyram, Netheril96, Glockenklang1, Anandteke, Dysrhythmia, Adwaele, APerson and Anonymous: 22

- **Entropy (statistical thermodynamics)** *Source:* https://en.wikipedia.org/wiki/Entropy_(statistical_thermodynamics)?oldid=673985065 *Contributors:* Tobias Bergemann, Jason Quinn, Pjacobi, Jheald, Count Iblis, Gene Nygaard, Nanite, Commander Nemet, Enormousdude, SmackBot, InverseHypercube, Pokipsy76, Silly rabbit, Mahanchian, Acepectif, Sadi Carnot, Archimerged, Signo, Scottring, Gregbard, FilipeS, Cydebot, Gtxfrance, Boardhead, Headbomb, Samkung, Process2, Wpegden, Algorithms, Dolphin51, Infiniteawe, VQuakr, Djr32, CohesionBot, Sun Creator, Rror, Addbot, Yobot, Ht686rg90, AnomieBOT, Jockocampbell, Constructive editor, Aircorn, Netheril96, Hhhippo, Quondum, Khazar2, Dark Silver Crow, DavRosen, Dorianluparu and Anonymous: 25

- **Entropy (information theory)** *Source:* https://en.wikipedia.org/wiki/Entropy_(information_theory)?oldid=681323553 *Contributors:* Tobias Hoevekamp, Derek Ross, Bryan Derksen, The Anome, Ap, PierreAbbat, Rade Kutil, Waveguy, B4hand, Youandme, Olivier, Stevertigo, Michael Hardy, Kku, Mkweise, Ahoerstemeier, Snoyes, AugPi, Rick.G, Ww, Sbwoodside, Dysprosia, Jitse Niesen, Fibonacci, Paul-L-enwiki, Omegatron, Jeffq, Noeckel, Robbot, Tomchiukc, Benwing, Netpilot43556, Rursus, Bkell, Tobias Bergemann, Stirling Newberry, Giftlite, Donvinzk, Boaz-enwiki, Peruvianllama, Brona, Romanpoet, Udo.bellack, Jabowery, Christopherlin, Neilc, Gubbubu, Beland, OverlordQ, MarkSweep, Karol Langner, Wiml, Sctfn, Zeman, Abdull, TheObtuseAngleOfDoom, Rich Farmbrough, ArnoldReinhold, ESkog, MisterSheik, Jough, Guettarda, Cretog8, Army1987, Foobaz, Franl, Flammifer, Sligocki, PAR, Cburnett, Jheald, Tomash, Oleg Alexandrov, Linas, Shreevatsa, LOL, Bkwillwm, Male1979, Ryan Reich, Btyner, Marudubshinki, Graham87, BD2412, Jetekus, Grammarbot, Nanite, Sjö, Rjwilmsi, Thomas Arelatensis, Nneonneo, Erkcan, Alejo2083, Mfeadler, Srleffler, Chobot, DVdm, Flashmorbid, Wavelength, Alpt, Kymacpherson, Ziddy, Kimchi.sg, Afelton, Buster79, Brandon, Hakeem.gadi, DmitriyV, GrinBot-enwiki, SmackBot, InverseHypercube, Fulldecent, Istvan-Wolf, Diegotorquemada, Mcld, Gilliam, Ohnoitsjamie, Dauto, Kurykh, Gutworth, Nbarth, DHN-bot-enwiki, Colonies Chris, Jdthood, Rkinch, Javalenok, CorbinSimpson, Wen D House, Radagast83, Cybercobra, Mrander, DMacks, FilippoSidoti, Daniel.Cardenas, Michael Rogers, Andrei Stroe, Ohconfucius, Snowgrouse, Dmh-enwiki, Ninjagecko, JoseREMY, Severoon, Nonsuch, Phancy Physicist, KeithWinstein, Seanmadsen, Shockem, Ryan256, Dan Gluck, Kencf0618, Dwmalone, AlainD, Ylloh, CmdrObot, Hanspi, CBM, Mcstrother, Citrus538, Neonleonb, FilipeS, Tkircher, Farzaneh, Blaisorblade, Ignoramibus, Michael C Price, Alexnye, SteveMcCluskey, Nearfar, Thijs!bot, WikiC-enwiki, Edchi, EdJohnston, D.H, Phy1729, Jvstone, Seaphoto, Heysan, Zylorian, Dougher, Husond, OhanaUnited, Time3000, Shaul1, Coffee2theorems, Magioladitis, RogierBrussee, VoABot II, Albmont, Swpb, First Harmonic, JaGa, Kestasjk, Tommy Herbert, Pax:Vobiscum, R'n'B, CommonsDelinker, Coppertwig, Policron, Jobonki, Jvpwiki, Ale2006, Idioma-bot, Cuzkatzimhut, Trevorgoodchild, Aelkiss, Trachten, Saigyo, Kjells, Go2slash, DragonLord, Mermanj, Spinningspark, PhysPhD, Bowsmand, Michel.machado, TimProof, Maxlittle2007, Hirstormandy, Neil Smithline, Dailyknowledge, Flyer22, Mdsam2-enwiki, EnOreg, Algorithms, Svick, AlanUS, Melcombe, Rinconsoleao, Alksentrs, Schuermann-enwiki, Vql, Djr32, Blueyeru, TedDunning, Musides, Ra2007, Qwfp, Johnuniq, Kace7, Porphyro, Addbot, Deepmath, Landon1980, Olli Niemitalo, Fgnievinski, Hans de Vries, Mv240, MrVanBot, Jill-Jênn, Favonian, ChenzwBot, Wikomidia, Numbo3-bot, Ehrenkater, Tide rolls, Lightbot, Fryed-peach, Eastereaster, Luckas-bot, Yobot, Sobec, Cassandra Cathcart, AnomieBOT, Jim1138, Zandr4, Mintrick, Informationtheory, Belkovich, ArthurBot, Xqbot, Gusshoekey, Br77rino, Almabot, GrouchoBot, Omnipaedista, RibotBOT, Ortvolute, Entropeter, Constructive editor, FrescoBot, Hobsonlane, GEBStgo, Mhadi.afrasiabi, Orubt, Rc3002, HRoestBot, Cesarth73, RedBot, Cfpcompte, Pmagrass, Mduteil, Lotje, BlackAce48, LilyKitty, Angelorf, 777sms, CobraBot, Duoduoduo, Aoidh, Spakin, Hoelzro, Jann.poppinga, Mitch.mcquoid, Born2bgratis, Lalaithion, Gopsy, Racerx11, Mo ainm, Hhhippo, Purplie, Quondum, SporkBot, Music Sorter, Erianna, Elsehow, Alexander Misel, ChuispastonBot, Sigma0 1, DASHBotAV, ClueBot NG, Tschijnmotschau, Matthiaspaul, Mesoderm, Sheriff KLap, Helpful Pixie Bot, Bibcode Bot, BG19bot, Guy vandegrift, Eli of Athens, Hushaohan, Trombonechamp, Manoguru, Muhammad Shuaib Nadwi, BattyBot, ChrisGualtieri, Marek marek, VI.Reeder77, Jrajniak89, Cerabot-enwiki, Fourshade, Frosty, SFK2, Szzoli, Chrislgarry, I am One of Many, Jamesmcmahon0, Altroware, OhGodItsSoAmazing, Tchanders, Suderpie, Orehet, Monkbot, Leegrc, Visme, Donen1937, WikiRambala, Oisguad, Boky90, Isambard Kingdom, Secvz, Tinyscienceeow, KasparBot, Radegast, UMD Xuechi and Anonymous: 310

- **Entropy (order and disorder)** *Source:* https://en.wikipedia.org/wiki/Entropy_(order_and_disorder)?oldid=664270819 *Contributors:* Stone, Tobias Bergemann, Chowbok, Dvavasour, Dave souza, Cedders, Jheald, Count Iblis, Dave.Dunford, Gene Nygaard, Carcharoth, Nanite, Fresheneesz, Sodin, SmackBot, InverseHypercube, Chris the speller, Droll, Sadi Carnot, JzG, Martian.knight, ChrisCork, FilipeS, Cydebot, Barticus88, Qwerty Binary, Coppertwig, Squids and Chips, Pleroma, ToaDjango, SchreiberBike, WizardOfKozz, RogGoetsch, Shirki, Daniele Pugliesi, Citation bot, Plumpurple, IVAN3MAN, Trappist the monk, Music Sorter, APerson, Atomsk1321, DavRosen, Monkbot, Mit0126, Nearwater, Dewitwhite, Alakzi, Karen2626 and Anonymous: 16

- **Entropy (energy dispersal)** *Source:* https://en.wikipedia.org/wiki/Entropy_(energy_dispersal)?oldid=637805059 *Contributors:* Chowbok, Chris Howard, Pjacobi, Dave souza, Bender235, Jheald, Gene Nygaard, Woohookitty, Linas, Carcharoth, DVdm, Buster79, Dhollm, Enormousdude, Arthur Rubin, Itub, SmackBot, Bduke, Can't sleep, clown will eat me, Sadi Carnot, John, JoseREMY, K, ChrisCork, CmdrObot, Jsd, Cydebot, JamesAM, Headbomb, VinnyTheFish, User A1, HEL, Martin451, Kbrose, Lisatwo, Fratrep, EconoPhysicist, Tassedethe, Cesiumfrog, Legobot, AnomieBOT, Daniele Pugliesi, Citation bot, Chjoaygame, Brianpetersn, Music Sorter, Mdann52, DavRosen, Monkbot and Anonymous: 17

- **Entropy production** *Source:* https://en.wikipedia.org/wiki/Entropy_production?oldid=674481555 *Contributors:* Officiallyover, Dirac66, Werquers, Sfan00 IMG, Dthomsen8, Yobot, Materialscientist, Citation bot, Jaborn, Chjoaygame, Steve Quinn, Claisam, Adwaele, APerson, Hillbillyholiday, Yllwhmmr, Prokaryotes, LVL_6, Monkbot, Mr.yannainglinn and Anonymous: 11

- **Entropy in thermodynamics and information theory** *Source:* https://en.wikipedia.org/wiki/Entropy_in_thermodynamics_and_information_theory?oldid=679565160 *Contributors:* The Anome, Edward, Michael Hardy, Kku, Xezbeth, Pmetzger, Army1987, Foobaz, Ryanmcdaniel, PAR, Jheald, Gene Nygaard, Kanenas, Nanite, Rjwilmsi, BradBeattie, DVdm, Alpt, Archelon, SmackBot, RDBury, InverseHypercube, Jackk, Bluebot, RDBrown, Vulcanstar6, Calbaer, Ohconfucius, Dave Carter, DavidConner, Thermochap, FilipeS, Cydebot, Headbomb, Bobblehead, Acasp, PotomacFever, VoABot II, First Harmonic, OttoMäkelä, Ale2006, Cuzkatzimhut, Julia Neumann, Antoni Barau, Kjells, Drojem, SchreiberBike, Nafis ru, L.smithfield, Johnuniq, Deepmath, Kmeisterling, Lightbot, Ht686rg90, Pcap, Armchair info guy, AnomieBOT, Citation bot, False vacuum, Entropeter, FrescoBot, Physics therapist, Thermoworld, Artoannila, Dewritech, GoingBatty, Quondum, Pwagle, ClueBot NG, Tschijnmotschau, Bibcode Bot, BG19bot, Rubmum, VLReeder77, JGTZ, Ajbilan, Monkbot, Vieque and Anonymous: 30

- **Entropic uncertainty** *Source:* https://en.wikipedia.org/wiki/Entropic_uncertainty?oldid=679391863 *Contributors:* Michael Hardy, Charles Matthews, Longhair, Remuel, PAR, Sheynhertz-Unbayg, GregorB, Salsb, Teply, Downwards, RomanSpa, A. Pichler, Myasuda, Michael C Price, Cuzkatzimhut, Pmigdal, Deepmath, Yobot, Citation bot, Colanderman, RjwilmsiBot, Slightsmile, Quondum, Dexbot and Anonymous: 5

- **Entropy (arrow of time)** *Source:* https://en.wikipedia.org/wiki/Entropy_(arrow_of_time)?oldid=678662954 *Contributors:* Michael Hardy, Gracefool, Elroch, Timothy Campbell, Rich Farmbrough, Pjacobi, Bender235, Pt, Mike Schwartz, PAR, Jheald, Woohookitty, Linas, BD2412, War, DVdm, JocK, Dhollm, Froth, Zero1328, Jules.LT, Ilmari Karonen, SmackBot, Hmains, Thumperward, Colonies Chris, Shalom Yechiel, Sadi Carnot, Plugimi, Gensemer, Dan Gluck, Esn, CmdrObot, Cydebot, Tkircher, Peterdjones, Boardhead, Michael C Price, DumbBOT, Headbomb, CharlotteWebb, Alphachimpbot, STBot, Minderbinder-enwiki, Chintu rohit, BostonRed, Richwil, JoeMP, Paradoctor, Djayjp, 718 Bot, Sun Creator, Frongle, Johnuniq, DumZiBoT, Clejan, Airplaneman, Addbot, Tibixe-enwiki, Eric Drexler, Ronhjones, Jockelinde, Lightbot, Drpickem, Yobot, AnomieBOT, Sophus Bie, Mumngb, Elockid, SkyMachine, Nickyus, RjwilmsiBot, Tuxedo junction, Wyhiugl, 497glbig, Sage321, Hmainsbot1, Thepasta, Monkbot, Seventhorbitday and Anonymous: 44

- **Entropy rate** *Source:* https://en.wikipedia.org/wiki/Entropy_rate?oldid=682243834 *Contributors:* Michael Hardy, Kku, Army1987, Linas, Emacsnw, Dvdsn, Skittleys, Mange01, Maxlittle2007, Wkboonec, Addbot, Fgnievinski, AnomieBOT, Elaquias, GrouchoBot, Klbrain, Alexander Misel, Helpful Pixie Bot, Mygskr, 5D2262B74 and Anonymous: 6

- **Residual entropy** *Source:* https://en.wikipedia.org/wiki/Residual_entropy?oldid=603096303 *Contributors:* CYD, Edsanville, Minghong, Jheald, Itsmine, Kelly Martin, Linas, Bluemoose, Mattopia, Conscious, Salsb, Sbyrnes321, SmackBot, Cydebot, Brichcja, AlleborgoBot, Arjayay, WikiDao, Addbot, Erik9bot, Steve Quinn, HRoestBot, Quondum, Alchemist314, Helpful Pixie Bot, Zedshort and Anonymous: 7

- **Second law of thermodynamics** *Source:* https://en.wikipedia.org/wiki/Second_law_of_thermodynamics?oldid=682155947 *Contributors:* The Anome, Jeronimo, XJaM, Roadrunner, Jdpipe, David spector, Lorenzarius, Michael Hardy, Ixfd64, Ahoerstemeier, Cyp, Theresa knott, Snoyes, Jebba, AugPi, Cherkash, Ilyanep, Tantalate, Reddi, Terse, Tb, Time, IceKarma, DJ Clayworth, Tpbradbury, Phys, Omegatron, Marc Girod-enwiki, Jeffq, ScienceGuy, ChrisO-enwiki, Fredrik, Romanm, Gandalf61, Postdlf, Ashley Y, Hadal, Robinh, Johnstone, Cutler, Tobias Bergemann, Stirling Newberry, Giftlite, ComaVN, N12345n, Karn, FunnyMan3595, Curps, FeloniousMonk, Chinasaur, Dav4is, Duncharris, Bobblewik, Edcolins, LiDaobing, Pearbonn, Antandrus, Eroica, Ravikiran r, Kaldari, Jossi, Karol Langner, Wikimol, Rdsmith4, Panzi, Sam Hocevar, Neutrality, Ratiocinate, Trevor MacInnis, Grstain, DanielCD, Brianhe, Rich Farmbrough, KillerChihuahua, Rhobite, Pjacobi, Vsmith, ArnoldReinhold, Dave souza, Ivan Bajlo, Number 0, Ignignot, Sietse Snel, Euyyn, SteveCoast, Bobo192, I9Q79oL78KiLOQTFHgyc, Aquillion, Nk, Maebmij, Helix84, AppleJuggler, Cpcjr, Jason One, Zenosparadox, Arthena, Mineralogy, PAR, Wtshymanski, Evil Monkey, Jheald, Count Iblis, Dominic, Pauli133, Alai, KTC, Oleg Alexandrov, Crosbiesmith, ChrisNoe, Madmardigan53, Miaow Miaow, Keta, Denevans, Funhistory, Christopher Thomas, Gerbrant, GSlicer, Kbdank71, Nanite, Rjwilmsi, Koavf, Eyu100, Yamamoto Ichiro, Dantecubed, Fresheneesz, Srleffler, Jittat-enwiki, Chobot, Flying Jazz, ChrisChiasson, Wavelength, Michaeladenner, RobotE, Hairy Dude, Bobby1011, Wavesmikey, Akamad, Stephenb, Gaius Cornelius, CambridgeBayWeather, Acusoes1, SCZenz, Ragesoss, Dhollm, Abb3w, Mgrierson, Dna-webmaster, WAS 4.250, Enormousdude, 2over0, Dieseldrinker, Arthur Rubin, BorgQueen, Ilmari Karonen, Fluent aphasia, Profero, Infinity0, Sbyrnes321, DVD R W, Knowledgeum, Luk, SmackBot, Cirejcon, Ashenai, ChXu, CarbonCopy, McGeddon, Palinurus, WebDrake, Jim62sch, David Shear, Neptunius, Gunnar.Kaestle, Lsommerer, Bmord, Jab843, Gilliam, The Gnome, ThorinMuglindir, Kmarinas86, Chris the speller, Bduke, Tisthammerw, MalafayaBot, Complexica, Bonaparte, Desp-enwiki, Zmanish, Verrai, Ben Rogers, Sholto Maud, Andyparkins, H-J-Niemann, EPM, Dreadstar, Henning Makholm, Sadi Carnot, Mikaduki, Zchenyu, AThing, Miftime, Rklawton, Doanison, JorisvS, Mgiganteus1, Nonsuch, IronGargoyle, AwesomeMachine, Stikonas, MrArt, Peyre, Xionbox, Astrobradley, Dan Gluck, Seqsea, K, Michaelbusch, CzarB, Kommando797, Spk ben, George100, Tubbyspencer, Josedaniele, Mikiemike, Ale_jrb, Wafulz, AlbertSM, Father Ignatius, Jucati, Emilio Juanatey, Myasuda, Cydebot, Rifleman 82, Meno25, Ring0, Miguel de Servet, Michael C Price, DumbBOT, JodyB, Spookpadda, Daa89563, LeBofSportif, DMZ, Headbomb, Marek69, John254, EdJohnston, AntiVandalBot, Widefox, Gökhan, Canadian-Bacon, Narssarssuaq, MER-C, Physical Chemist, Acroterion, Meeples, Magioladitis, VoABot II, Mbarbier, Hubbardaie, Daarznieks, Dirac66, Hbent, Heqwm, Tercer, Wkussmaul, Jtir, Aeternium, Hweimer, R'n'B, Mbweissman, Time traveller, J.delanoy, Pharaoh of the Wizards, Musaran, Ian.thomson, Bluecheese333,

Salih, LordAnubisBOT, Frisettes, Stootoon, Ppithermo, VolkovBot, Larryisgood, Joeoettinger, ABF, Speaker to Lampposts, JayEsJay, Reibot, Anonymous Dissident, Michael H 34, LeaveSleaves, Maxim, Antixt, Enviroboy, San Diablo, Zebas, Kbrose, Subh83, SieBot, YonaBot, BotMultichill, Dawn Bard, Caltas, Jewk, Crash Underride, Arjun r acharya, Wpegden, Nrsmith, Jdaloner, Barry Fruitman, Sunrise, Denisarona, Vanished user qkqknjitkcse45u3, ClueBot, PaulLowrance, The Thing That Should Not Be, Wisemove, Hjlim, Bbanerje, Lbrewer42, LizardJr8, LonelyBeacon, Manishearth, Jimbomonkey, Nymf, Simonmckenzie, Wndl42, Estirabot, Sun Creator, Laughitup2, Nafis ru, AC+79 3888, Crowsnest, DumZiBOT, Darkicebot, AP Shinobi, BodhisattvaBot, Jovianeye, Lilmy13, Gonfer, Subversive.sound, Aunt Entropy, MystBot, Addbot, Magus732, Jncraton, Ashanda, MrOllie, CarsracBot, Favonian, Tide rolls, Lightbot, Gatewayofintrigue, Teles, Arbitrarily0, Hartz, Luckas-bot, Yobot, Fraggle81, Sanyi4, Egbertus, AnomieBOT, Rubinbot, Jim1138, Jacob2718, Materialscientist, Citation bot, Chemeditor, Xqbot, Nanog, GrouchoBot, ChristopherKingChemist, Rhettballew, Waleswatcher, Sin.pecado, Chjoaygame, FrescoBot, Tobby72, JMS Old Al, D'ohBot, RWG00, Tomerbot, Vh mby, Citation bot 1, PigFlu Oink, Pinethicket, I dream of horses, AnandaDaldal, Alfredwongpuhk, كاشف عقيل, Howzeman, Klangenfurt, Naji Khaleel, Yappy2bhere, LoStrangolatore, RjwilmsiBot, Ptbptb, Aircorn, EmausBot, John of Reading, Lea phys, Da500063, Netheril96, Arjun S Ariyil, Evanh2008, JSquish, John Cline, Bollyjeff, Mattedia, Kenan82, Tls60, WikiPidi, Ems2715, BF6-NJITWILL, Spicemix, Rocketrod1960, ClueBot NG, Snoid Headly, Jj1236, Mormequill, Widr, WikiPuppies, Helpful Pixie Bot, Bibcode Bot, Lowercase sigmabot, BG19bot, Savarona1, Cdh1001, Ugncreative Usergname, Glevum, Rs2360, Crio, Rowan Adams, Pratyya Ghosh, LeeMcLoughlin1975, Adwaele, Mdkssner, Jchammel, Pterodactyloid, Lugia2453, Zmicier P., Jochen Burghardt, Reatlas, Nerlost, Nicksola, Glenn Tamblyn, Mre env, The-vegan-muser, Aspro89, Prokaryotes, Nakitu, PhoenixPub, Ammamaretu, Skr15081997, Burnandquiver, Monkbot, Douglas Cotton, Tylerleeredd, BiologicalMe, Crystallizedcarbon, Yusefgbouth, Captain Chesapeake and Anonymous: 543

- **Irreversible process** *Source:* https://en.wikipedia.org/wiki/Irreversible_process?oldid=679389824 *Contributors:* Michael Hardy, Charles Matthews, Fredrik, Hadal, Sundar, Antandrus, Vsmith, Pmetzger, Kjkolb, Sparksm4, PAR, Jheald, Linas, Marskell, Rjwilmsi, YurikBot, Gaius Cornelius, Dhollm, Bob Hu, Cazort, Sadi Carnot, 16@r, Olag, Bbold, Chetvorno, ChrisCork, Peterdjones, Michael C Price, Thijs!bot, Headbomb, MER-C, J.delanoy, M C Y 1008, Pdcook, TXiKiBoT, Rei-bot, SieBot, Paradoctor, Phe-bot, Flyer22, Wrpscott, Dolphin51, SuperHamster, Auntof6, Addbot, LatitudeBot, Duketlstout, Yobot, Amirobot, Alchimista, AnomieBOT, Xqbot, J04n, MuffledThud, Perceiverguy, Citation bot 2, Red-Bot, Jauhienij, RjwilmsiBot, EmausBot, Slawekb, Sultryshoebox, Gary Dee, PatrickJmahoney, Gareth Griffith-Jones, Davidcarfi, Helpful PixieBot, Bibcode Bot, Movinbutnotshakin, Gwickwire, Mdann52, Adwaele, Dexbot, Mr. Guye, Babitaarora, Ginsuloft and Anonymous: 38
- **Reversible process (thermodynamics)** *Source:* https://en.wikipedia.org/wiki/Reversible_process_(thermodynamics)?oldid=678557980 *Contributors:* Bryan Derksen, Heron, Michael Hardy, Lexor, Kku, Angela, Glenn, Reddi, DJ Clayworth, Robbot, Romanm, Saulisagenius, Polychrome, DavidCary, Mintleaf~enwiki, Jorend, CSTAR, John Vandenberg, Kjkolb, PAR, Jheald, Artur adib, Joriki, Xhin, Karam.Anthony.K, Grammarbot, MaltaGirl, Mpfrank, Dhollm, Katieh5584, Boris Barowski, Gilliam, Idamlaj, Sadi Carnot, Agapefan, 2T, Thijs!bot, Jtir, Cpiral, Izno, TXiKiBoT, Dolphin51, Alexbot, DumZiBOT, Ladsgroup, LatitudeBot, Jncraton, LaaknorBot, Materialscientist, AbigailAbernathy, Mnmngb, Dbjohn, Maghemite, AMSask, Jauhienij, Dinamik-bot, EmausBot, ZéroBot, Arnaugir, Paulmiko, ResearchRave, ClueBot NG, Widr, GoShow, Adwaele, Yurkows, Esdcesdcas and Anonymous: 41

- **Dissipation** *Source:* https://en.wikipedia.org/wiki/Dissipation?oldid=677973078 *Contributors:* Bryan Derksen, Michael Hardy, Wapcaplet, William M. Connolley, Phys, Riddley, Chuunen Baka, Karl Dickman, Vsmith, Dave.Dunford, Oleg Alexandrov, Woohookitty, Linas, Eleassar777, YurikBot, RobotE, Sir48, Mikeblas, Kain2396, Zwobot, SmackBot, Tarret, Melchoir, WebDrake, Stepa, Yamaguchi先生, PieRRoMaN, Anlace, FrozenMan, Breno, Benplowman, Wingman358, Thijs!bot, EdJohnston, Escarbot, Magioladitis, WolfmanSF, Afil, MartinBot, Schmloof, Lisamh, Idioma-bot, VolkovBot, Oshwah, A4bot, Bjo2314, Cuvette, ClueBot, PixelBot, Cenarium, SoxBot III, Crowsnest, Ovis23, Addbot, AndersBot, Zorrobot, Amirobot, Daniele Pugliesi, W.stanovsky, Xqbot, SassoBot, Bellerophon, Chjoaygame, FrescoBot, Pinethicket, WikitanvirBot, Marcreif, Wikipelli, Hhhippo, ClueBot NG, CocuBot, Widr, NuclearEnergy, Adwaele, HazeySpace, Sinnema313, Prokaryotes, 1:001100001101011100100, Jinxblau and Anonymous: 41

- **Internal energy** *Source:* https://en.wikipedia.org/wiki/Internal_energy?oldid=676721713 *Contributors:* Bryan Derksen, Peterlin~enwiki, Patrick, Michael Hardy, SebastianHelm, Cyan, Andres, Robbot, Hankwang, Fabiform, Giftlite, Andries, Dratman, Bensaccount, Bobblewik, H Padleckas, Icairns, Edsanville, Spiko-carpediem~enwiki, El C, Shanes, Euyyn, Kine, Nhandler, Haham hanuka, Lysdexia, PAR, Count Iblis, RainbowOfLight, Reaverdrop, GleasSpty, Isnow, BD2412, Qwertyus, Saperaud~enwiki, Rjwilmsi, Thechamelon, HappyCamper, Margosbot~enwiki, ChrisChiasson, DVdm, Bgwhite, YurikBot, RussBot, Stassats, Dhollm, 2over0, RG2, SmackBot, Oloumi, David Shear, KocjoBot~enwiki, Ddcampayo, BirdValiant, ThorinMuglindir, Zgyorfi~enwiki, Persian Poet Gal, MalafayaBot, Complexica, DHN-bot~enwiki, Sbharris, Rrburke, AFP~enwiki, Henning Makholm, Sadi Carnot, Vina-iwbot~enwiki, Stikonas, Vaughan Pratt, CmdrObot, Xanthoxyl, Cydebot, Christian75, Omicronpersei8, Barticus88, Headbomb, Bobblehead, Mr pand, Ste4k, Trakesht, JAnDbot, PhilKnight, Magioladitis, VoABot II, Cardamon, MartinBot, R'n'B, LedgendGamer, Pdcook, Lseixas, Squids and Chips, Sheliak, VolkovBot, TXiKiBoT, SQL, Riick, SHL-at-Sv, Kbrose, SieBot, Da Joe, The way, the truth, and the light, Andrewjlockley, Dolphin51, Atif.t2, ClueBot, The Thing That Should Not Be, Mild Bill Hiccup, SuperHamster, Djr32, CohesionBot, Jusdafax, DeltaQuad, Hans Adler, ChrisHodgesUK, Crowsnest, Avoided, Hess88, That-guy flint, Addbot, Xp54321, DOI bot, Arcturus87, Aboctok, Morning277, CarsracBot, PV=nRT, Luckas-bot, Yobot, Fraggle81, Becky Sayles, AnomieBOT, Daniele Pugliesi, Ipatrol, Materialscientist, The High Fin Sperm Whale, Citation bot, LilHelpa, Xqbot, J04n, GrouchoBot, Mnmngb, MLauba, Vatbey, Chjoaygame, Maghemite, RWG00, Cannolis, Citation bot 1, DrilBot, Pinethicket, MastiBot, RazielZero, FoxBot, Derild4921, Gosnap0, Artemis Fowl III, LcawteHuggle, John of Reading, WikitanvirBot, Max139, Dewritech, Faolin42, Sportgirl426, GoingBatty, Googamooga, Mmeijeri, Hhhippo, JSquish, Fæ, Timmytoddler, Qclijun, Vramasub, ClueBot NG, NuclearEnergy, Mariraja2007, Cky 2250, Aisteco, Acratta, Adwaele, Qsq, Eli4ph, Jamesx12345, Anaekh, SkateTier, The Last Arietta, Scipsycho, LuFangwen, Amangau-tam 1995 and Anonymous: 162

- **Thermodynamic system** *Source:* https://en.wikipedia.org/wiki/Thermodynamic_system?oldid=680278531 *Contributors:* Toby Bartels, Fxmastermind, Eric119, Stevenj, Smack, Filemon, Giftlite, Peruvianllama, Andycjp, Blazotron, Rdsmith4, Icairns, Jfraser, Helix84, Mdd, Alansohn, Rw63phi, PAR, Pion, Jheald, Dan100, BD2412, Chobot, Wavesmikey, Gaius Cornelius, Dhollm, Jpbowen, Bota47, Light current, E Wing, SmackBot, Bomac, MalafayaBot, Stepho-wrs, DinosaursLoveExistence, Sadi Carnot, 16@r, CmdrObot, Cydebot, Krauss, Sting, Headbomb, Stannered, Akradecki, JAnDbot, Athkalani~enwiki, MSBOT, .anacondabot, VoABot II, Rich257, KConWiki, An1MuS, Ac44ck, Pbroks13, Nev1, Trusilver, Maurice Carbonaro, Cmbankester, Usp, VolkovBot, Kbrose, PaddyLeahy, SieBot, Mercenario97, OKBot, ClueBot, Auntof6, Excirial, PixelBot, Wdford, Mikaey, SchreiberBike, Addbot, CarsracBot, Redheylin, Glane23, Ht686rg90, Fraggle81, Becky Sayles, AnomieBOT, ArthurBot, Xqbot, DSisyphBot, GrouchoBot, Chjoaygame, FrescoBot, Pshmell, Pinethicket, RedBot, Thinking of England, Artem Korzhimanov, AznFiddl3r, EmausBot, Abpk62, Glockenklang1, ClueBot NG, Gokulchandola, Loopy48, BZTMPS, BG19bot, Gryffon5147, Tutelary, ChrisGualtieri, Upsidedowntophat, Adwaele, Frosty, PhoenixPub, Eclipsis Proteo, Zortwort and Anonymous: 72

- **Disgregation** *Source:* https://en.wikipedia.org/wiki/Disgregation?oldid=640078271 *Contributors:* Jheald, Rjwilmsi, Dhollm, TimBentley, Sadi Carnot, JzG, Cydebot, N8sbug, PC78, Warut, Airplaneman, AnomieBOT, Daniele Pugliesi, Gryllida, Hij987 and Anonymous: 3

- **Boltzmann's entropy formula** *Source:* https://en.wikipedia.org/wiki/Boltzmann'{|s_entropy_formula?oldid=652703653 *Contributors:* Tobias Bergemann, DocWatson42, Jheald, Deltabeignet, ChrisChiasson, SmackBot, Complexica, Colonies Chris, Sadi Carnot, Jsd, Jikmo, FilipeS, Cydebot, Michael C Price, Headbomb, GuidoGer, Jorgenumata, Idioma-bot, DAK4Blizzard, Chphe, Agor153, Nafis ru, MystBot, Addbot, Luckas-bot, Citation bot, Thehelpfulbot, Aircorn, Libb Thims, Guy vandegrift, Dark Silver Crow, Hmainsbot1, Makecat-bot, Irises23, FLeupold, Kevin.dale.parrish, Thebeast707 and Anonymous: 12

- **Tsallis entropy** *Source:* https://en.wikipedia.org/wiki/Tsallis_entropy?oldid=679400291 *Contributors:* Michael Hardy, Kku, Charles Matthews, Tobias Bergemann, Giftlite, Kappa, Jeodesic, PAR, Jheald, Linas, HappyCamper, SmackBot, C.Fred, Colonies Chris, Tekhnofiend, Derek farn, Loodog, FilipeS, Cydebot, Headbomb, Liberio, Addbot, Yobot, Wireader, Citation bot, Omnipaedista, Julian Birdbath, Alph Bot, Bibcode Bot, Purple Post-its, Woland, Dexbot, Limit-theorem, Gatsekouras and Anonymous: 11

- **Standard molar entropy** *Source:* https://en.wikipedia.org/wiki/Standard_molar_entropy?oldid=682344189 *Contributors:* Michael Hardy, Tim Starling, Gentgeen, Robbot, Altenmann, Gtrmp, Eequor, Van Flamm~enwiki, Pearle, Benjah-bmm27, Jheald, Gene Nygaard, Coolgamer, Linas, StradivariusTV, FlaBot, Ffaarr, Bgwhite, YurikBot, Avalon, Afrotrance, SmackBot, PieRRoMaN, G N Frykman, SilverStar, Olin, Knights who say ni, Judgesurreal777, Cabria, Cydebot, Brad101, JamesAM, Zadeez, .anacondabot, Su-no-G, Jcwf, Jackfork, RAAckerman, Cerireid, Addbot, RobertHannah89, Fulcanelli, Citation bot, Chemeditor, Brane.Blokar, Astrorocks, Skyerise, Rlholden, Dg.de, Mn-imhotep, Robotsheepboy, Monkbot, Vieque and Anonymous: 12

- **Entropy of mixing** *Source:* https://en.wikipedia.org/wiki/Entropy_of_mixing?oldid=623208022 *Contributors:* William Avery, Glueball, PAR, Jheald, V8rik, Mortice, Ogro, Dhollm, Alfredo.correa, David Shear, Chris the speller, Bduke, Gpetty, Colonies Chris, Cubbi, Rezecib, Cydebot, Christian75, Headbomb, Pjvpjv, Dirac66, R'n'B, Jorgenumata, Locke9k, Lord Phat, Cerireid, Addbot, Potekhin, Yobot, Daniele Pugliesi, Xqbot, Nathanielvirgo, Listerineman, Pberndt, Chjoaygame, FrescoBot, Hhhippo, Widr, Helpful Pixie Bot, Scienxe and Anonymous: 19

- **Loop entropy** *Source:* https://en.wikipedia.org/wiki/Loop_entropy?oldid=569293930 *Contributors:* SebastianHelm, Jheald, Salix alba, Phantomsteve, Cydebot, WillowW, Calvero JP and Anonymous: 1

- **Conformational entropy** *Source:* https://en.wikipedia.org/wiki/Conformational_entropy?oldid=655131529 *Contributors:* Jheald, Linas, FilipeS, Cydebot, WillowW, Opabinia regalis, Hodja Nasreddin, My very best wishes, Jesse V. and Anonymous: 3

- **Entropic force** *Source:* https://en.wikipedia.org/wiki/Entropic_force?oldid=676400046 *Contributors:* Michael Hardy, Dratman, Jheald, Linas, Mouvement, Helvetius, Bubba73, Yahya Abdal-Aziz, JocK, Msuzen, Sbyrnes321, SmackBot, Hongooi, Pwjb, Sadi Carnot, Leo C Stein, Dr.saptarshi, Robofish, CmdrObot, Cydebot, Thijs!bot, Nick Number, BillGordon, Hooplehead, David Eppstein, Boghog, Lamro, Adabow, Nergaal, Webridge, Biggerj1, Edkarpov, Addbot, Numbo3-bot, Yobot, AnomieBOT, Götz, Erik9bot, Sanpitch, D'ohBot, My very best wishes, Reaper Eternal, Marie Poise, Fey, Grondilu, Sitar Physics, Bibcode Bot, BG19bot, Pfd1986 and Anonymous: 21

- **Free entropy** *Source:* https://en.wikipedia.org/wiki/Free_entropy?oldid=672080037 *Contributors:* Waltpohl, Keenan Pepper, PAR, Jheald, Jaxal1, Physchim62, ChrisChiasson, Retired username, Dhollm, SmackBot, Sprocedato, JorisvS, Nonsuch, CmdrObot, Vyznev Xnebara, Cydebot, Christian75, Headbomb, Katharineamy, VolkovBot, Dmcq, StevenDH, TimothyRias, Addbot, DOI bot, Yobot, Daniele Pugliesi, Citation bot, Citation bot 1, EmausBot, Quondum, Libb Thims, Antichristos, Creator666~enwiki, Bibcode Bot, 9p4gh9gkj, BattyBot, Jfizzix, Monkbot and Anonymous: 8

- **Entropic explosion** *Source:* https://en.wikipedia.org/wiki/Entropic_explosion?oldid=622407571 *Contributors:* –K, A2Kafir, Jheald, V8rik, N0YKG, Nonsuch, Eastlaw, Cydebot, Notreallydavid, Whoop whoop pull up and Anonymous: 3

- **Sackur–Tetrode equation** *Source:* https://en.wikipedia.org/wiki/Sackur%E2%80%93Tetrode_equation?oldid=662876971 *Contributors:* MichaelHardy, Giftlite, Hooperbloob, PAR, Jheald, Count Iblis, Gene Nygaard, Linas, Tabletop, Zzyzx11, Margosbot~enwiki, Physchim62, Salsb, Dhollm, Bo Jacoby, Kmarinas86, Sbharris, G716, Dicklyon, Mets501, Cydebot, Plasma g, Jorgenumata, Isaac Sanolnacov, Cwkmail, Winggundam, Laburke, Addbot, SPat, Yobot, Kotika98, Citation bot 1, John of Reading, Helpful Pixie Bot, Aglecer, Krebs49 and Anonymous:

[10] **Entropy (computing)** *Source:* https://en.wikipedia.org/wiki/Entropy_(computing)?oldid=667497148 *Contributors:* DavidCary, Robert Brockway, Urhixidur, Jhertel, BD2412, Zenten, Maxal, SmackBot, Thumperward, Lambiam, Nuwewsco, X96lee15, NapoliRoma, NerdyNSK, CWii, Sfan00 IMG, Gene93k, SF007, DumZiBoT, Addbot, Yobot, Bunnyhop11, AnomieBOT, J04n, Skyerise, Dewritech, Elandy2009, DennisIsMe, ClueBot NG, Doctor Ruud, Ferroin, Klilidiplomus, Laggingreflex and Anonymous: 27

- **Heat death of the universe** *Source:* https://en.wikipedia.org/wiki/Heat_death_of_the_universe?oldid=683254353 *Contributors:* Bryan Derksen, Andre Engels, Jdpipe, Anders Feder, Jeff Relf, Michael Shields, Evercat, Schneelocke, Robinh, Seth Ilys, Cutler, Giftlite, Anville, Alison, Arturus~enwiki, Gracefool, Golbez, Andycjp, Mike R, Zantolak, Tzarius, B.d.mills, SamSim, Burschik, Joyous!, Esperant, Discospinster, Brianhe, Dbachmann, Pavel Vozenilek, Spiralx, El C, Jomel, Marathoner, Viriditas, Foobaz, I9Q79oL78KiL0QTFHgyc, Giraffedata, Nk, VBGFscJUn3, Jonathunder, Knucmo2, Alansohn, JYolkowski, Yamla, Caesura, Hohum, Wtmitchell, Jheald, LukeSurl, Adrian.benko, Falcorian, Dmitrybrant, Bjones, Linas, LOL, Bkkbrad, BillC, Robert K S, Christopher Thomas, The Nameless, RichardWeiss, Ketiltrout, Rjwilmsi, Urbane Legend, SMC, Dougluce, Remurmur, Nihiltres, Bgwhite, Wjfox2005, YurikBot, Spacepotato, Phlip, Rsrikanth05, Mike411, Herve661, Thiseye, Expensivehat, Brandon, Lomn, Aaron Schulz, Mission9801, JasonAD, Wknight94, Fram, Tzepish, CIreland, Sardanaphalus, SmackBot, Ashill, Furry, Thumperward, Silly rabbit, SchfiftyThree, Latrosicarius, Rickythesk8r, Hippo43, Steve Pucci, Blake-, Sbluen, Sadi Carnot, RossF18, Byelf2007, Nishkid64, Richard L. Peterson, Korean alpha for knowledge, Beetstra, AdultSwim, Mdanh2002, Dan Gluck, Slammer111, ScottHolden, Tawkerbot2, Chetvorno, Rapidflash, JForget, Vyznev Xnebara, Gregbard, Lyml, Cydebot, Gogo Dodo, Ttiotsw, Michael C Price, Malleus Fatuorum, Thijs!bot, Epbr123, Coelacan, Mbell, Matny1989, Headbomb, Najro, Pmrobert49, Peter Gulutzan, Davidhorman, Jonny-mt, Greg L, Chillysnow, Astrokid, JAnDbot, Andonic, CrazySpas, VoABot II, Telanis, Swpb, Email4mobile, Beetfarm Louie, Zygar2k6, Skylights76, Thompson.matthew, Fastman99, Tgeairn, Mike.lifeguard, Jerry, SkaryMonk, Adamd1008, Jarry1250, Idioma-bot, Signalhead, Larryisgood, RingtailedFox, Indubitably, HowardFrampton, Zanardm, Anonymous Dissident, Reddoor, MearsMan, SwordSmurf, Wasted Sapience, Vector Potential, Kbrose, Comosabi, SieBot, BotMultichill, Paradoctor, Chazzlyy, Lightmouse, Crisis, Hamiltondaniel, Martarius,

ClueBot, Artichoker, Arakunem, Der Golem, Leopard850, Bubbles4sale, Zomno, M.O.X, Johnuniq, Matma Rex, MystBot, Truthnlove, Crack-Dragon, NonvocalScream, Osarius, Maldek, Addbot, Atethnekos, ThisIsMyWikipediaName, Dansk59, WFPM, Lightbot, Zorrobot, Mikey58, Luckas-bot, Amble, Maldek2, AnomieBOT, AngusCA, VX, Citation bot, ArthurBot, Hammersbach, Chjoaygame, Sanpitch, Citation bot 1, Pinethicket, Tom.Reding, Σ, Diannaa, Suryamp, Stephen Graetzel, RjwilmsiBot, Tanshai, EmausBot, WikitanvirBot, Slightsmile, Ornithikos, Serketan, Ὁ οἶστρος, K kisses, Donner60, ChuispastonBot, Whoop whoop pull up, ClueBot NG, HRO'Neill, 497glbig, Widr, Pennykohl, Bibcode Bot, Snow Rise, LeeMcLoughlin1975, ChrisGualtieri, Zeeyanwiki, Grown up atheist, Reatlas, Madreterra, Jwratner1, Spgoggles, 1bandsaw, Stamptrader, Crow, Thundergodz, Tetra quark and Anonymous: 202

34.6.2 Images

- **File:Adiabatic-demagnitization.svg** *Source:* https://upload.wikimedia.org/wikipedia/en/2/29/Adiabatic-demagnitization.svg *License:* PD *Contributors:*
  I (Dave.Dunford (talk)) created this work entirely by myself. *Original artist:*
  Dave.Dunford (talk)

- **File:Adiabatic-irrevisible-state-change.svg***Source:*https://upload.wikimedia.org/wikipedia/commons/9/94/Adiabatic-irrevisible-state-svg *License:* CC0 *Contributors:* Own work *Original artist:* Andlaus

- **File:Adjabatic-revisible-state-change.svg** *Source:* https://upload.wikimedia.org/wikipedia/commons/1/19/Adjabatic-revisible-state-change.svg *License:* CC0 *Contributors:* Own work *Original artist:* Andlaus

- **File:Ambox_contradict.svg** *Source:* https://upload.wikimedia.org/wikipedia/commons/2/2e/Ambox_contradict.svg *License:* Public domain *Contributors:* self-made using Image:Emblem-contradict.svg *Original artist:* penubag, Rugby471

- **File:Ambox_important.svg** *Source:* https://upload.wikimedia.org/wikipedia/commons/b/b4/Ambox_important.svg *License:* Public domain *Contributors:* Own work, based off of Image:Ambox scales.svg *Original artist:* Dsmurat (talk · contribs)

- **File:Ambox_question.svg** *Source:* https://upload.wikimedia.org/wikipedia/commons/1/1b/Ambox_question.svg *License:* Public domain *Contributors:* Based on Image:Ambox important.svg *Original artist:* Mysid, Dsmurat, penubag

- **File:Binary_entropy_plot.svg** *Source:* https://upload.wikimedia.org/wikipedia/commons/2/22/Binary_entropy_plot.svg *License:* CC-BY-SA-3.0 *Contributors:* original work by Brona, published on Commons at Image:Binary entropy plot.png. Converted to SVG by Alessio Damato *Original artist:* Brona and Alessio Damato

- **File:Boltzmann-molecules.svg** *Source:* https://upload.wikimedia.org/wikipedia/commons/0/05/Boltzmann-molecules.svg *License:* Public domain *Contributors:* Own work *Original artist:* Dave.Dunford

- **File:Boltzmann_equation.JPG** *Source:* https://upload.wikimedia.org/wikipedia/commons/f/fa/Boltzmann_equation.JPG *License:* CC-BY-SA-3.0 *Contributors:* http://en.wikipedia.org/wiki/Image:Zentralfriedhof_Vienna_-_Boltzmann.JPG *Original artist:* User:Daderot http://en.wikipedia.org/wiki/User:Daderot

- **File:Carnot_heat_engine_2.svg** *Source:* https://upload.wikimedia.org/wikipedia/commons/2/22/Carnot_heat_engine_2.svg *License:* Public domain *Contributors:* Based upon Image:Carnot-engine.png *Original artist:* Eric Gaba (Sting - fr:Sting)

- **File:Clausius-1.jpg** *Source:* https://upload.wikimedia.org/wikipedia/commons/3/34/Clausius-1.jpg *License:* Public domain *Contributors:* unknown *Original artist:* unknown

- **File:Clausius.jpg** *Source:* https://upload.wikimedia.org/wikipedia/commons/4/40/Clausius.jpg *License:* Public domain *Contributors:* http://www-history.mcs.st-andrews.ac.uk/history/Posters2/Clausius.html *Original artist:* Original uploader was user:Sadi Carnot at en.wikipedia

- **File:Crab_Nebula.jpg** *Source:* https://upload.wikimedia.org/wikipedia/commons/0/00/Crab_Nebula.jpg *License:* Public domain *Contributors:* HubbleSite: gallery, release. *Original artist:* NASA, ESA, J. Hester and A. Loll (Arizona State University)

- **File:Crypto_key.svg** *Source:* https://upload.wikimedia.org/wikipedia/commons/6/65/Crypto_key.svg *License:* CC-BY-SA-3.0 *Contributors:* Own work based on image:Key-crypto-sideways.png by MisterMatt originally from English Wikipedia *Original artist:* MesserWoland

- **File:Deriving_Kelvin_Statement_from_Clausius_Statement.svg***Source:*https://upload.wikimedia.org/wikipedia/commons/8/83/Kelvin_Statement_from_Clausius_Statement.svg *License:* CC BY-SA 3.0 *Contributors:* Own work *Original artist:* Netheril96

- **File:DiffusionMicroMacro.gif** *Source:* https://upload.wikimedia.org/wikipedia/commons/4/4d/DiffusionMicroMacro.gif *License:* Public domain *Contributors:* Own work *Original artist:* Sbyrnes321

- **File:Drops_I.jpg** *Source:* https://upload.wikimedia.org/wikipedia/commons/f/f4/Drops_I.jpg *License:* CC BY 2.0 *Contributors:* Flickr.com - image description page *Original artist:* Staffan Enbom from Finland

- **File:Earth-moon.jpg** *Source:* https://upload.wikimedia.org/wikipedia/commons/5/5c/Earth-moon.jpg *License:* Public domain *Contributors:* NASA [1] *Original artist:* Apollo 8 crewmember Bill Anders

- **File:Edit-clear.svg** *Source:* https://upload.wikimedia.org/wikipedia/en/f/f2/Edit-clear.svg *License:* Public domain *Contributors:* The *Tango! Desktop Project*. *Original artist:*
  The people from the Tango! project. And according to the meta-data in the file, specifically: "Andreas Nilsson, and Jakub Steiner (although minimally)."

- **File:Entropy-diagram.png** *Source:* https://upload.wikimedia.org/wikipedia/en/d/dc/Entropy-diagram.png *License:* PD *Contributors:* ? *Original artist:* ?

- **File:Entropy_flip_2_coins.jpg** *Source:* https://upload.wikimedia.org/wikipedia/commons/d/d4/Entropy_flip_2_coins.jpg *License:* CC BY-SA 3.0 *Contributors:* File:Ephesos_620-600_BC.jpg *Original artist:* http://www.cngcoins.com/

- **File:First_law_open_system.svg** *Source:* https://upload.wikimedia.org/wikipedia/commons/8/86/First_law_open_system.svg *License:* Public domain *Contributors:*
- First_law_open_system.png *Original artist:*
- derivative work: Pbroks13 (talk)
- **File:Fisher_iris_versicolor_sepalwidth.svg** *Source:* https://upload.wikimedia.org/wikipedia/commons/4/40/Fisher_iris_versicolor_sepalwidth.svg *License:* CC BY-SA 3.0 *Contributors:* en:Image:Fisher iris versicolor sepalwidth.png *Original artist:* en:User:Qwfp(original); Pbroks13(talk) (redraw)
- **File:Folder_Hexagonal_Icon.svg** *Source:* https://upload.wikimedia.org/wikipedia/en/4/48/Folder_Hexagonal_Icon.svg *License:* Cc-by-sa-3.0 *Contributors:* ? *Original artist:* ?
- **File:Gnome-searchtool.svg** *Source:* https://upload.wikimedia.org/wikipedia/commons/1/1e/Gnome-searchtool.svg *License:* LGPL *Contributors:* http://ftp.gnome.org/pub/GNOME/sources/gnome-themes-extras/0.9/gnome-themes-extras-0.9.0.tar.gz *Original artist:* David Vignoni
- **File:Green_check.svg** *Source:* https://upload.wikimedia.org/wikipedia/commons/0/03/Green_check.svg *License:* Public domain *Contributors:* Derived from Image:Yes check.svg by Gregory Maxwell *Original artist:* gmaxwell
- **File:Heat_engine_and_refrigerator01.jpg** *Source:* https://upload.wikimedia.org/wikipedia/en/5/56/Heat_engine_and_refrigerator01.jpg *License:* CC-BY-SA-3.0 *Contributors:*
Made by SliteWrite
Original artist:
Adwaele
- **File:IceBlockNearJoekullsarlon.jpg** *Source:* https://upload.wikimedia.org/wikipedia/commons/7/71/IceBlockNearJoekullsarlon.jpg *License:* CC BY-SA 4.0 *Contributors:* Own work *Original artist:* Andreas Tille
- **File:Ice_water.jpg** *Source:* https://upload.wikimedia.org/wikipedia/commons/0/0c/Ice_water.jpg *License:* Public domain *Contributors:* ? *Original artist:* ?
- **File:Ilc_9yr_moll4096.png** *Source:* https://upload.wikimedia.org/wikipedia/commons/3/3c/Ilc_9yr_moll4096.png *License:* Public domain *Contributors:* http://map.gsfc.nasa.gov/media/121238/ilc_9yr_moll4096.png *Original artist:* NASA / WMAP Science Team
- **File:James-clerk-maxwell3.jpg** *Source:* https://upload.wikimedia.org/wikipedia/commons/6/6f/James-clerk-maxwell3.jpg *License:* Public domain *Contributors:* ? *Original artist:* ?
- **File:Lord_Kelvin_photograph.jpg** *Source:* https://upload.wikimedia.org/wikipedia/commons/a/a0/Lord_Kelvin_photograph.jpg *License:* Public domain *Contributors:* http://www.sil.si.edu/digitalcollections/hst/scientific-identity/CF/by_scientist_display_results.cfm?scientist=kelvin *Original artist:* ?
- **File:Merge-arrow.svg** *Source:* https://upload.wikimedia.org/wikipedia/commons/a/aa/Merge-arrow.svg *License:* Public domain *Contributors:* ? *Original artist:* ?
- **File:Merge-arrows.svg** *Source:* https://upload.wikimedia.org/wikipedia/commons/5/52/Merge-arrows.svg *License:* Public domain *Contributors:* ? *Original artist:* ?
- **File:NAtomVacuumPumpMemory.png** *Source:* https://upload.wikimedia.org/wikipedia/commons/4/48/NAtomVacuumPumpMemory.png *License:* CC BY-SA 3.0 *Contributors:* Own work *Original artist:* P. Fraundorf
- **File:OpenSystemRepresentation.svg** *Source:* https://upload.wikimedia.org/wikipedia/commons/7/77/OpenSystemRepresentation.svg *License:* CC BY-SA 4.0 *Contributors:* Own work *Original artist:* Krauss
- **File:Operation_Upshot-Knothole_-_Badger_001.jpg** *Source:* https://upload.wikimedia.org/wikipedia/commons/7/79/Operation_Upshot_-_Badger_001.jpg *License:* Public domain *Contributors:* This image is available from the National Nuclear Security Administration Nevada Site Office Photo Library under number XX-34. *Original artist:* Federal Government of the United States
- **File:Polymers_Logo_svg.svg** *Source:* https://upload.wikimedia.org/wikipedia/commons/2/23/Polymers_Logo_svg.svg *License:* Public domain *Contributors:* Own work *Original artist:* Anon126
- **File:Question_book-new.svg** *Source:* https://upload.wikimedia.org/wikipedia/en/9/99/Question_book-new.svg *License:* Cc-by-sa-3.0 *Contributors:*
Created from scratch in Adobe Illustrator. Based on Image:Question book.png created by User:Equazcion *Original artist:* Tkgd2007
- **File:Red_x.svg** *Source:* https://upload.wikimedia.org/wikipedia/en/b/ba/Red_x.svg *License:* PD *Contributors:* ? *Original artist:* ?
- **File:ST_diagram_of_N2_01.jpg** *Source:* https://upload.wikimedia.org/wikipedia/en/8/80/ST_diagram_of_N2_01.jpg *License:* CC-BY-SA-3.0 *Contributors:*
created by slidewrite
Original artist:
Adwaele
- **File:Sadi_Carnot.jpeg** *Source:* https://upload.wikimedia.org/wikipedia/commons/8/80/Sadi_Carnot.jpeg *License:* Public domain *Contributors:* http://www-history.mcs.st-and.ac.uk/history/PictDisplay/Carnot_Sadi.html *Original artist:* Louis-Léopold Boilly
- **File:Schematic_diagram_of_a_heat_engine02.jpg** *Source:* https://upload.wikimedia.org/wikipedia/en/0/01/Schematic_diagram_of_a_heat_engine02.jpg *License:* CC-BY-SA-3.0 *Contributors:*
Made by SliteWrite
Original artist:
Adwaele

- **File:Solar_system.jpg** *Source:* https://upload.wikimedia.org/wikipedia/commons/8/83/Solar_system.jpg *License:* Public domain *Contributors:* ? *Original artist:* ?

- **File:Solid-liquid-gas.svg** *Source:* https://upload.wikimedia.org/wikipedia/commons/9/9b/Solid-liquid-gas.svg *License:* Public domain *Contributors:*

- Solid-liquid-gas.jpg *Original artist:* Solid-liquid-gas.jpg: Sadi Carnot

- **File:Symbol_template_class.svg** *Source:* https://upload.wikimedia.org/wikipedia/en/5/5c/Symbol_template_class.svg *License:* Public domain *Contributors:* ? *Original artist:* ?

- **File:System_boundary.svg** *Source:* https://upload.wikimedia.org/wikipedia/commons/b/b6/System_boundary.svg *License:* Public domain *Contributors:* en:Image:System-boundary.jpg *Original artist:* en:User:Wavesmikey, traced by User:Stannered

- **File:System_boundary2.svg** *Source:* https://upload.wikimedia.org/wikipedia/commons/6/63/System_boundary2.svg *License:* CC BY-SA 4.0 *Contributors:* Own work *Original artist:* Krauss

- **File:Temperature-entropy_chart_for_steam,_US_units.svg** *Source:* https://upload.wikimedia.org/wikipedia/commons/6/63/Temperature-chart_for_steam%2C_US_units.svg *License:* CC BY-SA 3.0 *Contributors:* Own work Data retrieved from: E.W. Lemmon, M.O. McLinden and D.G. Friend, "Thermophysical Properties of Fluid Systems" in NIST Chemistry WebBook, NIST Standard Reference Database Number 69, Eds. P.J. Linstrom and W.G. Mallard, National Institute of Standards and Technology, Gaithersburg MD, 20899, http://webbook.nist.gov, (retrieved November 2, 2010).) *Original artist:* Emok

- **File:Thermodynamic_system01b.jpg** *Source:* https://upload.wikimedia.org/wikipedia/en/e/e7/Thermodynamic_system01b.jpg *License:* CC-BY-SA-3.0 *Contributors:*

  Made by Slitewrite

  Original artist:

  Adwaele

- **File:Unbalanced_scales.svg** *Source:* https://upload.wikimedia.org/wikipedia/commons/f/fe/Unbalanced_scales.svg *License:* Public domain *Contributors:* ? *Original artist:* ?

- **File:Wiki_letter_w_cropped.svg** *Source:* https://upload.wikimedia.org/wikipedia/commons/1/1c/Wiki_letter_w_cropped.svg *License:* CC-BY-SA-3.0 *Contributors:*

- Wiki_letter_w.svg *Original artist:* Wiki_letter_w.svg: Jarkko Piiroinen

- **File:Wikiquote-logo.svg** *Source:* https://upload.wikimedia.org/wikipedia/commons/f/fa/Wikiquote-logo.svg *License:* Public domain *Contributors:* ? *Original artist:* ?

- **File:Wiktionary-logo-en.svg** *Source:* https://upload.wikimedia.org/wikipedia/commons/f/f8/Wiktionary-logo-en.svg *License:* Public domain *Contributors:* Vector version of Image:Wiktionary-logo-en.png. *Original artist:* Vectorized by Fvasconcellos (talk · contribs), based on original logo tossed together by Brion Vibber

- **File:Zentralfriedhof_Vienna_-_Boltzmann.JPG** *Source:* https://upload.wikimedia.org/wikipedia/commons/6/63/Zentralfriedhof_Vienna_-_Boltzmann.JPG *License:* CC-BY-SA-3.0 *Contributors:* transferred from the English language Wikipedia *Original artist:* Own work by Daderot

34.6.3   Content license

- Creative Commons Attribution-Share Alike 3.0

Chapter 35

Negentropy

The **negentropy**, also **negative entropy, syntropy, extropy, ectropy** or **entaxy**,[1] of a living system is the entropy that it exports to keep its own entropy low; it lies at the intersection of entropy and life. The concept and phrase "negative entropy" was introduced by Erwin Schrödinger in his 1944 popular-science book *What is Life?*[2] Later, Léon Brillouin shortened the phrase to *negentropy*,[3][4] to express it in a more "positive" way: a living system imports negentropy and stores it.[5] In 1974, Albert Szent-Györgyi proposed replacing the term *negentropy* with *syntropy*. That term may have originated in the 1940s with the Italian mathematician Luigi Fantappiè, who tried to construct a unified theory of biology and physics. Buckminster Fuller tried to popularize this usage, but *negentropy* remains common.

In a note to What is Life? Schrödinger explained his use of this phrase.

Indeed, negentropy has been used by biologists as the basis for purpose or direction in life, namely cooperative or moral instincts.[6]

In 2009, Mahulikar & Herwig redefined negentropy of a dynamically ordered sub-system as the specific entropy deficit of the ordered sub-system relative to its surrounding chaos.[7] Thus, negentropy has units [J/kg-K] when defined based on specific entropy per unit mass, and [K^{-1}] when defined based on specific entropy per unit energy. This definition enabled: *i*) scale-invariant thermodynamic representation of dynamic order existence, *ii*) formulation of physical principles exclusively for dynamic order existence and evolution, and *iii*) mathematical interpretation of Schrödinger's negentropy debt.

35.1 Information theory

In information theory and statistics, negentropy is used as a measure of distance to normality.[8][9][10] Out of all distributions with a given mean and variance, the normal or Gaussian distribution is the one with the highest entropy. Negentropy measures the difference in entropy between a given distribution and the Gaussian distribution with the same mean and variance. Thus, negentropy is always nonnegative, is invariant by any linear invertible change of coordinates, and vanishes if and only if the signal is Gaussian.

Negentropy is defined as

$$J(p_x) = S(\phi_x) - S(p_x)$$

where $S(\phi_x)$ is the differential entropy of the Gaussian density with the same mean and variance as p_x and $S(p_x)$ is the differential entropy of p_x :

$$S(p_x) = -\int p_x(u) \log p_x(u) du$$

Negentropy is used in statistics and signal processing. It is related to network entropy, which is used in Independent Component Analysis.[11][12]

35.2 Correlation between statistical negentropy and Gibbs' free energy

There is a physical quantity closely linked to free energy (free enthalpy), with a unit of entropy and isomorphic to negentropy known in statistics and information theory. In 1873, Willard Gibbs created a diagram illustrating the concept of free energy corresponding to free enthalpy. On the diagram one can see the quantity called capacity for entropy. The said quantity is the amount of entropy that may be increased without changing an internal energy or increasing its volume.[13] In other words, it is a difference between maximum possible, under assumed conditions, entropy and its actual entropy. It corresponds exactly to the definition of negentropy adopted in statistics and information theory. A similar physical quantity was introduced in 1869 by Massieu for the isothermal process[14][15][16] (both quantities differs just with a figure sign) and then Planck for the isothermal-isobaric process.[17] More recently, the Massieu-Planck thermodynamic potential, known also as *free entropy*, has been shown to play a great role in the so-called entropic formulation of statistical mechanics,[18] applied among the others in molecular biology[19] and thermodynamic non-equilibrium processes.[20]

$$J = S_{max} - S = -\Phi = -k \ln Z$$

where:

J - negentropy (Gibbs "capacity for entropy")

Φ – Massieu potential

Z - partition function

k - Boltzmann constant

35.3 Risk management

In risk management, negentropy is the force that seeks to achieve effective organizational behavior and lead to a steady predictable state.[21]

35.4 Brillouin's negentropy principle of information

In 1953, Léon Brillouin derived a general equation[22] stating that the changing of an information bit value requires at least kT ln(2) energy. This is the same energy as the work Leó Szilárd's engine produces in the idealistic case. In his book,[23] he further explored this problem concluding that any cause of this bit value change (measurement, decision about a yes/no question, erasure, display, etc.) will require the same amount of energy.

35.5 See also

- Exergy

- Extropy

- Free entropy

- Entropy in thermodynamics and information theory

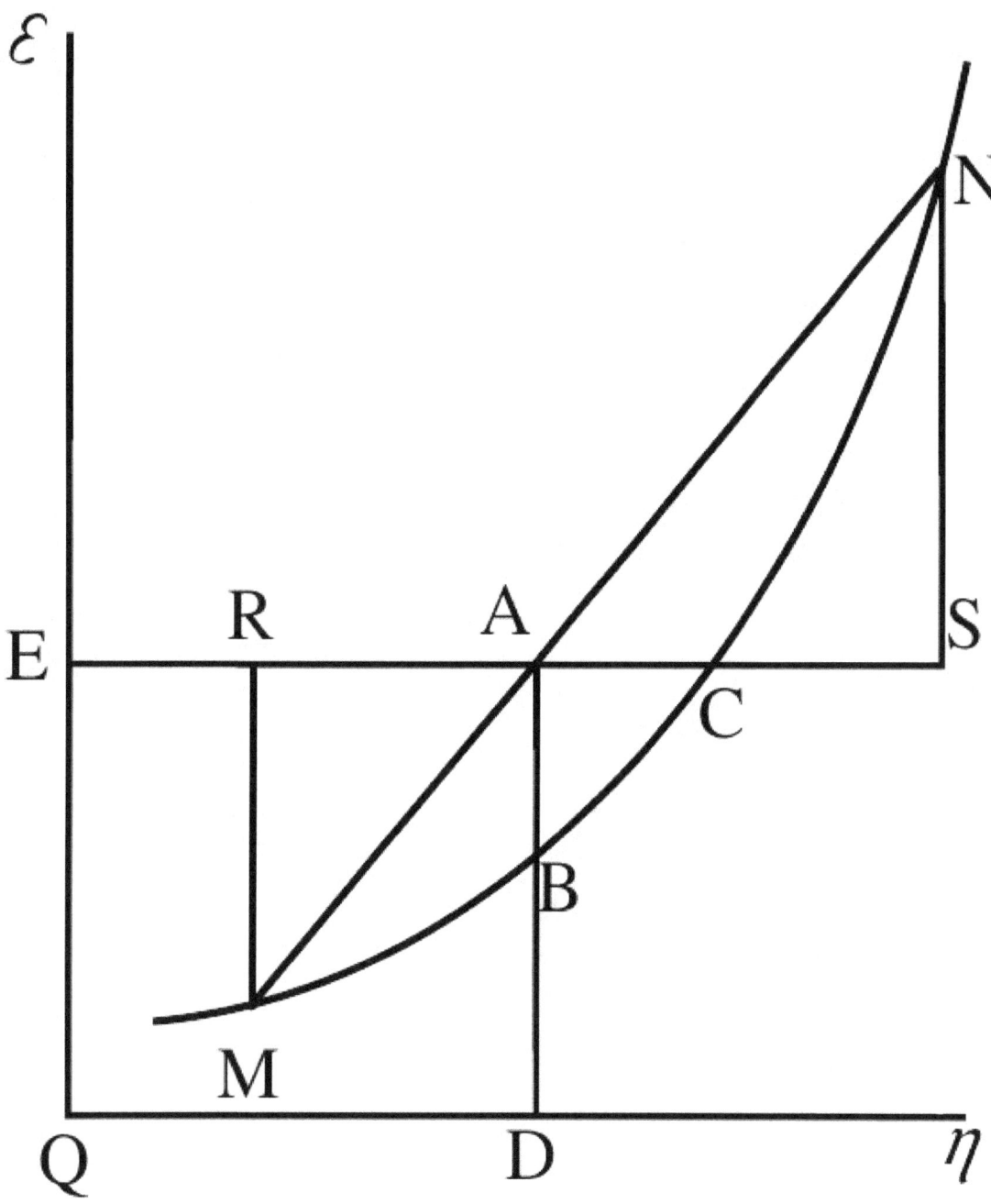

*Willard Gibbs' 1873 **available energy** (free energy) graph, which shows a plane perpendicular to the axis of v (volume) and passing through point A, which represents the initial state of the body. MN is the section of the surface of dissipated energy. Qε and Qη are sections of the planes η = 0 and ε = 0, and therefore parallel to the axes of ε (internal energy) and η (entropy) respectively. AD and AE are the energy and entropy of the body in its initial state, AB and AC its available energy (Gibbs free energy) and its capacity for entropy (the amount by which the entropy of the body can be increased without changing the energy of the body or increasing its volume) respectively.*

35.6 Notes

[1] Wiener, Norbert

[2] Schrödinger, Erwin, *What is Life - the Physical Aspect of the Living Cell*, Cambridge University Press, 1944

[3] Brillouin, Leon: (1953) "Negentropy Principle of Information", *J. of Applied Physics*, v. **24(9)**, pp. 1152-1163

[4] Léon Brillouin, *La science et la théorie de l'information*, Masson, 1959

[5] Mae-Wan Ho, What is (Schrödinger's) Negentropy?, Bioelectrodynamics Laboratory, Open university Walton Hall, Milton Keynes

[6] Jeremy Griffith. 2011. *What is the Meaning of Life?*. In *The Book of Real Answers to Everything!* ISBN 9781741290073. From http://www.worldtransformation.com/what-is-the-meaning-of-life/

[7] Mahulikar, S.P. & Herwig, H.: (2009) "Exact thermodynamic principles for dynamic order existence and evolution in chaos", *Chaos, Solitons & Fractals*, v. **41(4)**, pp. 1939-1948

[8] Aapo Hyvärinen, Survey on Independent Component Analysis, node32: Negentropy, Helsinki University of Technology Laboratory of Computer and Information Science

[9] Aapo Hyvärinen and Erkki Oja, Independent Component Analysis: A Tutorial, node14: Negentropy, Helsinki University of Technology Laboratory of Computer and Information Science

[10] Ruye Wang, Independent Component Analysis, node4: Measures of Non-Gaussianity

[11] P. Comon. Independent Component Analysis - a new concept?, *Signal Processing*, **36** 287-314, 1994.

[12] Didier G. Leibovici and Christian Beckmann, An introduction to Multiway Methods for Multi-Subject fMRI experiment, FM-RIB Technical Report 2001, Oxford Centre for Functional Magnetic Resonance Imaging of the Brain (FMRIB), Department of Clinical Neurology, University of Oxford, John Radcliffe Hospital, Headley Way, Headington, Oxford, UK.

[13] Willard Gibbs, A Method of Geometrical Representation of the Thermodynamic Properties of Substances by Means of Surfaces, *Transactions of the Connecticut Academy*, 382-404 (1873)

[14] Massieu, M. F. (1869a). Sur les fonctions caractéristiques des divers fluides. *C. R. Acad. Sci.* LXIX:858-862.

[15] Massieu, M. F. (1869b). Addition au precedent memoire sur les fonctions caractéristiques. *C. R. Acad. Sci.* LXIX:1057-1061.

[16] Massieu, M. F. (1869), *Compt. Rend.* **69** (858): 1057.

[17] Planck, M. (1945). *Treatise on Thermodynamics*. Dover, New York.

[18] Antoni Planes, Eduard Vives, Entropic Formulation of Statistical Mechanics, Entropic variables and Massieu-Planck functions 2000-10-24 Universitat de Barcelona

[19] John A. Scheilman, Temperature, Stability, and the Hydrophobic Interaction, *Biophysical Journal* **73** (December 1997), 2960-2964, Institute of Molecular Biology, University of Oregon, Eugene, Oregon 97403 USA

[20] Z. Hens and X. de Hemptinne, Non-equilibrium Thermodynamics approach to Transport Processes in Gas Mixtures, Department of Chemistry, Catholic University of Leuven, Celestijnenlaan 200 F, B-3001 Heverlee, Belgium

[21] Pedagogical Risk and Governmentality: Shantytowns in Argentina in the 21st Century (see p. 4).

[22] Leon Brillouin, The negentropy principle of information, *J. Applied Physics* **24**, 1152-1163 1953

[23] Leon Brillouin, *Science and Information theory*, Dover, 1956

Chapter 36

Extropianism

Extropianism, also referred to as the philosophy of *Extropy*, is an evolving framework of values and standards for continuously improving the human condition. Extropians believe that advances in science and technology will some day let people live indefinitely. An extropian may wish to contribute to this goal, e.g. by doing research and development or volunteering to test new technology.

Extropianism describes a pragmatic consilience of transhumanist thought guided by a proactionary approach to human evolution and progress.

Originated by a set of principles developed by Dr. Max More, *The Principles of Extropy*,[1] extropian thinking places strong emphasis on rational thinking and practical optimism. According to More, these principles "do not specify particular beliefs, technologies, or policies". Extropians share an optimistic view of the future, expecting considerable advances in computational power, life extension, nanotechnology and the like. Many extropians foresee the eventual realization of indefinite lifespans, and the recovery, thanks to future advances in biomedical technology or mind uploading, of those whose bodies/brains have been preserved by means of cryonics.

36.1 Extropy

The term 'extropy', as an antonym to 'entropy' was used in a 1967 academic volume discussing cryogenics[2] and in a 1978 academic volume of cybernetics.[3] Diane Duane was the first to use the term "extropy" to signify a potential transhuman destiny for humanity.[4] 'Extropy' as coined by Tom Bell (T.O. Morrow) and defined by Max More in 1988, is "the extent of a living or organizational system's intelligence, functional order, vitality, energy, life, experience, and capacity and drive for improvement and growth." Extropy is not a rigorously defined technical term in philosophy or science; in a metaphorical sense, it simply expresses the opposite of entropy.

A more recent definition of Extropy has been provided by Kevin Kelly, senior maverick at Wired magazine.[5] "Extropy is neither wave, nor particle, nor pure energy. It is a non-material force that is very much like information. Since Extropy is defined as negative entropy-the reversal of disorder-it is, by definition, an increase in order." Kelly gives this definition of extropy in his research on the evolution of technology.

In the philosophy of digital probabilistic physics, the extropy of a physical system is defined to be the self-information of the Markov chain probability of the physical system at a moment in time. This was to distinguish the probability of the Markov state of the physical system from the probability defined by entropy which creates ensembles of equivalent microstates.

36.2 The Extropy Institute

In 1987, Max More moved to Los Angeles from Oxford University in England, where he had helped to establish (along with Michael Price, Garret Smyth and Luigi Warren) the first European cryonics organization, known as Mizar Limited (later Alcor UK), to work on his Ph.D. in philosophy at the University of Southern California.

In 1988, *Extropy: The Journal of Transhumanist Thought* was first published. (For the first few issues, it was "Extropy: Vaccine for Future Shock.") This brought together thinkers with interests in artificial intelligence, nanotechnology, genetic engineering, life extension, mind uploading, idea futures, robotics, space exploration, memetics, and the politics and economics of transhumanism. Alternative media organizations soon began reviewing the magazine, and it attracted interest from like-minded thinkers. Later, More and Bell co-founded the Extropy Institute, a non-profit 501(c)(3) educational organization. "ExI" was formed as a transhumanist networking and information center to use current scientific understanding along with critical and creative thinking to define a small set of principles or values that could help make sense of new capabilities opening up to humanity.

The Extropy Institute's email list was launched in 1991 (and, as of April 2015, continues to exist as "Extropy-Chat"), and in 1992 the institute began producing the first conferences on transhumanism. Affiliate members throughout the world began organizing their own transhumanist groups. Extro Conferences, meetings, parties, on-line debates, and documentaries continue to spread transhumanism to the public.

The Internet soon became the most fertile breeding ground for people interested in exploring transhumanist ideas, with the availability of websites for such organizations that have joined the Extropy Institute in developing and advocating transhumanist (and related) ideas. These include Humanity+, the Alcor Life Extension Foundation, the Life Extension Foundation, Foresight Institute, Transhumanist Arts & Culture, Betterhumans, the Singularity Institute for Artificial Intelligence, and the Institute for Ethics and Emerging Technologies.

In 2006, the board of directors of the Extropy Institute made a decision to close the organisation, stating that its mission was "essentially completed."[6]

36.3 Extropism

Extropism is a modern derivative of the transhumanist philosophy of Extropianism. It follows in the same tradition, hence the similarity of name, but has been revised to better suit the paradigms of the 21st century. As introduced in *The Extropist Manifesto*,[7] it promotes an optimistic futuristic philosophy that can be summed up in the following five phrases, which spell out the word "EXTROPISM":

- Endless eXtension

- Transcending Restriction

- Overcoming Property

- Intelligence

- Smart Machines

These five key points, when taken together, formulate a philosophy and world view which embraces bio-ethical abolitionism, life extension, singularitarianism, technogaianism, freedom of information and several other related disciplines and philosophies. While it does not make a firm political stance, it is most closely related to libertarian socialism (given that it supports the abolition of money and property). Philosophically, it draws from the philosophy of Jeremy Bentham and utilitarianism.

Extropists desire to prolong their life span to a near-immortal state and exist in a world where artificial intelligence and robotics have made work irrelevant. As in utilitarianism, the purpose of one's life should be to increase the overall happiness of all creatures on Earth through cooperation.

The Extropist Manifesto, written by Breki Tomasson and Hank Hyena of The Extropist Examiner in January 2010 (site since discontinued), details the ways in which Extropism has evolved away from, while building upon the original tenets of Extropianism. For example, it moves away from the original Extropian Principles[8] by placing a significant focus on the need to abolish and/or restrict the current use of surveillance, copyright and patent laws. This philosophy, inspired in part by the philosophy of the International Pirate Party, is one of the five basic tenets of the Extropist philosophy, falling under the category "Overcoming Property". Other noteworthy topics that appear frequently in Extropist writings is the focus on equal rights for LGBT couples and individuals and a general distaste for organized religiosity.

36.4 See also

- Cyborg anthropology

- Democratic transhumanism

- Digital probabilistic physics

- Eclipse Phase (role-playing game), a tabletop game which uses the philosophy in its futuristic setting.

- Futures studies

- Holism

- Law of Complexity/Consciousness

- Negentropy

- Proactionary Principle

- Sustainability

- Systems philosophy

- Systems thinking

- Transhumanism

36.5 References

[1] Max More (2003). "Principles of Extropy (Version 3.11) : An evolving framework of values and standards for continuously improving the human condition". Extropy Institute. Archived from the original on 2013-10-15

[2] *Cryogenics*, IPC Science and Technology Press, vol. 7, pg. 225 (1967)

[3] Proceedings of the Fourth International Congress of Cybernetics & Systems: "Current Topics in Cybernetics and Systems", pg. 258 (1978)

[4] Duane, Diane. "The Wounded Sky" (1983)

[5] Kelly, Kevin (April 2011). "Understanding Technological Evolution and Diversity". *The Futurist* 45 (2): 44–48. ISSN 0016-3317.

[6] Extropy Institute (2006). "Next Steps". Retrieved 2006-05-05.

[7] The Extropist Manifesto. *The Extropist Examiner* (blog).

[8] Max More (1998). "The Extropian Principles (Version 3.0) : A Transhumanist Declaration". Extropy Institute.

36.6 External links

- Kevin Kelly on Extropy - Kevin Kelly at The Technium, August 29, 2009

- "Transhumanism's Extropy Institute - Transhumanism for a better future". Retrieved 1 August 2013.

www.ingramcontent.com/pod-product-compliance
Lightning Source LLC
Chambersburg PA
CBHW080802180526
45168CB00006B/2301

9781519182494